Studies in Systems, Decision and Control

Volume 144

Series editor

Janusz Kacprzyk, Polish Academy of Sciences, Warsaw, Poland
e-mail: kacprzyk@ibspan.waw.pl

The series "Studies in Systems, Decision and Control" (SSDC) covers both new developments and advances, as well as the state of the art, in the various areas of broadly perceived systems, decision making and control- quickly, up to date and with a high quality. The intent is to cover the theory, applications, and perspectives on the state of the art and future developments relevant to systems, decision making, control, complex processes and related areas, as embedded in the fields of engineering, computer science, physics, economics, social and life sciences, as well as the paradigms and methodologies behind them. The series contains monographs, textbooks, lecture notes and edited volumes in systems, decision making and control spanning the areas of Cyber-Physical Systems, Autonomous Systems, Sensor Networks, Control Systems, Energy Systems, Automotive Systems, Biological Systems, Vehicular Networking and Connected Vehicles, Aerospace Systems, Automation, Manufacturing, Smart Grids, Nonlinear Systems, Power Systems, Robotics, Social Systems, Economic Systems and other. Of particular value to both the contributors and the readership are the short publication timeframe and the world-wide distribution and exposure which enable both a wide and rapid dissemination of research output.

More information about this series at http://www.springer.com/series/13304

Fernando Lopes · Helder Coelho
Editors

Electricity Markets with Increasing Levels of Renewable Generation: Structure, Operation, Agent-based Simulation, and Emerging Designs

Springer

Editors
Fernando Lopes
National Laboratory of Energy and Geology
Lisbon
Portugal

Helder Coelho
University of Lisbon
Lisbon
Portugal

ISSN 2198-4182 ISSN 2198-4190 (electronic)
Studies in Systems, Decision and Control
ISBN 978-3-319-74261-8 ISBN 978-3-319-74263-2 (eBook)
https://doi.org/10.1007/978-3-319-74263-2

Library of Congress Control Number: 2017963843

Printed on acid-free paper

This Springer imprint is published by the registered company Springer International Publishing AG part of Springer Nature
The registered company address is: Gewerbestrasse 11, 6330 Cham, Switzerland

To my son, Nuno Lopes,
and my parents,
Apolinário Lopes and Maria Lopes,
who taught me two valuable lessons in life,
which guided me on this journey:
to work with people from different
backgrounds
and to respect humanity and nature.

F. L.

Foreword I

The power sector is undeniably going through deep structural changes. The impact of these changes is amplified by the speed at which they are occurring and by the diversity of their nature. In the European Union, the decarbonization agenda is the biggest driver of change. It has direct impact on generation given the required quick pace of investments in low-carbon generation. It also impacts electricity demand as it becomes the clean energy carrier of choice for other sectors such as transport and heating and cooling.

Technological advancements are also impacting the sector. The rise of diverse and increasingly cheap decentralized resources, such as distributed generation and storage, are challenging the past logic of fully centralized systems. Digitalization of the sector is also enabling customers to become increasingly involved in electricity markets. This involvement allows customers to be at the center of power systems and reap benefits from demand-side participation in markets and energy efficiency.

All of the changes that the sector is undergoing have one common underlying economic feature. We are evolving toward a sector increasingly based on fixed costs along the whole value chain. In the upstream, low-carbon investments are mostly based on capital expenditure, e.g., renewables, nuclear or carbon capture and store, in stark contrast with traditional thermal generation. In the midstream, intermittency management is achieved through storage, interconnections, and under-utilized thermal backup thus, all fixed costs. At the downstream level, energy efficiency is also achieved through a replacement of variable costs (e.g., burning gas to heat homes) with fixed costs (e.g., deploying capital in homes like insulation or more efficient appliances).

However, the current electricity market design is based on an energy-only pool with marginal pricing which was conceived for the 1990s liberalization of the power sector. This period coincided with low commodity prices and a thermal technology investment cycle (namely the dash for gas using combined cycle gas turbines, a variable cost technology). However, this market design is clearly not adequate anymore in the current context of investments in capital-intensive technologies (with zero or very low marginal prices) and volatile commodities.

Exposing technologies with these characteristics to market risk, in a marginalist system, makes therefore little sense: it can either lead to overcompensation if the marginal fuel price is very high or to stranded costs if it is too low. In any case, it is a (price) risk these infra-marginal technologies cannot manage since they are composed of fixed (sunk) costs. The current market arrangements therefore make it difficult to invest because investors are forced to take on too many risks (regulatory, price, policy, economic cycle, etc.), far beyond the ones it can and should manage (development, construction, operation, financing).

This book provides very insightful contributions on how electricity markets will behave in the future and how they should be modeled. It is a welcome technical and economic contribution to the long-term policy discussions that are continuously held across the European Union.

Lisbon, Portugal
June 2017

Eduardo Catroga
President
EDP—Energias de Portugal, S.A.

Ana Quelhas
Head of Energy Planning
EDP—Energias de Portugal, S.A.

Foreword II

The liberalization of electricity markets was promoted with the assumption that the competition of different actors should bring an improvement on the efficiency and security of electricity supply, and, as a consequence, better prices and quality of service for customers. This process has been enforced in Europe during the last decade, not only at national level, but increasing the interconnectedness of European energy markets toward building a common market. However, contrary to other liberalized markets, such as the telecommunications, both private and industrial consumers have experienced rather steep price increases, and the market stays highly concentrated on the same players. Furthermore, the progressive installation of new sources of renewable energy, which are very dependent on changing climate conditions, introduces new variables for consideration. There is also an impact of the evolution of prices on other markets, such as those of oil, gas, and carbon, which are still prevalent for electricity generation.

All these factors make a difficult task for the regulator to establish the appropriate directives for the electricity market. Tools are required to test the consequences and validate the norms that control the interactions among stakeholders. Pure mathematical models are difficult to implement because of the evolutionary and distributed nature of the electricity markets, as well as challenging scalability requirements when analyzing real cases. This book promotes agent-based modeling as a tool to simulate the complexity and dynamics of electricity markets. Agents facilitate modeling of market competition where each actor is an agent with its own goals and strategies. They also allow to model the consequences of misbehaviors of individual or groups of actors, disruptive events, such as abrupt variations on the renewable generation, or unexpected events such as failures and accidents on key elements of the supply network. Their simulation facilitates the characterization of emergent behaviors that result from the interactions of the agents on particular scenarios. Nevertheless, the specification of the agents and their interactions, and the setup of the simulation configurations, require a methodical design, which has been the subject of research during the last decade. This book shows relevant contributions from leading experts in the field.

One of the merits of this book is the ability to integrate works from selected research groups in a coherent progression, which allows the reader to move from the basics of agent-based modeling of electricity markets toward more complex scenarios that result from the addition of more heterogeneous electricity generation systems. The application to real markets, which is illustrated in some chapters, and recommendations derived from the corresponding analysis, show the potential of the agent-based approach in this context.

Works on this concrete application of agent-based modeling and simulation have been appearing in conferences and scientific journals during the last decade. This book comes in the right moment, as the tools and results have got maturity, and there was a need to put together the major contributions to get a comprehensive view of the state of the art. We have to acknowledge the great effort of the editors to involve the most representative researchers in this field, as well as their ability to cooperate in producing a coherent ensemble. It is important to note also that, given the multidisciplinary nature of the subject, the editors and the authors have succeeded to make the text quite accessible to a variety of audiences, so it is not required to have a very specific expertise to go through the working of agent models and electricity markets. The result is a reference book for those interested on a better understanding of the complex interactions of the different actors in electricity markets.

Madrid, Spain
June 2017

Juan Pavón
Universidad Complutense de Madrid

Preface

The electricity industry was traditionally heavily regulated, with extensive public ownership, federalized organizational structures, and inefficient competition. In the past, most electric power companies operated as vertically integrated systems, having complete control over power production, transmission, and distribution, and were therefore considered natural monopolies. Typically, customers paid a tariff that reflected all associated costs plus a reasonable rate of return that was controlled by specific regulation. At present, the electricity industry has evolved into a distributed and competitive industry in which market forces drive the price of energy and reduce the net cost through increased competition among suppliers. The economic operation of most power systems is now managed by market operators responsible for balancing supply and demand and for setting energy prices. The security of the systems is normally assigned to independent system operators. Market participants have open access to transmission networks and can freely engage in electricity trades between any two points in a specific network, subject only to the laws of physics and the capacity of transmission lines.

Most existing electricity markets (EMs) were designed according to the principles proposed in the standard market design. The "common" design framework reflects a pool-based market in which there exists a two-settlement system for day-ahead and balancing markets, with ancillary services, and a financial transmission rights market for financial hedging. This framework was, however, set out when the vast majority of power plants were controllable and fired with fossil fuels.

Today, a significant part of the traded power comes from renewable energy sources that are variable and uncertain, largely due to the inability to precisely forecast their output. In fact, renewable generation or variable generation (VG), such as wind and photovoltaic solar power, has increased substantially in recent years. The European Union has been one of the major drivers of the development of renewable energies. The energy policies of most European countries have involved subsidized tariffs—such as, the feed-in tariff in Portugal, the regulated tariff and the market price plus premium in Spain, and the Renewables Obligation in UK. In the United States, many states have also incentives or requirements that will provide for a further increase in variable generation in the coming years.

VG has several unique characteristics compared to the traditional technologies that supply energy in electricity markets. Specifically, VG has significant fixed capital costs but near-zero or zero variable production costs, increases the variability and uncertainty of the net load, and has unique diurnal and seasonal patterns. Together, these characteristics may significantly influence the outcomes of EMs. In particular, large penetrations of VG may reduce market-clearing prices due to their low-bid costs, and increase price volatility because of their increased variability.

As noted earlier, most existing market designs are unique in their complex relationships between economics and the physics of electricity, but were created without the notion that large penetrations of VG would be part of the supply mix. Accordingly, the potential impacts of VG should be monitored to determine if the original designs are still effective. If existing market designs lead to inefficiency, reduced competition or increased market power, improvements to these designs may be required or newer designs may be needed. Simply put, there is a growing need to accurately model, analyze in detail, and fully understand the behavior of today's evolving electricity markets and how market participants may act and react to the changing economic and regulatory environments in which they operate.

Multi-agent systems (MAS) represent a relatively new and rapidly expanding area of research and development. MAS are essentially systems composed of software agents that interact to solve problems that are beyond the individual capabilities of each agent. Software agents are elements situated in some environment and capable of flexible autonomous action in order to meet their design objectives. The major motivations for the increasing interest in MAS include the ability to solve problems in which data, expertise, or control is distributed, and the ability to enhance performance along the dimensions of computational efficiency, reliability, and robustness. Conceptually, a multi-agent approach presents itself as an advanced modeling approach to simulate the behavior of power markets over time. Software agents can be designed to act in an open and distributed environment, with incomplete and uncertain information, limited resources, and may efficiently manage cooperative and competitive interactions with other agents.

This book is about the common ground between two fields of inquiry: electricity markets and multi-agent systems (or artificial intelligence generally). The field of electricity markets has grown significantly in the past few years resulting in a substantial body of work and well-established technical literature. There are several journals that focus on research in this area (e.g., IEEE Transactions on Power Systems and Applied Energy) and several books have been presented in the literature. Also, research on multi-agent systems has a vigorous, exciting tradition, and has generated many useful ideas and concepts, leading to important theories and relevant computing systems. Various journals and forums have been dedicated almost exclusively to the study of intelligent agents, such as the Autonomous Agents and Multi-Agent Systems journal and the AAMAS Conference series. And development has occurred on the practitioner side as well. This book lets these different strands come together—it includes methods and techniques from energy and power systems, economics, artificial intelligence, and the social sciences.

Agent-based simulation has been an important approach to model and analyze electricity markets over the past decade. Several agent-based energy management tools have emerged, including the Electricity Market Complex Adaptive System (EMCAS), developed by the Argonne National Laboratory, and the Simulator for the Electric Power Industry Agents (SEPIA), developed by the Honeywell Technology Center and the University of Minnesota. However, despite the power and elegance of these and other relevant tools, they were arguably developed to simulate traditional market mechanisms.

At present, the study of existing and emerging market designs to manage the potential challenges of VG, making use of software agents and methods from artificial intelligence (AI), has received only selective attention from both scholars and practitioners. Although some valuable journal articles and technical reports exist, there is not an up-to-date introduction to the area nor a comprehensive presentation of the research progress and achievements. Also, efforts to integrate research contributions from different fields into a broader understanding of electricity markets to meet the variability and uncertainty of VG were only beginning to occur. The main purpose of this book is to fulfill these needs.

The book has 11 chapters organized into three major parts: Electricity Markets and Autonomous Computational Agents (Part I), Electricity Markets with Large Penetrations of Variable Generation: Current and Emerging Designs (Part II), and Agent-based Simulation of Electricity Markets with Increasing Levels of Variable Generation: Traditional and New Design Elements (Part III). A comprehensive overview of the book is as follows.

Part I introduces the reader to the essentials of electricity markets and software agents. This part contains three chapters. Chapter 1 focuses on EMs and introduces the various markets for the different electrical related products: energy, reserves, transmission rights, and capacity. The chapter ends with a list of the potential impacts of VG on market outcomes. Chapters 2 and 3 introduce a generic framework for agent-based simulation of EMs. The framework provides a coherent set of concepts related to electricity markets and software agents, helps to compare disparate research efforts, and facilitates the development of future models and systems. It includes three groups of dimensions: market architecture, market structure, and software agents.

In particular, Chap. 2 deals with EMs and discusses, in considerable detail, the architecture and core structure of power markets. The chapter introduces the three key market sectors: wholesaling, retailing, and central coordination and transmission. It also describes some important market types (notably, pool, and bilateral) and discusses the role of the main entities operating in EMs. Chapter 3 deals with intelligent agents and presents some important features of agency. The chapter introduces the concepts of “agent architecture” and “agent capability”, and discusses six key types of agents: purely reactive, model-based, goal-based, utility-based, and learning agents. It also presents a core set of capabilities central to the definition and development of agents for EMs, including autonomy, proactiveness, social ability, and adaptability.

Part II discusses existing electricity markets and evolving market designs, notably potential improvements to current market designs to manage the challenges of VG. This part contains four chapters. Chapter 4 looks at the current design of the Nordic power market (Nord Pool). The chapter analyzes the hourly market data for Western Denmark in the period from 2004 to 2014, particularly the occurrence of extreme events (e.g., 100 < price < 5 €/MWh or 100% < wind < 1% of the hourly demand). The authors conclude that the current market organization has been able to handle the amount of wind power installed so far (in 2014, wind power provided 51.7% of the electricity consumption in Western Denmark). They point out, however, that the hydro power capacity is limited and larger penetrations of wind power will require additional measures. Chapters 5 and 6 focus on two key issues related to market design: incentivizing flexibility in short-term operations and revenue sufficiency for long-term reliability.

Specifically, Chap. 5 discusses whether existing market designs provide adequate incentives for suppliers to offer their flexibility into markets to meet the increased levels of variability and uncertainty introduced by VG. The chapter provides a definition of power system flexibility and examines how the introduction of VG may increase the need for flexibility. It analyzes five existing market design elements to incentivize flexibility: centralized scheduling and efficient dispatch, frequent scheduling and frequent settlement intervals, existing ancillary service markets, make-whole payment guarantees, and day-ahead profit guarantees. It also discusses a number of emerging market design elements that impact flexibility incentives, including pay-for-performance regulation, primary frequency control, convex hull pricing, and explicit products for flexible ramping provision.

Chapter 6 discusses whether suppliers who are needed to ensure a reliable system in the long run have sufficient opportunity to recover their variable and fixed costs to remain in the market. The focus is mainly on the investment time horizon and the installation of sufficient generation capability (operational issues, which are closely related, are discussed in Chap. 5). The chapter examines how increasing penetrations of VG may exacerbate the missing-money problem. It describes the two primary market mechanisms traditionally adopted by EMs to mitigate the issues of resource adequacy and revenue sufficiency: scarcity pricing (both through administrative prices as well as offered prices) and forward capacity markets. It also discusses the most recent market design changes to address these issues, with a focus on how they are evolving to meet the needs due to increased VG. Significant changes include scarcity pricing through dynamic demand curves for operating reserve and forward flexible capacity requirements.

As in part of Chap. 6, Chap. 7 looks at the impact of significant levels of VG on reliability requirements. The focus is on capacity markets to ensure the long-term viability of suppliers. The chapter considers three situations differing mainly in the mix of generation technologies: open cycle gas turbines (OCGTs) only, OCGTs and wind power plants, and OCGTs and nuclear plants. The author uses data from Sweden (e.g., the Swedish load and real Swedish wind power production data) to perform a detailed analysis of the influence of VG on capacity adequacy requirements, and makes a systematic comparison with the results of the nuclear case.

Part III is devoted to agent-based simulation of electricity markets with large penetrations of renewable generation. This part analyzes the potential impacts of VG on EMs and discusses the advantages of specific market design elements. Also, it explores new opportunities to bridge EMs and emerging technologies—such as demand response (DR) and distributed generation (DG)—and examines specific market designs that are inclusive of such technologies. It contains four chapters.

Chapter 8 introduces the agent-based simulation tool MATREM (for Multi-Agent TRading in Electricity Markets), which allows the user to simulate the behavior and outcomes of EMs, including markets with large penetrations of VG. The chapter begins by describing the two exchanges supported by the tool: a power exchange, comprising a day-ahead market and an intra-day market, and a derivatives exchange, comprising a futures market for trading standardized bilateral contracts. Next, it describes the marketplace for negotiating the details of tailored (or customized) long-term bilateral contracts and presents the various market entities currently being implemented (e.g., generating companies, retailers, consumers and market operators). Following this material, the chapter presents the paradigm of human-computer interaction (involving both direct manipulation interface techniques and intelligent assistant agents). The final part of the chapter delves into the technical details of the agent model: a belief-desire-intention (BDI) model.

Chapter 9 focuses on variable generation, support policies, and the merit order effect (MOE). The first part of the chapter analyzes the sustained growth of VG worldwide and discusses the global policy landscape. The second part describes in detail the principles underlying the MOE. Following this introductory material, the chapter investigates the reduction in the Portuguese day-ahead prices achieved by wind power as a result of the MOE in the first half of 2016. The results generated by MATREM indicate an average price reduction of about 17 €/MWh. The net cost of the wind energy support policy was −8.248 million € in January 2016, indicating that a net profit has occurred in the month. The net cost for the entire study period reached, however, the value of 69.011 million €. Although considerable, this cost should be interpreted carefully, since it takes into account the feed-in tariff for wind energy, the market value of the wind electricity, and the financial volume of the MOE, but does not account for the carbon price effect on the electricity market.

Chapter 10 looks at demand response (DR) in electricity markets to incentivize system flexibility and help managing the variability and uncertainty introduced by VG. The authors introduce two key categories of DR programs and present a brief overview of DR in Spain and Portugal. Following this introductory material, the authors investigate the price effect of DR on the Iberian market (MIBEL) during the period 2014–2017. The results generated by MATREM are striking. They indicate that modest amounts of DR—modeled as load reductions between 1 and 5% when prices rise above a threshold between 80 and 100 €/MWh—have a relatively large effect on market prices, creating substantial benefits to market participants (and most retail customers). For instance, in 2017, a load reduction of 5% when prices rose above 80 €/MWh yielded the benefit of 76.62 million €. The chapter concludes with recommendations—for consideration by state institutions, system operators, electric utilities, and other market participants—to foster DR in Portugal.

Chapter 11 brings an additional impact to agent-based modeling and simulation of power and energy systems, by combining the simulation of electricity markets and smart grids with the physical emulation of a laboratory micro-grid. The chapter introduces the Multi-Agent Simulator of Competitive Electricity Markets (MASCEM) and the Multi-Agent Smart grid Platform (MASGriP). It also presents a case study based on real data, which involves a smart grid (SG) composed by a simulated distribution network with several real loads, including eight residential houses, eight residential buildings, and one commercial building, and also accommodating distributed generation (photovoltaic and wind power generation) and storage units. The case study illustrates the potentialities of integrating the two agent-based systems into a unified platform. The authors conclude that the cooperation between MASCEM and MASGriP opens important studying opportunities under different perspectives, resulting in an important contribution to the fields of transactive energy, electricity markets, and SGs.

Overall, the book is a confluence of a comprehensive exploration and a deep exposition of the common ground between electricity markets and multi-agent systems. While no single volume could cover the entire rich terrain at the intersection between these two areas of inquiry, the book gives the reader an insightful view of a landscape of stimulating ideas and offers a number of features, notably:

- Scope. The text is organized into three major parts. The book covers the fundamentals of electricity markets and software agents (Part I), discusses both traditional and emerging market designs to accommodate the variability and uncertainty of VG (Part II), and deals with agent-based simulation of electricity markets with increasing levels of renewable generation (Part III).
- Theory. The book gives a clear and careful presentation of the key concepts and methods from the two aforementioned areas as well as techniques from the common ground between these areas. We try to avoid excessive formality in the text while retaining precision. Several examples and illustrative case studies are provided.
- Practice. The emphasis is not only on theory but also on practice. The methods and techniques presented in the book are supplemented with actual cases involving real-world electricity markets and applications drawn from real-world situations.
- Expertise. The chapters have been written by leading and outstanding researchers, who have helped shaped the two areas of inquiry. The book is thus built on a diverse basis of knowledge and experience.

An explanatory and cautionary note is in order here. Broadly speaking, any book prepared by just a few authors is likely to be more coherent than any book in which several authors are involved. But as the reader will see, the editors have invested a considerable effort in ensuring the coherence of this book. The order of the chapters—and the chapters' topics—was done carefully to produce a highly organized text. Also, the contributors had the chance to review specific chapters, helping to significantly improve the quality of the book.

The intended audience of the book includes professionals associated mainly with the electric industry and electricity markets, including utility business leaders, engineers (notably, electrical, industrial, software, computer, power, and systems engineers), market operators, market players, energy economics, and investors related to energy projects. Also, the book is intended to be very valuable to researchers and academics who wish to better understand the areas of energy markets (with increasing levels of VG) and software agents, and mainly to investigate the common ground between these two fascinating areas—the book successfully integrates theory, scientific research, and real-world applications, and is sufficiently informative to earn the respect of specialists. Given the scope and the depth of the chapters—and since the book is written in a highly accessible style, the concepts and methods are carefully explained, and the text is liberally supported with practical applications—we are confident that the content of the book should also provide a coherent foundation for several different graduate courses.

This book could not have been completed without the help of many people. We are most grateful to:

- All authors of the book for participating in this challenging project.
- The organizations that have supported the authors and the editors.
- Many of our colleagues working on energy markets and software agents, who have given helpful feedback about earlier versions of the text.
- All Ph.D. and M.Sc. students from the NOVA University of Lisbon, University of Lisbon and ISCTE–University Institute of Lisbon, who have been involved in the MAN-REM project[1] and/or the IRPWind project.[2]
- The Springer team for their encouragement, tolerance, and full support, especially the editors Janusz Kacprzyk and Thomas Ditzinger, as well as Jeyashree Kumar.
- Our families, who have provided us with the time and the personal support required to finish this book.

In conclusion, this book is very much a team effort of different people, whose credentials as researchers are excellent, and whose research efforts have made the growth of the two areas of inquiry possible.

Lisbon, Portugal
June 2017

Fernando Lopes
Helder Coelho

[1]Project MAN-REM (http://www.lneg.pt/iedt/projectos/473/), supported by FEDER funds through the programme COMPETE ("Programa Operacional Temático Factores de Competividade"), and National funds through FCT ("Fundação para a Ciência e a Tecnologia").

[2]Project IRPWind (http://www.irpwind.eu/), funded by the European Union's seventh programme for research, technological development and demonstration, under grant agreement 609795.

Contents

Contributors

Hugo Algarvio Instituto Superior Técnico, Lisbon, Portugal

Aaron Bloom National Renewable Energy Laboratory, Golden, CO, USA

Audun Botterud Argonne National Laboratory, Argonne, IL, USA

Jaquelin Cochran National Renewable Energy Laboratory, Golden, CO, USA

Helder Coelho University of Lisbon, Lisbon, Portugal

Erik Ela Electric Power Research Institute, Palo Alto, CA, USA

Pedro Faria GECAD-Research Group on Intelligent Engineering and Computing for Advanced Innovation and Development, Institute of Engineering, Polytechnic of Porto, Porto, Portugal

Luis Gomes GECAD-Research Group on Intelligent Engineering and Computing for Advanced Innovation and Development, Institute of Engineering, Polytechnic of Porto, Porto, Portugal

Poul Erik Grohnheit DTU Management Engineering, Technical University of Denmark, Kongens Lyngby, Denmark

Todd Levin Argonne National Laboratory, Argonne, IL, USA

Fernando Lopes LNEG–National Laboratory of Energy and Geology, Lisbon, Portugal

Michael Milligan National Renewable Energy Laboratory, Golden, CO, USA

Tiago Pinto GECAD-Research Group on Intelligent Engineering and Computing for Advanced Innovation and Development, Institute of Engineering, Polytechnic of Porto, Porto, Portugal

Isabel Praça GECAD-Research Group on Intelligent Engineering and Computing for Advanced Innovation and Development, Institute of Engineering, Polytechnic of Porto, Porto, Portugal

João Santana Instituto Superior Técnico, Lisbon, Portugal

Klaus Skytte DTU Management Engineering, Technical University of Denmark, Kongens Lyngby, Denmark

João Sá Instituto Superior Técnico, Lisbon, Portugal

Lennart Söder KTH, the Royal Institute of Technology, Stockholm, Sweden

Aaron Townsend National Renewable Energy Laboratory, Golden, CO, USA

Zita Vale GECAD-Research Group on Intelligent Engineering and Computing for Advanced Innovation and Development, Institute of Engineering, Polytechnic of Porto, Porto, Portugal

Acronyms and Abbreviations

AAMAS	Autonomous agents and multi-agent systems
ABS	Agent-based simulation
AC	Alternating current
ACE	Agent-based computational economics
ACL	Agent communication language
AI	Artificial intelligence
AID	Agent IDentifier
ALBidS	Adaptive Learning Strategic Bidding System
AMS	Agent management system
API	Application programming interface
AS	Ancillary service provider
ASCP	Ancillary service clearing price
BDI	Belief–desire–intention
CAISO	California Independent System Operator
CBA	Coordinated balancing activity
CBL	Customer base load
CCP	Central counter-party
CFD	Contract for difference
CHP	Combined heat and power
CIC	Curtailment initiation cost
CONE	Cost of new entry
CPP	Critical peak pricing
DAM	Day-ahead market
DAMAP	Day-ahead margin assurance payment
DF	Directory facilitator
DG	Distributed Generation
DistCo	Distribution company
dMARS	distributed Multi-Agent Reasoning System
DR	Demand response
DRA	Demand response aggregator

DSV	Delivery settlement value
ECB	European Central Bank
EEX	European electricity exchange (Germany)
ElBAS	Electricity Balance Adjustment Service
ELCC	Effective load-carrying capability
ELDC	Equivalent load duration curve
ELMP	Extended LMP
EM	Electricity market
EMCAS	Electricity Market Complex Adaptive System
EPSA	Electric Power Supply Association
ERCOT	Electric Reliability Council of Texas
ERM	Energy resource management
ERSE	Portuguese Energy Services Regulatory Authority
ETS	Emission trading system
EU	European Union
EUA	European Union Allowance (CO_2 emissions)
EUE	Expected unserved energy
EV	Electric vehicle
FERC	Federal Energy Regulatory Commission
FIPA	Foundation for Intelligent Physical Agents
FIT	Feed-in tariff
FOR	Forced outage rate
FTD	First trading day
FTR	Financial transmission rights
GECAD	Knowledge Engineering and Decision Support Research Group
GenCo	Generating company
GWEC	Global Wind Energy Council
HVAC	Heating, ventilation, and air conditioning
HVDC	High-voltage direct current
ICAP	Installed capacity
IEA	International Energy Agency
IMF	International Monetary Fund
ISO	Independent system operator
ISO-NE	Independent System Operator of New England
JADE	Java Agent DEvelopment (framework)
KB	Knowledge base
KQML	Knowledge query and manipulation language
LDC	Load duration curve
LMP	Locational marginal pricing
LOLE	Loss-of-load expectancy
LOLH	Loss-of-load hours
LOLP	Loss-of-load probability
LRT	Last resort trader
LSE	Load-serving entity
LTD	Last trading day

MAS	Multi-agent systems
MASCEM	Multi-Agent Simulator of Competitive Electricity Markets
MASGriP	Multi-Agent Smart Grid Platform
MATREM	Multi-Agent TRading in Electricity Markets
MIBEL	Iberian Electricity Market (or "Mercado IBérico de ELectricidade)"
MISO	Midcontinent Independent System Operator
MO	Market operator
MOE	Merit order effect
MRC	Multiregional coupling
MTS	Message transport service
NAPRE	National Action Plan for Renewable Energy
NEMO	Nominated Electricity Market Operator
NERC	North American Electric Reliability Corporation
NYISO	New York Independent System Operator
OCGT	Open cycle gas turbine
OMIE	Spanish electricity market operator (or "Operador del Mercado Ibérico de electricidade—polo Espanol, S.A.")
OMIP	Portuguese electricity market operator (or "Operador do Mercado Ibérico de electricidade—pólo Português, S.A.")
OPF	Optimal power flow (procedure)
ORDC	Operating reserve demand curve
OTC	Over-the-counter
OWL	Web ontology language
PCR	Price coupling of regions
PFC	Primary frequency control
PJM	Pennsylvania-New Jersey-Maryland Independent System Operator
PRM	Planning reserve margin
PRS	Procedural reasoning system
PV	Photovoltaic
PX	Power exchange
RE	Renewable energy
REE	Spanish electrical grid (or "Red Eléctrica de Espana")
REN	Portuguese electrical grid (or "Redes Energéticas Nacionais")
RES	Renewable energy sources
RetailCo	Retailer
RTM	Real-time market
RTO	Regional transmission organization
RTP	Real-time pricing
RUC	Reliability unit commitment
SC	Scheduling coordinator
SCADA	Supervisory Control and Data Acquisition
SCUC	Security-constrained unit commitment
SEDC	Smart Energy Demand Coalition
SEG	Solar electricity generation (or electricity generation from solar power)

SEK	Swedish krona
SEPIA	Simulator for the Electric Power Industry Agents
SG	Smart grid
SMP	System marginal pricing
SO	System operator
SOICAM	SCADA Office Intelligent Context Awareness Management
SP	Settlement price
SPP	Southwest Power Pool
SQL	Structured Query Language
SRP	Spot reference price
TO	Transmission owner
TOU	Time-of-use
TransCo	Transmission company
TSO	Transmission system operator
UCAP	Unforced capacity
UML	Unified modeling language
US	United States
VG	Variable generation
VOLL	Value of lost load
VPP	Virtual power player
VPSC	Voluntary Price for Small Consumers
WWSIS-2	Western Wind and Solar Integration Study Phase 2
XML	Extensible markup language

Units Used to Measure Energy (and Costs)

A	amp (the unit of electrical current)
h	hour (time)
k	kilo (used, e.g., in kW, kWh and kV)
M	mega (million) (used, e.g., in MW and M€)
G	giga (used, e.g., in GW and GWh)
T	tera (used, e.g., in TWh)
V	volt (the unit of electrical pressure)
W	watt (power)
Wh	watt-hour (energy)

Part I
Electricity Markets and Autonomous Computational Agents

Chapter 1
Overview of Wholesale Electricity Markets

Erik Ela, Michael Milligan, Aaron Bloom, Jaquelin Cochran, Audun Botterud, Aaron Townsend and Todd Levin

Abstract This chapter provides a comprehensive review of four key electricity markets:

- Energy markets (day-ahead and real-time markets).
- Ancillary service markets.
- Financial transmission rights markets.
- Capacity markets.

It also discusses how the outcomes of each of these markets may be impacted by the introduction of high penetrations of variable generation. Furthermore, the chapter examines considerations needed to ensure that wholesale market designs are inclusive of emerging technologies, such as demand response, distributed generation, and distributed storage.

This chapter is based on the overview of wholesale electricity market designs presented by Ela et al. [1, Sect. 2].

E. Ela (✉)
Electric Power Research Institute, Palo Alto, CA 94304, USA
e-mail: eela@epri.com

M. Milligan · A. Bloom · J. Cochran · A. Townsend
National Renewable Energy Laboratory, Golden, CO 80401, USA
e-mail: Michael.Milligan@nrel.gov

A. Bloom
e-mail: Aaron.Bloom@nrel.gov

J. Cochran
e-mail: jaquelin.cochran@nrel.gov

A. Botterud · T. Levin
Argonne National Laboratory, Argonne, IL 60439, USA
e-mail: abotterud@anl.gov

T. Levin
e-mail: tlevin@anl.gov

F. Lopes and H. Coelho (eds.), *Electricity Markets with Increasing Levels of Renewable Generation: Structure, Operation, Agent-based Simulation, and Emerging Designs*, Studies in Systems, Decision and Control 144, https://doi.org/10.1007/978-3-319-74263-2_1

1.1 Introduction

The goal of all electricity systems, whether they are operated by regulated monopolies or centrally administered by an independent system operator (ISO) or regional transmission organization (RTO), is to ensure the reliable delivery of electricity at the lowest cost to consumers. These goals are rooted in a long history of regulatory principles that influence the entry of new market participants, set prices, prescribe the quality and condition of entry, and obligate a utility to provide service. The rationale for this regulation emerges from the physical constraints of the electric grid. This chapter is not intended to explain the intricacies of the grid and electric utility regulation, but a brief review is important to understanding the challenges of electricity markets.

Three fundamental components comprise the wholesale electricity supply: generation, transmission, and coordination services. Each of these has a financial and physical component that must accommodate for the lack of a consumer response inherent to electricity markets while ensuring the constant balance of generation and load. Because of the extreme cost associated with failure of the power system, the physical requirements of the system must be ensured, even though market and operational inefficiencies are introduced to do so. Assuring this reliability requires procuring adequate generation, transmission, and coordination services. In short, resource adequacy—i.e., having enough available capacity in the system—is required to reliably meet load at all times. This includes adequate transmission capacity which is also required to ensure energy can be delivered to where it is needed. Because electricity demand is relatively inelastic, variable in time, and uncertain in quantity, both generation and transmission must be constantly coordinated to meet load in a reliable manner.

To gain the system requirements necessary to support the security and reliability of the electric grid, adequate market policies must be crafted that address the financial implications of these requirements. Ideally, these policies will provide sufficient opportunity for generators to recover both fixed and variable costs if they contribute to resource adequacy; promote the construction and upkeep of a viable transmission network; and incentivize generators to coordinate scheduling of resources to meet the variable and uncertain load while maintaining the reliability of the transmission network. Simultaneously, these policies must avoid incentivizing an overbuilt system or overcompensating inefficient units.

Electricity Industry in North America. Historically the electricity industry in North America has been operated as a natural monopoly, regulated by a combination of state commissions and Federal oversight for some aspects of interstate trade. The legal justification for electric utility regulation in the United States can be traced through British common law and a series of Supreme Court cases.[1] Generally speaking, these cases have found that utilities, such as those in the electricity and natural gas

[1] Munn v. Illinois (94 U.S. 113, 1887), Smyth v. Ames (169 U.S. 466), and Federal Power Commission v. Hope Natural Gas Company (320 U.S. 591).

industries, provide services that are in the "public interest" and are necessary for the common welfare of the people. The economic justification for regulation has been focused on the inherently noncompetitive nature of the market. The market can be uncompetitive for a variety of reasons, including: (1) technology that allows a limited number of companies to provide adequate capacity to supply all demand; (2) the unique position of a principle buyer; and (3) conditions in the market that do not produce competitive results [2]. Because of these characteristics, regulators and policy makers have adopted certain regulatory frameworks to meet the essential needs of society and to ensure that utilities are capable of earning a fair return on their investments. However, the potential benefits of competitive generation instigated the restructuring of the electricity markets in the late 1990s. Currently, more than two thirds of the electricity consumption in the United States is purchased within restructured electricity markets. The restructured markets have been designed to support the financial constraints of generation, transmission, and coordination that are necessary to secure a stable and reliable physical power system while addressing the problems of inefficient pricing, investment risk, and market power.

In the United States, RTO/ISO administered markets have evolved in similar directions to a large extent following the principles proposed in the standard market design [3]. This design reflects a pool-based market in which there exists a two-settlement system for day-ahead markets (DAMs) and balancing/real-time markets (RTMs), with co-optimized energy and ancillary services, locational marginal pricing (LMP) for energy, and financial transmission rights markets (FTRs) in place for financial hedging. Energy is sold in forward (e.g., day-ahead, hourly) markets and balanced in 5-min RTMs with LMPs. Locational energy markets in the United States are cleared once a day for hourly trading intervals for day-ahead markets and every 5-min for real-time markets. At present, ancillary service markets are in place in all markets, including those for spinning contingency reserve, nonspinning contingency reserve, and regulating reserve. The ancillary service markets operate in a similar manner to energy markets and are cleared using the same model, with day-ahead and real-time prices and schedules for the capacity reservation of the ancillary service. FTRs are cleared in forward markets and are an instrument put in place to hedge against locational differences in energy prices.

Each regional transmission organization/independent system operator also has a process for procuring sufficient resources to meet the peak load requirements. In the Pennsylvania-New-Jersey-Maryland Independent System Operator (PJM), New York Independent System Operator (NYISO), and the Independent System Operator of New England (ISO-NE), mandatory capacity markets have been designed to incentivize investment in installed capacity and to allow peaking units to recover fixed costs. At present, the Electric Reliability Council of Texas (ERCOT), California Independent System Operator (CAISO), Midcontinent Independent System Operator (MISO), and Southwest Power Pool (SPP) do not have mandatory capacity markets available and utilize various administrative processes and spot (scarcity) prices to provide fixed-cost recovery for resources.

Europe. In Europe, most markets offer day-ahead and intraday markets. The power systems and energy markets are operated separately; the market clears a dispatch order, which then can be adjusted to accommodate transmission constraints. Germany, for example, with its extensive bilateral market contracts, requires longer gate closures to allow the transmission system operator (TSO) to conduct load-flow calculations and coordinate with neighboring TSOs, which in turn requires significant re-dispatch to resolve transmission constraints [4].

In the Nord Pool Spot, there is a day-ahead market followed by an intraday market, which matches bids continuously until one hour before the hour of delivery. This decreases liquidity in comparison to the Iberian intraday market, which has sessions that concentrate the trades. The Iberian intraday market, however, has a longer delay between the trade and delivery. Consequently, in Nord Pool there is no need for a market between the intraday and tertiary regulation market, which is called the regulating power market in Nord Pool (and the real-time market in the two-step markets). Nord Pool's regulating power market requires activation in 15 min and also is used to meet operating reserves.

1.2 Energy Markets

As discussed in the previous section, energy is bought and sold in most US and European markets through a two-settlement system. A forward market sells energy to load-serving entities (LSEs) and buys from sellers in advance of the time when the energy is produced and consumed. This is typically through the day-ahead market (DAM). The DAM clears to meet bid-in load demand for the entire day, one day in advance. Schedules and prices are calculated from the market-clearing engine, and this price-quantity pair is settled for all market participants regardless of their actual performance. The DAM is important because it provides a hedge against price volatility in the real-time markets caused by load forecast errors, generator outages, or other imbalances. The DAM also allows for make-whole payments when resources do not recover their costs, and it provides price incentives in advance toward reliable operation when resources may need ample notification time to be able to start their generating resources [5]. To reflect changes that may occur between the day-ahead market and real-time operations, a second market clearing is used by RTOs/ISOs to re-dispatch resources and commit new resources to meet system requirements. This is generally referred to as the RTM. Variability and uncertainty is present throughout the power system including changes in weather that can cause unexpected deviations in load and variable resource output, and forced outages that can take resources and network facilities offline unexpectedly. The RTM is in place to set prices and schedules to match the imbalances caused by such events. It reflects the actual operation of the resources participating in the market. Many markets also have intermediate scheduling procedures on the hour ahead or a few hours ahead to facilitate this process in advance of real time when the differing conditions from the

DAM are apparent. These markets typically have advisory prices and schedules, but they may have binding commitment directions.

In both day-ahead and real-time markets, suppliers will offer energy bids as a price and quantity pair. In US markets, there is further complexity in supplier offers, which are designed as three-part bids. Due to the non-convexity of costs of many generating resources, the generators submit a bid for (1) incremental energy, (2) no-load cost—i.e., a cost just to be online, or at its minimum generation level, and (3) a cost of starting up the generating unit and synchronizing it to the grid. The generators also submit to the ISO their unit constraints, including how fast they can ramp, how long they must stay online if committed, and other constraints. The market operator will select the least-cost set of suppliers to meet the demand based on these three-part bids and generating unit constraints while also obeying many of the physical power system constraints.

It is important that the average prices of the day-ahead and real-time markets converge, so that market participants should not have a strong preference to be in either market. Virtual trading, or convergence bidding, is used in most RTO/ISO markets to ensure that the prices of the DAM and RTM converge to the same price on average [6]. Virtual traders will sell or buy energy in the DAM and buy or sell it back in the RTM. They have no requirement to have physical assets to supply or consume energy. By taking advantage in either market when there is a premium in one, they will drive down the difference in prices between these markets. This design feature of the market recognizes the natural tendency of traders to arbitrage across different markets. In the absence of virtual trading, there is potential for a premium in one market that can lead to uncompetitive and inefficient behavior.

In addition to the day-ahead market process, a subsequent process is used, generally referred to as the reliability unit commitment (RUC) process. The day before the operating day, an initial security-constrained unit commitment (SCUC) will solve to meet the bid-in load with bid-in generation and create the schedules and prices for the DAM. These bid-in quantities, in particular the bid-in load or bid-in variable generation capacity, may or may not be close to reality. To ensure the system has sufficient capacity available, a subsequent SCUC will be solved to meet the RTO/ISO forecasted load. The exact practices vary by region, but generally the RUC will only commit additional resources and will not decommit any resources needed in the DAM. For example, while most markets solve the RUC subsequent to the DAM, the NYISO solves the DAM and RUC iteratively, so that resources committed by the RUC can affect the DAM prices and schedules [7]. Most ISOs are now also using the RTO/ISO forecasted variable generation as part of the RUC process as well. Energy markets that consist of short-dispatch intervals (e.g., 5-min dispatch intervals), which already have been adopted in many restructured markets, improve system flexibility by more closely matching the changes in variable generation (VG) and load ("net load") economically. As net load changes, the dispatch optimization responds as well-cost-effectively optimizing generation. Short-dispatch interval markets also reduce the required levels of regulating reserves needed, which are the automatic resources that can respond to minute-to-minute fluctuations and are the most expensive ancillary service [8]. High energy prices during the ramp periods also could provide an incentive for flexible supply. All generation receives the energy market clearing price in an energy market, as opposed to markets with ramp products, described below. A two-step market with unit commitment in the day-ahead timescale will leave significant forecast errors to be resolved during real-time balancing. The balancing resources acting on the timescale of a few minutes can be relatively expensive [9]. An alternative is to have some form of intraday market that enables participation from power plants with intermediate lead/start-up times [10].

For example, the Iberian market already has a considerable share of variable generation. The market structure consists of a day-ahead market followed by six sessions in the intraday market. The gate closure in the intraday market is 3 h and 15 min. The intraday market is at times followed by a deviation management market, which is used when a deviation of more than 300 MWh is expected to last several hours. A tertiary regulation market is used to recover secondary regulation reserves in the intra-hour timescale. In the Nord Pool Spot, there is a day-ahead market followed by an intraday market, which matches bids continuously until one hour before the hour of delivery. This decreases liquidity in comparison to the Iberian intraday market, which has sessions that concentrate the trades. The Iberian intraday market, however, has a longer delay between the trade and delivery. Consequently, in Nord Pool there is no need for a market between the intraday and tertiary regulation market, which is called the regulating power market in Nord Pool (and the real-time market in the two-step markets). Nord Pool's regulating power market requires activation in 15 min and also is used to meet operating reserves.

Ramp products, akin to proposals for flexible ramping and ramp capability products in the CAISO and MISO markets, respectively, are designed to periodically complement the fast energy market by providing for operational flexibility to meet load more reliably and efficiently, as well as incentivizing the specific resources that provide the flexibility to do so. The ramp product market price can have supplemental payments that are provided only to those resources providing the ramping support. Ramp products therefore reward only the flexible generation and, during these flexibility-scarce periods, do not reward inflexible resources. The ramp capability price would be zero during most hours, when ramping capacity in the energy dispatch mix is sufficient to follow load [11]. When ramping is needed whether due to expected variability, or uncertainty in meeting the net load in future intervals and not provided by the energy market, the price would reflect the marginal cost of providing that ramping capability, incentivizing flexible resources.

To add ramp capability and ensure sufficiently fast response, the Spanish TSO in May 2012 implemented a new market for the management of additional upwards reserves [12]. EirGrid, the TSO in Ireland, also has proposed a new ramping product to respond to imbalances that occur over the minutes-to-hours timeframe, such as from changes in demand, wind generation, and interconnector flows. The TSO anticipates a broad range of resources to supply this service, including wind and photovoltaic (PV) plants that have been dispatched down, conventional generators, storage, and demand [13]. Negative pricing can occur when serving the next increment of demand would actually save the system money; that is, the marginal cost to serve load is negative. For example, negative pricing can occur due to a lack of flexibility within the system.

This might be due to limited transmission capacity creating location-specific negative pricing, minimum generation periods during which resources cannot be shut down, and other reasons. Negative prices also can occur during periods of high variable renewable energy generation and low loads. In general, this can happen either due to resources setting the price with negative cost offers (e.g., due to production credits), or because of reduced capability to reduce generation and increase load (e.g., due to self-scheduled resources). Incorporating negative pricing into market design facilitates balancing and provides a financial incentive to increase system flexibility for several reasons:

- Negative pricing can discourage generators, such as wind (unless tax incentives encourage production), nuclear, and coal from providing too much power when demand is low.
- Negative pricing sends a strong signal to generators to be more flexible and reduce constraints on flexibility. In Denmark, the minimum running capacity of some older coal-fired power plants has been reduced from 30 to 10% of maximum capacity due to dynamic and negative pricing [14].
- Negative pricing can encourage greater diversification in the location and types of variable renewable energy, especially in transmission-constrained areas.
- Negative pricing can encourage the use of storage to absorb excess production, and load to increase demand.
- Negative pricing can provide a transparent mechanism for curtailment of renewable resources via market means rather than out-of-market procedures.

One concern about negative pricing in the United States is that with the production tax credit—which in 2013 offers wind generators a $0.023 subsidy for each kilowatt-hour of energy produced—wind energy can still generate revenue when prices have become negative. They then can offer negative prices representing this"effective"cost of generating. This subsidized bidding can distort the clearing price and impact the rest of the generation fleet. A second concern with negative pricing is that it makes revenue streams more difficult to calculate, and therefore can deter investors from participating in energy markets.

When implementing negative prices, it is important for markets to coordinate with neighbors with respect to the use of administratively defined minimum price levels. At present these minimum price levels differ, for example, between Germany and Denmark, where flows from Germany to Denmark have been observed when Danish prices were negative and extra power was not needed, but German prices were even more negative. For example, this occurred in December 2012, when Danish bids were curtailed to achieve market equilibrium above the minimum price level, but even cheaper German power was imported anyway. Currently, measures are under consideration to avoid this occurrence in future. As already occurs in Denmark, individually negotiated compensation for offshore plants could be designed to eliminate fixed feed-in compensation during hours of negative prices to relieve stress on the power system.

1.3 Ancillary Service Markets

Ancillary services are used to support power system reliability and perform the necessary services that the energy market cannot provide [15]. In the United States, all transmission providers are required to procure ancillary services. The six required ancillary services were defined by the Federal Energy Regulatory Commission's (FERC) landmark rule, Order No. 888 on "Promoting Wholesale Competition Through Open Access Non-discriminatory Transmission Services by Public Utilities: Recovery of Stranded Costs by Public Utilities and Transmitting Utilities" [16]. The functional unbundling of these services was deemed necessary by FERC to ensure that transmission access could be provided in an open and transparent manner. Although the requirement to procure and provide these services is consistent across all wholesale markets, the method of acquisition varies greatly. In non-RTO/ISO regions, these services are obtained and paid for according to a series of FERC-approved rate schedules. In the RTOs/ISOs, most of these services are procured in a competitive manner that is co-optimized with energy markets.

Although much research has focused on how variable renewable resources could increase the need for ancillary services, variable renewable resources also can be used to provide these ancillary services [17–19]. Currently, rules do not allow this provision in most of the ancillary services markets. In Germany, auctions for frequency control reserves occur six days in advance, which effectively precludes wind energy from bidding due to forecasting uncertainties [20]. Variable generation, however, can provide great flexibility. Variable renewable generators can have fast electronically controlled ramp rates, zero minimum generation levels, and no start-up time needs. With increased penetrations, it might be more economical to utilize variable renewable resources to provide these services for both consumers (in terms of reduced production costs) and for variable renewable generators (in terms of increased profits). Kirby et al. [21] describe the provision of ancillary services in some markets by demand response.

Demand-side resources increasingly are providing ancillary services to the grid, in roles that require faster and more verifiable performance than traditional uses of energy efficiency. Demand-side resources long have been employed in ways that only require several hours of lead time, such as "interruptible load" for emergency peak shaving [22] or to increase nighttime load during off-peak price periods. Yet, provision of ancillary services occurs on much shorter timescales, typically seconds to minutes. Such fast-acting demand response is employed in several US wholesale markets including ERCOT, PJM, and MISO [22]. System security requires that such systems ensure rigorous performance characteristics (response time and minimum load size), special contractual and compensation mechanisms, robust measurement and verification methodology, and high-speed communications interface to enable automatic control. As such, industrial sources have predominated in providing ancillary services. Pilot and demonstration projects are underway to aggregate residential and commercial resources to provide ancillary services [23], but significant legal and technical barriers remain to ensure adequate performance characteristics.

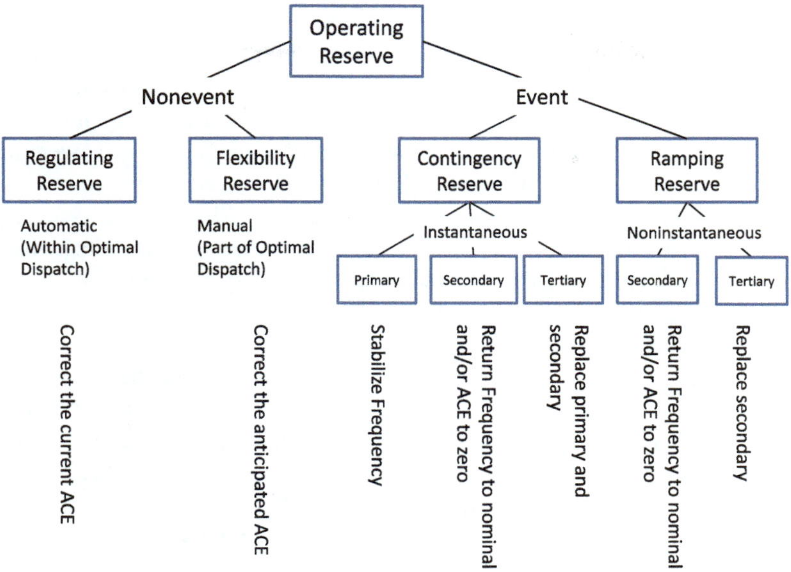

Fig. 1.1 Operating reserve types and their uses [15]

In previous literature, we categorized all active power control services that are ancillary to energy scheduling, also defined as operating reserve, as shown in Fig. 1.1 [15]. The operating reserve types that have existing dynamically priced markets—including synchronized reserve (contingency reserve-secondary), supplemental reserve (contingency reserve-tertiary), and regulation (regulating reserve)—are bought and sold in day-ahead and real-time markets in a similar manner to energy markets. In fact, the US markets that have ancillary service markets currently co-optimize energy and operating reserve when clearing DAM and RTM markets. This means that the markets are cleared simultaneously so that costs and requirements of both markets are considered when clearing the entire market.

Other ancillary services are not sold through dynamic markets. For example, reactive supply and voltage control are needed services both during steady state and disturbances. Reactive power, which supports voltage control, does not travel far due to high inductive impedances. It therefore is very localized which, in turn, inhibits a broad competitive market. In general, all generators except wind plants are required to be able of providing reactive power within a power factor range defined in their interconnection agreement, although in Spain new operating procedures are being studied to require wind turbines to provide voltage control [24]. Compensation for provision of this service varies by transmission provider. In US, there is no requirement to compensate generators for reactive power within the power factor range unless the transmission provider is compensating its own generators. Generators typically are paid for fixed costs as well as opportunity costs; that is, any costs it foregoes in the markets because of constraints on providing reactive power [25].

Other services are much more long term and are cost based. For example, black-start service is needed from generators for system restoration following blackout events. These resources must be capable of starting without outside power supply, able to maintain frequency and voltage under varying load, and able to maintain rated output for a significant period of time (e.g. 16 h) [26, 27]. Many markets will request black-start service proposals and will then have cost-based recovery mechanisms in place for these resources. Other services such as primary frequency response and inertial response currently lack markets or cost-based recovery mechanisms in many markets, which was detailed in [11].

Variable renewable energy lacks inherent inertial response, which helps the system remain stable in the initial moments after a disturbance, before the automatic response by governors. Simulations by the Western Electricity Coordinating Council have shown that frequency response degrades during periods of high wind and low load, when conventional generators comprise a small share of the dispatch mix [11]. The simulations also show that it is technically possible for wind to sufficiently emulate this inertial response by connecting to a power electronic converter; some load and storage also can supply similar capability. Inertia is an inherent part of synchronous generation, therefore it has no added cost other than being online, and so a market similar to the other ancillary service markets, with changing schedules and prices, might not be the best approach. If some resources do provide the service, and others do not, however, then some sort of compensation might be required.

Flexibility reserve, for additional ramping requirements to meet increasing levels of variability and uncertainty, have historically not been an ancillary service market either, but are garnering more interest in some markets.

1.4 Pricing Energy and Ancillary Services

Prices for energy and ancillary services are calculated in similar ways throughout all of the restructured regions in the United States. In US markets, these prices are based on the marginal pricing concept, in which the prices are equal to the bid-based marginal cost to provide each service. Market participant bids are meant to reflect true variable costs, and the marginal pricing design theoretically drives resources to bid their true variable costs. We refer to these prices as LMP and ancillary service clearing prices (ASCPs) for energy and ancillary services, respectively.

Ancillary service markets will also typically follow a pricing hierarchy [28]. The hierarchy will price higher quality reserve services that share the same capacity to be greater than or equal to the lower quality service. This is because some ancillary services are more critical than others, and the incentives provide transparency to market participants on which service they should provide. ASCP may also have locational differences when deliverability issues arise.

Most ASCP payments go to market participants for the provision of capacity to provide ancillary services. The payments usually are not modified based on how the market participant performs the ancillary service, or if the unit was even asked to respond, as long as its performance is satisfactory and the capacity reservation is held (although if deployed, the resource will be paid for the energy deployed with additional energy payments). Recently, there has been motivation to incentivize market participants based on how they performed. FERC Order 755 directs a pay-for-performance scheme for regulating reserve. Resources that provide greater movement and accuracy when providing regulating reserve are compensated more. This is an advantage for participants that can provide regulating reserve faster or more accurately.

Suppliers will be paid the DAM LMP at the DAM energy schedule and the DAM ASCP at the DAM ancillary service schedule. When asked to provide energy or ancillary services differently from the DAM schedule in the RTM, the suppliers will be paid the RTM LMP and RTM ASCP for the difference between the RTM—and DAM—scheduled energy and ancillary services, respectively. In both markets, load pays the LMP and generation is paid the LMP at their corresponding locations. The prices in RTM can change because of changing load, changing VG output, change in committed resource, or change in network topology (i.e., due to transmission outage). The change in RTM prices should incentivize suppliers to adjust schedules accordingly. The introduction of virtual trading (i.e., convergence bidding) should result in the average prices between DAM and RTM to converge, thereby not leading to suppliers, or consumers, to prefer one market than another.

Another important factor to the pricing of energy and ancillary service prices is the administratively-set scarcity prices. Scarcity pricing implies that when demand is very high, the supply may be insufficient and/or costly to deploy to meet the load [5, p. 70]. These price spikes reflect the relative inelasticity of supply (and demand) at high load levels or due to other sources of capacity constraints. Scarcity pricing can be designed to encourage investments in flexible response, such as storage and price-responsive load, because these resources can respond quickly to brief periods of scarcity. Scarcity pricing is favored in some markets on the basis that policy interference in pricing mechanisms, such as through a capacity market, would jeopardize market participants' trust in the market and discourage investors from investing in new capacity.

These pricing methods are designed to incentivize resources to offer their true costs for energy and true capabilities for ancillary services. The RTO/ISO is responsible for solving an optimization problem to minimize the total costs to meet the energy and ancillary service demands while also meeting numerous generation and reliability constraints. This schedule should place each market participant in a position to make the most amount of profit given the prices generated by the market-clearing engine. However, because of issues such as non-convex costs and commitment constraints, it is possible for the RTO/ISO to direct a market participant to provide energy and ancillary services that cause that market participant to lose money. When this happens, the RTO/ISO provides a make-whole payment to ensure that the market participant does not receive a negative profit. After actual power data is measured, resources are paid this make-whole payment in addition to the scheduled payments. Sometimes penalties are in place for market participants that stray too far from their directed energy or ancillary service schedules. These vary depending on the market region but give further incentives to ensure reliable operation.

Scarcity pricing predominantly is found in the European Union, where the policy goal in several states has been to combine scarcity pricing with carbon prices to increase the competitiveness of low-carbon flexible units and use extensive interconnections to balance integrated regions. Nevertheless, the European Union reflects different policy approaches to adequacy, and member state policy actions have yet to create a coordinated market-based approach. The differing approaches to adequacy have complicated cross-border trades, such as those between countries with and without capacity payments [29].

1.5 Financial Transmission Rights Markets

FTR markets, also called transmission congestion contracts and financial congestion rights, are markets designed to hedge the volatility in locational differences of energy pricing [30]. When the transmission system is congested, the load at the receiving side of the constraint would typically pay more for energy than the generators supplying energy at the sending side. This difference is allocated to the FTR holders between the two locations. These FTRs are not part of system operation, because they are purely financial and do not affect the objective of the system operator to dispatch the supply at least cost. Bilateral agreements between supply and demand at different locations can avoid the volatility of pricing between their locations with the purchase of FTRs.

Market participants can obtain FTRs through an RTO-specific allocation process and auctions. Initial FTR allocations are based on historical usage and entities that fund the construction of new facilities. FTRs are typically auctioned at annual, seasonal, and monthly periods. They can also be traded bilaterally. Each auction can include new potential buyers and sellers of FTRs, and it will include a market-clearing engine similar to the one used in the energy market, in which the objective is to minimize the cost of all FTR bids while incorporating the network security constraints. The pricing that results from the FTR auction is performed in a very similar manner to the prices of energy, where in this case the marginal cost of transmission is paid to the seller and taken from the new buyer. Many other characteristics can be included in the FTR market [31]. FTR options are rights in which the owner earns only the locational difference in energy prices if that difference is in their favor. Some markets will have FTRs that are different for on-peak and off-peak periods to signify the differences in transmission flows between these periods. Other areas also have multi-round auctions, in which each round will sell only a portion of the available transmission capacity to FTR purchasers. This is said to make the FTR market more flexible and competitive and allows for the market participants to adjust the bids each round after learning the results from the previous round.

The revenues that FTR holders receive when they own the rights are typically through the congestion costs that occur in the DAM rather than the RTM. The more the prices differ between the DAM and RTM, the more that FTRs may not reflect the true cost of congestion. The congestion patterns are well understood in most markets, although on-peak and off-peak times and transmission outages can certainly affect the outcomes differently than anticipated during the auction periods. Also, at the onset of FTRs it was thought that they could promote future investment in new transmission, but there is a lot of argument about whether FTRs provide sufficient incentives for transmission investment [32]. How these markets may evolve in the future is still very unclear, as is the impact that higher penetrations of VG have on them. However, the scope of this chapter (and book) has only marginal relevance to FTR markets and so we provide little focus on this market product.

1.6 Capacity Markets

Capacity markets are motivated by the desire to employ a market mechanism to ensure that new generation is developed on time to meet resource adequacy targets and help these resources recover their capital costs. Power plants are large, capital-intensive resources that take considerable time to permit and build. The decision to build a power plant must be made well before the plant is needed. Some RTO/ISO regions rely on high and volatile energy prices that are sometimes constrained by administratively-set scarcity prices or price caps. Other RTO/ISOs operate explicit capacity markets to ensure that sufficient generation will be available to meet the expected load. In vertically integrated systems, resource adequacy assessments are carried out by the utility, and any needed additional capacity could be acquired internally or via contract, subject to regulatory oversight. The costs for procuring that capacity are typically subject to rate-making proceedings with state public utility commissions.

Mandatory capacity markets are intended to address long-term reliability needs and ensure that resources have adequate opportunity to recover their variable and fixed costs over time. Capacity markets are often backstop mechanisms that evaluate potential capacity shortfalls after considering bilateral contracts or other power purchase agreements [33].

In Europe, the question of capacity remuneration mechanisms is discussed very differently among the Member States. Conventional power plants (even new flexible gas plants) are being closed or are threatening to close not only because some are at the end of their lifetimes, but in some cases because of changes in fuel prices. As a result, generation adequacy regionally is becoming a matter of concern [4, 29, 34]. Also, limited interconnection capacity, for example in countries such as Spain, has increased interest in capacity payments. In Europe, security of supply is a national question, but over-capacities would occur if solved strictly nationally. Thus, European organizations and associations strongly recommend international coordination [35–38].

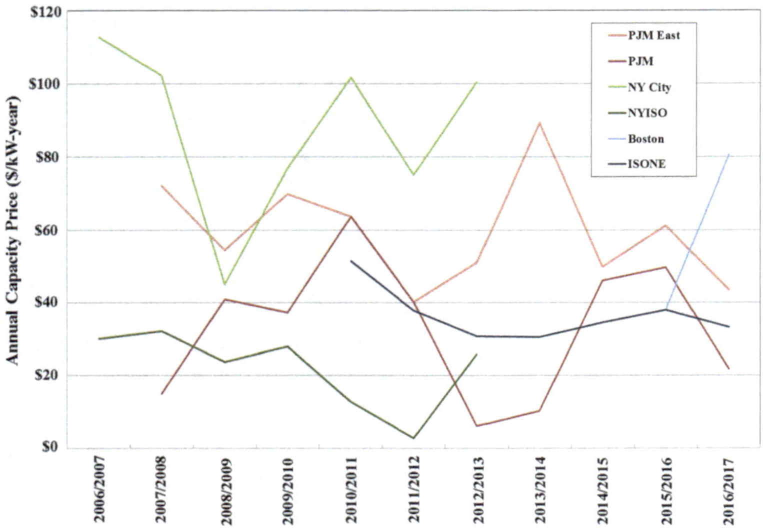

Fig. 1.2 Capacity prices in some RTO/ISOs (adapted from [40])

In the United States, the methods for calculating capacity prices in each of the RTO/ISOs are based on the market design choices of each region. In general, regions with capacity markets find that the capacity prices tend to be limited to the capital cost of a new gas-fired plant that can be sited and built within three years [39]. As shown in Fig. 1.2, prices generated by mandatory capacity markets have been considerably volatile [40]. These results are driven by a variety of market considerations that vary from one region to another.

The demand for capacity is based on an administrative process that determines the total amount of capacity necessary to meet peak load requirements. NYISO, PJM, and ISO-NE all use a downward-sloping demand curve for capacity rather than a fixed target. The downward-sloping demand curve is constructed to reflect the marginal value of capacity to load, and it serves to reduce the potential exercise of market power in capacity auctions. Although the specific demand curve parameters vary between the markets, the main principles are illustrated in Fig. 1.3. The curve is constructed around a target for new capacity at which the price is set equal to the cost of new entry (CONE). The cost of new entry is typically set equal to the annualized capital cost of a new peaking plant (e.g., a combustion turbine), and it may be adjusted for the expected revenue from the energy market (i.e., net CONE). Administered price caps are common and are designed to protect against potential market power and provide a backstop mechanism in case insufficient bids are received from the market.

Resources participating in the capacity markets must verify their capabilities to determine the total capacity they can bid into the market. Each of the mandatory capacity markets has a process for qualifying as a capacity resource. Generally speaking, resources interested in participating in capacity markets must verify their

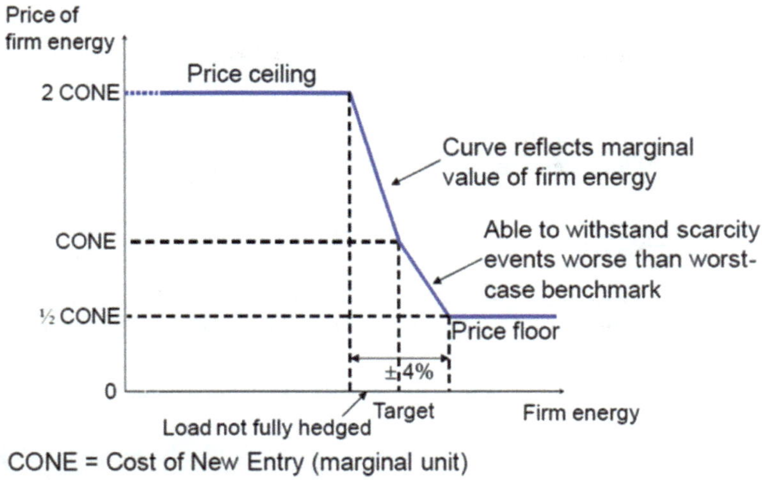

Fig. 1.3 Illustration of demand curve for capacity (based on [41])

operating capability in MW for a specified time period, usually the winter or summer peak. Each organized market has different capacity qualification rules for existing resources, new resources, external resources, demand response, and renewables. Many of the markets will require capacity market resources to offer their capacity in the day-ahead market. Current capacity markets typically do not require capacity resources to have specific attributes other than the provision of capacity during periods of peak demand.

The physical location of a resource is also important for capacity markets. Transmission limitations can limit the ability of a load to access a resource. Local capacity obligations are enforced in each of the markets to ensure that load-serving entities have adequate supply and transmission capacity to deliver energy to an area. The issue is most prevalent in regions with constrained export and import capabilities. Accurately identifying zones that have deliverability constraints is critical to developing efficient capacity markets.

There is no widespread agreement on the need for a capacity mechanism to supplement energy-only markets—and, if the need exists, how best to do it. There also is little, if any, evidence regarding whether scarcity pricing would result in revenue sufficiency for capacity, as illustrated by the current review of options in ERCOT [42]. Because most retail consumers do not see real-time prices that reflect cost, the demand curve for electricity is muted [5, 43]. Proponents for capacity mechanisms argue that this malfunction of the market for electricity, coupled with the lack of ability to differentiate reliability among customers on a widespread basis, renders an energy-only market incapable of providing sufficient forward capacity [41]. This debate is not new, and began long before variable renewable energy sources were significant in the electricity supply.

1.7 The Impacts of Variable Generation on Market Outcomes

The outcomes of each of the markets discussed above may be impacted by the introduction of high penetrations of VG. Possible impacts are briefly discussed in the list below.

- Energy markets:
  - VG can reduce average LMPs because of its low variable costs.
  - VG can cause more occurrences of zero or negative LMP periods because of its variable cost and zero or negative bid-in costs.
  - VG's increased variability can cause LMPs to be more volatile from one time period to another.
  - VG's increased uncertainty can cause greater differences between DAM and RTM LMPs (although on average they are likely to remain converged as a result of virtual trading).
  - VG can cause a greater need for flexible resources in the energy market, and the energy market may or may not provide sufficient incentive for this flexibility.
- Ancillary service markets:
  - VG can increase the requirements for normal balancing reserve, such as regulating reserve, which can increase the ASCP for those services.
  - With higher balancing reserve demands and increased variability and uncertainty, administratively-set scarcity ASCP may be triggered more often, resulting in more frequent extreme price spikes.
  - VG can displace synchronous, frequency-responsive resources, and when not equipped with technology to provide a comparable response, it can cause the need for supplemental actions or market designs to ensure that sufficient frequency response and/or system inertia is available.
  - VG can cause the ancillary service requirements to change from one day to another and from DAM to RTM, if the requirements are based on correcting the variability and uncertainty of VG, which can cause uncertainty in ancillary service demands and changing demands for the same time periods between DAM and RTM, similar to load.
  - VG can cause a need for greater flexibility from the resources that correct for its variability and uncertainty. Certain forms of flexibility may or may not be built into the current ancillary service markets.
- FTR markets:
  - VG's increased variability and uncertainty can cause greater variation on power flow, which causes FTR holders to be uncertain about expected congestion patterns.
  - VG's increased uncertainty can cause greater deviations of power flows between DAM and RTM. Because FTR revenues are typically based on the DAM, there

could be greater divergence between FTR revenues and actual congestion patterns.

- Capacity markets:
  - The reduction in LMP and energy schedules from conventional resources will result in reduced revenues in the energy market. If these resources are still required to be available for short periods of time, more resources become capacity-based rather than energy-based.
  - VG's variability and uncertainty can cause the need for different types of resources to be built and available. In other words, it might require the need to plan and build more flexible resources to prepare for future needs and not to focus on the need for MW capacity alone.
  - VG's variability and uncertainty can cause the need for existing resources to modify their flexible capability potential. Market designs may need to incentivize the existing resources to spend the capital on retrofits to increase the flexible capability that it can provide.
  - Must-offer price rules, designed to limit the ability of buyers to suppress capacity prices by subsidizing relatively higher-cost new capacity to replace lower-cost existing capacity, may increase risk that a resource built to satisfy a state renewable portfolio standard will not clear the capacity market at the applicable minimum offer floor.

References

1. Ela, E., Milligan, M., Bloom, A., Botterud, A., Townsend, A., Levin, T.: Evolution of wholesale electricity market design with increasing levels of renewable generation. Technical Report NREL/TP-5D00-61765, Golden, Colorado (2014)
2. Phillips Jr., C.: The Regulation of Public Utilities: Theory and Practice. Public Utilities Reports, Arlington (1993)
3. Hogan, W.: Competitive Electricity Market Design: A Wholesale Primer. Harvard University, Cambridge. http://www.hks.harvard.edu/fs/whogan/empr1298.pdf (1998). Accessed 25 Oct 2016
4. Miller, M., Bird, L., Cochran, J., Milligan, M., Bazilian, M., Denny, E., Dillon, J., Bialek, J., O'Malley, M., Neuhoff, K.: RES-E-NEXT: Next Generation of RES-E Policy Instruments. International Energy Agency's Implementing Agreement on Renewable Energy Technology Deployment (IEA-RETD) Report, Utrecht, The Netherlands (2013)
5. Stoft, S.: Power System Economics: Designing Markets for Electricity. IEEE Press and Wiley Interscience (2002)
6. ISO New England: Impact of Virtual Transactions on New England's Energy Market. Holyoke, MA (2004)
7. NYISO Technical Bulletin 049: Multi-Pass Methodology of Security Constrained Unit Commitment. The New York Independent System Operator, Rensselaer, New York (2009)
8. Smith, J., Beuning, S., Durrwachter, H., Ela, E., Hawkins, D., Kirby, B., Lasher, W., Lowell, J., Porter, K., Schuyler, K., Sotkiewicz, P.: Impact of Variable Renewable Energy on U.S. Electricity Markets. In: IEEE Power and Energy Society General Meeting, pp. 1–12. IEEE (2010)

9. Kirby, B.: Ancillary Services: Technical and Commercial Insights. Wärtsilä North America Inc., USA (2007)
10. Kiviluoma, J., Meibom, P., Tuohy, A., Troy, N., Milligan, M., Lange, B., Gibescu, M., O'Malley, M.: Short term energy balancing with increasing levels of wind energy. IEEE Trans. Sustain. Energy **3**(4), 769–776 (2012)
11. Ela, E., Tuohy, A., Milligan, M., Kirby, B., Daniel, B.: Alternative approaches for incentivizing the PFR ancillary service. Electr. J. **25**(4), 88–102 (2012)
12. Ministerio de Industria Energía y Turismo: P.O. 3.9: Contratación y Gestión de Reserva de Potencia Adicional a Subir. Boletín Oficial del Estado, **60**, 22513–22520 (2012)
13. EirGrid and SONI: DS3: System Services Consultation - New Products and Contractual Arrangements. EirGrid GRoup, Dublin, Ireland (2012)
14. Blum, R.: Experience with high flexible power plants in denmark, consequences for the power producers and coal-fired power plants. In: EU-China Workshop on Future Flexible Power System for Renewable Energy Grid Integration, Beijing, DONG Energy (2013)
15. Ela, E., Milligan, M., Kirby, B.: Operating reserves and variable generation: a comprehensive review of current strategies, studies, and fundamental research on the impact that increased penetration of variable renewable generation has on power system operating reserves. Technical report NREL/TP-5500-51978, Golden, Colorado. http://www.nrel.gov/docs/fy11osti/51978.pdf (2011). Accessed 25 Oct 2016
16. Federal Energy Regulatory Commission (FERC): Promoting Wholesale Competition Through Open Access Non-discriminatory Transmission Services by Public Utilities: Recovery of Stranded Costs by Public Utilities and Transmitting Utilities. Order No. 888, Washington, D.C., USA. http://www.ferc.gov/legal/maj-ord-reg/land-docs/order888.asp (1997). Accessed 25 Oct 2016
17. Miller, N., Clark, K.: Advanced Controls Enable Wind Plants to Provide Ancillary Services. In: Power and Energy Society General Meeting, pp. 1–6. IEEE (2010)
18. Miller, N., Clark, K., Shao, M.: Frequency responsive wind plant controls: impacts on grid performance. In: Power and Energy Society General Meeting, pp. 1–8. IEEE (2011)
19. Ruttledge, L.; Flynn, D.: System-wide Inertial Response from Fixed Speed and Variable Speed Wind Turbines. In: Power and Energy Society General Meeting, pp. 1–7. IEEE (2011)
20. Holttinen, H., Cutululis, N., Gubina, A., Keane, A., Van Hulle, F.: Ancillary Services: Technical Specifications, System Needs and Costs. Deliverable Report D 2.2, REserviceS, Agreement IEE/11/814/SI2.616374, Technical University of Denmark (2012)
21. Kirby, B., Milligan, M., Ela, E.: Providing minute-to-minute regulation from wind plants. In: The 9th International Workshop on Large-Scale Integration of Wind Power into Power Systems and Transmission Networks for Offshore Wind Power Plants, Energynautics GmbH, Langen, Germany (2010)
22. Pfeifenberger, J., Hajos, A.: Demand Response Review. Presentation to AESO. The Brattle Group, Cambridge, MA (2011)
23. Navigant: Smart Electric Meters, Advanced Metering Infrastructure, and Meter Communications: Global Market Analysis and Forecasts. Boulder, Colorado, USA. https://www.navigantresearch.com/research/smart-meters (2016). Accessed 25 Oct 2016
24. Ministerio de Industria Energía y Turismo: P.O. 7.5: Servicio Complementario de Control de Tensión en el Sistema Eléctrico Espanõl Aplicable al Régimen Especial. draft, pp. 1–12 (2010)
25. Federal Energy Regulatory Commission (FERC): Principles for Efficient and Reliable Reactive Power Supply and Consumption. Staff Report, Docket No. AD05-1-000, Washington, D.C., USA (2005)
26. PJM Interconnection: RTO-Wide Five-Year Selection Process Request for Proposal for Black Start Service. Request for Proposal For Black Start Service, Audubon, PA (2013)
27. Saraf, N., McIntyre, K., Dumas, J., Santoso, S.: The annual black start service selection analysis of ERCOT grid. IEEE Trans. Power Syst. **24**(4), 1867–1874 (2009)
28. Oren, S.S.: Design of ancillary service markets. In: 34th Annual Hawaii International Conference on Systems Sciences Proceedings, IEEE (2001)

29. European Commission: Generation Adequacy. Capacity Mechanisms and the Internal Market in Electricity. Consultation Paper, Brussels, Belgium (2012)
30. Lyons, K., Fraser, H., Parmesano, H.: An introduction to financial transmission rights. Electr. J. **13**(10), 31–37 (2000)
31. Sarkar, V., Khaparde, S.: A comprehensive assessment of the evolution of financial transmission rights. IEEE Trans. Power Syst. **23**(4), 1783–1795 (2008)
32. Oren, S., Spiller, P., Varaiya, P., Wu, F.: Nodal prices and transmission rights: a critical appraisal. Electr. J. **8**(3), 24–35 (1995)
33. Rose, K.: An examination of RTO capacity markets. IPU Working Paper n. 2011-4, Institute of Public Utilities Regulatory Research and Education, Michigan State University (2011)
34. Council of European Energy Regulators: CEER Response to the European Commission Consultation Paper on Generation Adequacy, Capacity Mechanisms and the Internal Market in Electricity. Register number: 65470797015-89, Brussels (2013). Accessed 7 Feb 2013
35. European Union: Making the internal energy market work. Communication from the Commission to the European Parliament, the Council, the European Economic And Social Committee and the Committee of the regions, COM/2012/0663 (2012). Accessed 15 Nov 2012
36. European Wind Energy Association (EWEA): Creating the Internal Energy Market in Europe. A Report by the European Wind Energy Association (2012)
37. Agency for the cooperation of energy regulators (ACER): capacity remuneration mechanisms and the internal market for electricity. Technical report, Pursuant to Article 11 of Regulation (EC) No 713/2009 (2013). Accessed 30 July 2013
38. European Network of Transmission System Operators for Electricity (ENSO-E): Response to the European Commission Public Consultation on Generation Adequacy, Capacity Mechanisms and the Internal Market in Electricity. ENSO-E REport (2013)
39. Federal Energy Regulatory Commission (FERC): Energy Primer: A Handbook of Energy Market Basics. Staff Report, Washington, D.C., USA. http://www.ferc.gov/market-oversight/guide/energy-primer.pdf (2012). Accessed 25 Oct 2016
40. Federal Energy Regulatory Commission (FERC): Centralized Capacity Market Design Elements. Staff Report AD13-7-000, Washington, D.C., USA (2013)
41. Cramton, P., Stoft, S.: Forward reliability markets: less risk, less market power, more efficiency. Util. Policy **16**(3), 194–201 (2008)
42. Newell, S., Spees, K., Pfeifenberger, J., Mudge, R., DeLucia, M., Carlton, R.: ERCOT Investment Incentives and Resource Adequacy, A Report prepared for the Electric Reliability Council of Texas. ERCOT), The Brattle Group (2012)
43. Kirschen, D., Strbac, G.: Fundamentals of Power System Economics. Wiley, Chichester (2004)

Chapter 2
Electricity Markets and Intelligent Agents Part I: Market Architecture and Structure

Fernando Lopes

Abstract The electric power industry has undergone a sweep restructuring resulting in the emergence of electricity markets (EMs) worldwide. The trend towards EMs has led to extensive efforts by the research community to develop optimization and equilibrium models adapted to the new competitive industry. The complexity of EMs calls, however, for richer and more flexible modeling techniques. Agent-based simulation is a relatively new approach relying on advanced social science methods as well as established engineering modeling techniques. The agent-based approach presents itself as a promising approach to accurately model and analyze in detail the behavior of EMs over time. Agent-based simulation of EMs is, at the time of writing, an active area of research and a number of prominent models and systems have been proposed in energy-related journals. These high-quality scientific contributions exhibit fairly different features and make use of a diverse range of concepts. Currently, there seems to be no agreed framework to compare the usage of specific concepts in one contribution with usage in other contributions, nor to compare disparate research efforts. This chapter and its companion (Chap. 3) claim that the development of such a framework can be an important step to provide a coherent set of concepts related to the area, to assess progress in the area, and to facilitate the development of future models and systems. Accordingly, this chapter (Part I) and Chap. 3 (Part II) introduce a generic framework for agent-based simulation of EMs. The complete framework includes three groups (or categories) of dimensions: market architecture, market structure and software agents. This chapter describes, in considerable detail, the components of the first two groups of dimensions, notably the architecture and core structure of power markets. The third and last group of dimensions is the subject of Chap. 3.

F. Lopes (✉)
LNEG–National Laboratory of Energy and Geology,
Estrada do Paço do Lumiar 22, 1649-038 Lisbon, Portugal
e-mail: fernando.lopes@lneg.pt

F. Lopes and H. Coelho (eds.), *Electricity Markets with Increasing Levels of Renewable Generation: Structure, Operation, Agent-based Simulation, and Emerging Designs*,
Studies in Systems, Decision and Control 144, https://doi.org/10.1007/978-3-319-74263-2_2

2.1 Introduction

The electric power industry around the world has evolved to open markets that promote competition among their participants and provide consumers with a choice of services. Restructuring has enabled a paradigm shift away from "cost-of-service" to market-based pricing. Markets for different electricity related products—energy, capacity, reserves, regulation, and transmission rights—have emerged worldwide. Such markets are distributed in nature and may involve a variety of complex transactive techniques, including bidding, auctions, centralized and bilateral market clearing, and electricity pricing.

Typically, competition exists on each side of any market—among suppliers or among demanders, but not between suppliers and demanders—and may result in a lack of market power and "price taking" behavior. Specifically, competition can provide full strength cost-minimizing incentives, force average prices down toward their marginal costs, and encourage real-time pricing for retail customers. Also, competitive forces may drive suppliers to make cost-saving innovations quickly (e.g., efficient repairs and labor saving techniques) and thus operate in more efficient ways [1].

The trend towards electricity markets (EMs) has led to extensive efforts by the research community to develop optimization and equilibrium models adapted to the new competitive industry (see [2] for a comprehensive review). Most optimization models were developed under the implicit assumption of a centralized decision-making process. They consider a single profit maximization program involving an objective function subject to a set of technical and economic constraints (see, e.g., [3]). In contrast, equilibrium models represent the overall market behavior taking into account competition among several participants. They are formulated as a simultaneous profit maximization program involving multiple objective functions (see, e.g., [4, 5]). Two types of equilibrium are often considered. The commonest one is based on the Cournot competition concept, in which participants compete in quantity, whereas the other focuses on the supply function equilibrium approach, in which participants compete in both quantity and price. Both types of equilibrium are based on the well-known concept of Nash equilibrium [6].

The complexity of EMs calls, however, for richer and more flexible modeling techniques [7]. In fact, traditional models are often considered a poor fit to liberalized EMs—operation decisions are decentralized and strategically taken by each market participant to maximize individual profit. Also, equilibrium models are based on a formal definition of equilibrium—typically expressed in terms of a system of algebraic and/or differential equations, often considered very hard to solve. Equilibrium models also impose rigid limitations on the representation of competition between participants and disregard the consequences of learning effects from daily repeated interactions [8]. Simply put, the constant need of market participants to repeatedly probe EMs and adapt their strategies adds complexity that is difficult to represent with conventional techniques, such as standard optimization methods and game theory [9].

Agent-based simulation (ABS) presents itself as a promising approach to accurately model and analyze in detail the behavior of EMs over time and the complex interactions and inter-dependencies among market participants. ABS is a relatively new approach relying on advanced social science methods as well as established engineering modeling techniques (see, e.g., [10]). It does not postulate a single decision maker with a specific objective for the entire system—rather, software agents are allowed to establish their own objectives and apply their own strategies. Conceptually, software agents are computer systems situated in some environment and capable of flexible autonomous action to satisfy their design objectives [11]. Multi-agent systems are essentially loosely coupled networks of agents that interact to solve problems that are beyond the individual capabilities of each agent. Examples of common types of interactions include cooperation (working together towards a common objective) and negotiation (coming to a mutually acceptable agreement) [10, 12].

The agent-based approach is indeed an ideal fit to the naturally distributed domain of a deregulated electricity market. Accordingly, several researchers have paid attention to agent-based simulation of EMs over the last years and a number of prominent models and systems have been proposed in the technical literature. These high-quality scientific contributions exhibit fairly different features and make use of a diverse range of concepts (e.g., electricity pricing and agent architecture), showing performance characteristics that vary significantly depending on the modeling approach. There is a large degree of heterogeneity in describing details of existing models and systems, and also in representing software agents operating in EMs and in performance evaluation techniques. It follows that there seems to be no agreed framework to compare the usage of key concepts in one scientific contribution with usage in other contributions, nor to compare disparate research efforts. We believe that such a framework can be very important and instructive, providing a coherent set of concepts, helping to understand the interrelationships of disparate research efforts, and facilitating the development of future models and systems.

Against this background, this chapter (Part I) and the next chapter (Part II) introduce a generic framework for agent-based simulation of EMs. The complete framework includes the following three groups (or categories) of dimensions:

1. Market architecture.
2. Market structure.
3. Software agents.

The first group, labeled "market architecture", contains three key dimensions: sub-markets (e.g., day-ahead, ancillary service and forward), market types (e.g., bilateral, exchange and pool) and market linkages (e.g., arbitrage, timing and location). The group "market structure" is composed by two distinct, yet interrelated dimensions: market sector (e.g., wholesale, retail, and central coordination and transmission) and market participants (e.g., generating companies, retailers, and large and small consumers). The last group, labeled "software agents", includes two broad dimensions: agent architectures (e.g., model-based, goal-based and learning) and agent capabilities (e.g., autonomy, pro-activeness, social ability and adaptability).

Thus, this chapter and the next are about the common ground between two fields of inquiry: electricity markets and software agents. Both chapters are not meant as a survey of the area of agent-based simulation of electricity markets.[1] Rather, the description of the various components of the conceptual framework is generally undertaken with particular reference to work from power systems engineering, economics, artificial intelligence, and computer science generally.[2] Hence, the two chapters do not present new theorems nor important experimental results, but, instead, they aim at providing both a coherent set of concepts and a comprehensive and systematic basis for objectively comparing and contrasting disparate research efforts. Such a basis can be, we believe, an important step for the development of more sophisticated models and systems because it helps to define the core elements and features required by software agents able to simulate EMs in a realistic way. Given the diversity of approaches and the heterogeneity between existing pieces of work, we also believe that the conceptual framework can help the area to reach a higher level of stability and maturity.

In this chapter, we discuss the first two groups of dimensions of the framework, labeled "market architecture" and "market structure". Specifically, the purpose of this chapter is threefold:

1. to examine the literature on power markets and to identify the main strands of work in this research field;
2. to introduce the first two groups of dimensions of a generic framework for agent-based simulation of EMs;
3. to describe, in considerable detail, the various components of these two groups of dimensions and to discuss the core elements of competitive electricity markets.

In a companion chapter—that is, the next chapter—we introduce the third and last group of dimensions of the framework, labeled "software agents".

The remainder of this chapter is structured so that each of the two groups of dimensions is presented in a separate section. Specifically, Sect. 2.2 discusses the architecture of power markets and Sect. 2.3 deals with the core structure of power markets. Section 2.2 begins by introducing the concept of "market architecture", presenting the key market types and describing some important market linkages. Next, Sect. 2.2.1 discusses the key market model of pool trading, and Sect. 2.2.2 presents the main bilateral contracts and the markets for trading them. Following this, Sect. 2.3 introduces the concept of "market structure", which encompasses both market sectors (Sect. 2.3.1) and market entities (Sect. 2.3.2). Finally, Sect. 2.4 presents some concluding remarks.[3]

---

[1]The agent-based analysis of economic systems, including power systems and electricity markets, is often referred to as agent-based computational economics (ACE). However, throughout the book, we will use the broader term "agent-based simulation".

[2]We will draw from several different research traditions, but our focus will always be on promoting a deeper understanding of existing and emerging market designs to reliably and efficiently manage the potential challenges of variable generation (VG).

[3]Sections 2.2 and 2.3 refine and extend our previous work on a conceptual framework for agent-based electricity markets, presented in [13, 14], respectively.

2.2 Market Architecture

Market architecture includes the list of all submarkets, the type of each submarket and the different linkages between them (see Table 2.1).[4] A power market typically comprises several components or submarkets—it may comprise a few or many submarkets depending on its degree of vertical integration or unbundling. Representative examples of *submarkets* include the day-ahead and real-time energy markets, ancillary-service markets, futures markets and swap markets.[5]

The problem of which particular submarkets to include in a power market is surrounded by many controversies. Perhaps the most fundamental controversy concerns the question of whether the system operator (SO) should operate an energy market or a financial transmission rights market [1]. An interesting point of view holds that the SO is only needed to operate the grid and sell rights to its use, but should minimize the role in the market and refrain from trading or pricing electrical energy. The extreme version of this perspective eliminates both the day-ahead and real-time energy markets, but considers a centralized transmission-rights market. The opposing point of view holds that such an architecture, plausible in a simplified theoretical world, is wholly impractical. At least in real time, the SO needs to buy and sell energy directly and needs to set different prices for energy provided at different locations. Interestingly, the extreme version of this perspective eliminates both the day-ahead and a real-time transmission-rights markets, but states that a real-time locational energy market should be on the list of submarkets.

A power market often includes a mixture of market types, such as a decentralized bilateral market and a centralized exchange or pool. *Market type* classifies markets as bilateral or mediated, with the latter usually more organized.[6] *Bilateral markets* can be, in order of increasing centralization, search, bulletin-board, or brokered markets, while *mediated markets* can be dealer markets, exchanges or pools (see Table 2.1). Generally speaking, there are two basic ways to arrange trades between buyers and sellers. They can trade directly—a single buyer and a single seller can negotiate the terms and conditions of a bilateral contract in such a way that other market participants do not observe the trading—or a supplier can contact an intermediary who may negotiate on behalf of him and possibly sell his product to end-use customers. Public centralized markets tend to have certain advantages over decentralized private markets: lower transaction costs, quicker transactions, greater transparency of price, and easier monitoring. Trading in decentralized private markets can be more expensive but these markets typically provide more flexibility [1].

---

[4]Both the classification presented in Table 2.1 and the description of the several markets presented in the first part of this section are based on [1, Chaps. 1–8], though the section is not intended as a summary of their perspectives on market architecture.

[5]For simplicity and clarity of presentation, and since submarkets are themselves markets, they will often be referred to simply as markets.

[6]The distinction between bilateral and mediated markets is not absolute and the reader may find some overlap between them. Section 2.2.2 examines in detail several types of markets to make the distinctions between them clearer.

Table 2.1 Conceptual framework: the category "market architecture"

Group (or category)	Dimension	Element (or characteristic)
Market architecture	(Organized) submarkets	• Energy market
		o Day-ahead market
		o Intra-day market
		o Real-time market
		• Ancillary-service market
		• Transmission-rights market
		• Capacity market
		• Forward market
		• Futures market
		• Options market
		• Swap market
		• Other markets
	Market types	• Bilateral market
		o Direct-search market
		o Bulletin-board market
		o Brokered market
		• Mediated market
		o Dealer market
		o Exchange (or auction) market
		o Pool market
	Market linkages	• Implicit
		o Arbitrage linkages
		o Spacial linkages
		o Temporal linkages
		• Explicit

In bilateral markets, buyers and sellers often trade directly, although this can be facilitated by brokers. *Direct search markets* are the least organized bilateral markets—here, buyers and sellers should seek each other out directly. An example of a transaction taking place in such markets is the sale of a product in which the seller/buyer advertises for buyers/sellers in specific places. *Bulletin-board markets* are a partially centralized variety of direct-search markets. The next level of organization are *brokered markets*—brokers do not actually buy or sell in a market but are paid a commission for arranging trades. Brokered markets arise when the trading activity in a particular product increases. More specifically, in markets where trading in a product is sufficiently active, brokers can find it profitable to offer search services to buyers and sellers. Put another way: economies of scale in searching for buyers/sellers may make it worthwhile for market participants to pay brokers to help them to conduct some searches. To this end, brokers often develop specialized knowledge on valuing assets traded in such markets [15].

Generally speaking, the economic function of a *broker* is to connect, and possibly negotiate on behalf of, ultimate buyers and sellers. Specifically, a potential seller S notifies a broker of their willing to sell a product at price P_s, and then the broker contacts a potential buyer B and notifies him of the possibility of buying at P_s or inquires about their willingness to buy at some other price. Once the broker negotiates a single price at which S is willing to sell and B is willing to buy, then the transaction can be executed [16].

Bilateral markets are flexible, since the trading parties may specify some contract terms. However, this flexibility comes frequently at a price—negotiating and writing contracts could be expensive. Accordingly, it is often advantageous to move toward more standardized and centralized trading when this is made possible by the volume of trade. *Dealer markets* are the most rudimentary type of mediated markets. Dealers, unlike brokers, trade for their own account, and usually maintain an inventory—they buy a product and hold it before reselling. There are no brokerage fees, but at any point in time dealers buy for a price that is lower than the price they sell for. The difference is called the spread [1]. Thus, the spread between the buy (or "bid") and sell (or "ask") prices is a source of profit.

Dealer markets save traders on search costs because market participants can easily look up the prices at which they can buy from or sell to dealers. The economic function of a *dealer* is to make-a-market so that buyers and sellers can readily buy and sell in the liquid market. A potential buyer B or seller S contacts a dealer (e.g. by using an electronic trading platform) and inquires him about a specific bid or offer quote. If the quote is acceptable, B can execute a trade at the offer price (called lifting an offer) or S can sell at the bid price (hitting a bid). In either case, the dealer is the counter party [16].

Energy exchanges utilize auctions and are sometimes called *auction markets*—all traders converge at one place (either physically or "electronically") to buy or sell a product (e.g., electricity). Auction markets are the most integrated type of markets. An advantage of these markets over dealer markets is that one need not search across dealers to find the best price for a product. If all participants converge, they can arrive at a mutually agreeable price and thus save the spread [15]. Also, exchanges have a number of advantages over bilateral markets [1]: they can reduce trading costs, increase competition, and produce publicly observable prices. Furthermore, exchanges can operate much nearer to real time than bilateral markets, making them the obvious choice for both day-ahead and real-time energy markets. However, bilateral markets often provide more flexibility and, weeks in advance, may play a larger role than exchanges.

Pools are exchanges in which the supply bids are complex, and the system operator carries out complex calculations to select and pay the winners. Complex bids often attempt to comprise a complete economic description of the generation process—that is, generators can include in bids their marginal costs, expected start-up and no-load costs, etc. Thus, pools are designed to accept multipart bids—generators bid their marginal costs and other costs and limitations into a pool which computes the energy prices and the accepted bids.

Energy exchanges may accept simple bids expressing only energy quantities and prices, meaning that generators could not account for their start-up and no-load costs directly, i.e., within the bid format. As a result, they may need to manipulate their bids and adopt gaming strategies to avoid losses. Gaming strategies are inherent in market designs requiring bidders to manipulate their bids to account for factors that cannot be expressed directly within the bid format. To avoid this problem, two-part bids, three-part bids or even more complex bids may be necessary. Alternatively, pools can be used. As noted, pools typically accept complex, multipart bids (though exchanges can also use multipart bids). Nevertheless, the competitive forces present in energy markets can induce traders to represent true costs as accurately as possible within the bid format. Accordingly, even in exchanges with one-part bids, where considerable manipulation may be necessary, truly competitive markets may do a remarkable efficient job of dispatching generation [1].

Furthermore, energy exchanges often accept bids that, according to their bid-in values, at least break even. In contrast, pools may accept some apparently losing bids—that is, bids that are found to lose money because the pool prices are not enough higher than the marginal costs to cover other bid-in costs. Some accepted bids that would otherwise lose money are compensated with a "make-whole" side payment. Thus, pools are defined by the existence of side payments (exchanges do not make side payments).

Market linkages are very important to the functioning of power markets. Linkages can be either implicit or explicit [1]. Arbitrage, timing and location are the keys to most naturally arising implicit linkages. Specifically, implicit linkages are often produced by arbitrage, the most important example being the approximation between a forward price for delivery at a specific period and the expected spot price at that period (i.e., the price in a forward energy market approximates the expected price in the spot market). Also, because energy markets are geographically distributed, many of their submarkets are multi-product markets and contain a vast array of internal (spacial) linkages. Furthermore, the market architecture establishes a temporal order for submarkets, and that order causes implicit (temporal) linkages to develop between them. Explicit linkages are often limited only by the imagination of market designers (representative examples are explicit rules linking rights purchased in one submarket to activities in another).

At this stage, the author wishes to note that three key market models (or types) have emerged [17]: (i) pool trading, (ii) bilateral trades or contracts, and (iii) a hybrid model. *Pool trading* is carried out through a centrally operated entity that determines generation levels and prices based on submitted generation bids and load offers. *Bilateral trades* are defined by privately negotiated bilateral contracts that can be either physical or financial obligations. *The hybrid model* combines several features of pools and bilateral contracts. All models have their merits and shortcomings, and their application has been extensively discussed and analyzed in the literature (see, e.g., [18–22]). Practically all real-world electricity markets have one of the first two basic models as the predominant structure and thus pools and bilateral contracts will receive the preponderance of our attention in the next two subsections.

2.2.1 Pool Trading

Electricity is typically bought and sold through a two-settlement system involving a *day-ahead market* (DAM) and a *real-time market* (RTM), also known as a balancing market.[7] The DAM clears to meet bid-in load demand for an entire day divided in periods of an hour or half an hour, one day in advance. To maintain reliability, the production and consumption of electric power must be balanced in real-time. Accordingly, the RTM sets prices and schedules to match the imbalances caused by the variability and uncertainties present in power systems. Many power markets also have intermediate scheduling and pricing procedures on the hour ahead or a few hours ahead to facilitate balancing in advance of real time.[8]

The pricing mechanism of most day-ahead markets is founded on the marginal pricing theory. There are two main variations of marginal pricing [23, 24]:

- System marginal pricing (SMP).
- Locational marginal pricing (LMP).

In markets operating on the basis of the *system marginal pricing*, generators compete to supply demand by submitting bids in the form of price and quantity pairs, for example. These bids are ranked in increasing order of price, leading to a supply curve. Similarly, retailers and possibly other market participants submit offers to buy certain amounts of energy at specific prices. These purchase offers are ranked in order of decreasing price, leading to a demand curve. The market clearing price (or system marginal price) is defined by the intersection of the supply curve with the cumulative demand curve. This price is normally determined on an hourly or half-hourly basis and applied to all generators uniformly, regardless of their bids or location. Generators are instructed to produce the amount of energy corresponding to their accepted bids and buyers are informed of the amount of energy that they are allowed to draw from the system [19]. SMP does not explicitly take into consideration transmission constraints, nor does it explicitly accounts for ancillary services.

Locational marginal pricing is a more complex variation of marginal pricing—as in SMP, the system operator collects generator bids and load offers, and then determines the optimal generation dispatch.[9] The difference, however, is that the optimization process is now subject to several system constraints, such as voltage limits, and can even include the supply of losses and other ancillary services necessary to support system operation. Typically, the SO runs an optimal power flow procedure that defines the energy prices at different locations in the system—that is, the marginal cost depends on the location where the electrical energy is produced or consumed.

[7]Chapter 1 presents an introduction to energy markets, and also ancillary service markets, financial transmission rights markets, and capacity markets. The reader is therefore referred to it for details.

[8]These markets are often referred to as *intra-day markets*.

[9]The *dispatch* represents essentially a set of instructions from the SO regarding the operation and control of a power system, especially with respect to defining the generators that provide power at any point in time and their output levels [1].

The format of costs and bids is a topic of central importance in electricity markets. As mentioned above, generators can submit simple bids consisting of price/quantity pairs. However, they often also submit bids in the form of curves of $/h versus MW output.[10] Most literature on EMs considers piecewise linear or quadratic functions, relating money and electric power, to represent costs and bids (see, e.g., [19]). Piecewise linear curves are usually considered compatible with the physical characteristics of electricity generators. Specifically, generators with multiple units are well approximated by these curves, since there is a jump in cost each time a unit is turned on or off, and then a gradual increase as individual units are ramped up or down. However, jumps make calculus-based analysis difficult and can result in inconsistencies. Quadratic curves provide smooth dispatch, revenue, and profit curves that facilitate calculus-based analysis. Yet they are not a perfect characterization of generators' cost structure. Clearly, the implications of the supply bid format for the operation of EMs is a topic that requires further research [25].

Market power is the ability to alter profitably prices away from competitive levels [1]. It is also a central topic in EMs. Generators can exercise market power by withholding their output as well as by manipulating production so as to cause network congestion. Specifically, a high concentration of ownership in a specific region can enable generators to exercise market power by restricting production and raising prices—they can profitably maintain prices above competitive levels by restricting output below competitive levels. Also, generators can exercise market power by increasing production, lowering prices, and exploiting feasibility constraints in electrical networks to foreclose competition and increase profits [26].

The problem of market power can be addressed in several ways, including [1]: (i) forward contracts and obligations of suppliers, and (ii) uncertainty of demand which causes supply-curve bidding. Long-term forward contracting is often considered an effective form to reduce market share. In contrast, medium-term contracts work only to the extent that suppliers do not believe forward contract prices equal the average level of recent spot prices. Demand uncertainty causes suppliers to enter elastic supply-curve bids, which can increase the elasticity of residual demand as seen by other suppliers and decrease their market power.

Yet another topic of central importance in EMs is *price volatility*. The need to maintain a real-time balancing between generation and demand at every bus, on a second to second basis, and also the lack of practical ways to effectively store electricity, mean that energy prices are more volatile than those of other commodities. As a result, market participants are normally exposed to a significant price risk. To hedge against price uncertainty, generators and loads may choose to enter into bilateral contracts that typically consider prices valid over a longer time horizon than the spot-prices.

---

[10]Generating companies can also submit more complex bids for each of their units, reflecting the cost characteristics of each unit (including marginal, start-up and no-load costs) as well as some technical parameters (e.g., minimum and maximum output). Rather than simply stacking the bids, the market performs complex calculations (e.g., unit commitment calculations) to determine the production schedule and the prices for an entire day.

2.2.2 Bilateral Trading

A *bilateral market* is a market in which private parties, sellers and buyers, trade directly at negotiated prices and conditions. Some markets are mediated or conducted by brokers and dealers., i.e., trades may also be arranged by brokers and dealers. A *bilateral contract* is an agreement between two parties for the exchange of electricity under mutually acceptable terms, including starting date, ending date, price, amount of traded energy, and any other terms which may be deemed applicable. Simply put, bilateral contracts are contracts used to make trades between two parties [1].

Two major types of bilateral contracts can be distinguished: physical and financial. *Physical bilateral contracts* mean that all the power transacted bilaterally should be self-generated and self-consumed at a pair of specific network buses—that is, they specify the parties that generate and consume the power, the buses of injection and consumption, as well as the amount of power agreed to. Selling generators have the obligation to produce the power to supply at least all of its physical contracts, while loads are expected to consume at least all of its contracts. *Financial bilateral contracts*, on the other hand, mean that the power transacted need not be self-generated nor self-consumed but could be transferred up to the short-term market-clearing time to another entity such as the pool. In these contracts, the points of injection and consumption may or may not be defined, and if known, they are not binding. The selling side of a contract is free to appoint any market participant willing to supply the electrical energy, while the buyer can also resell the contract further.

Bilateral contracts can also be firm or non-firm. *Firm bilateral contracts* are contracts in which delivery is unconditional—that is, sellers unable the fulfill their contractual obligations must buy the missing amount of electrical energy from the system (or buyers who cannot take full delivery must sell the amount in excess to the system). Typically, imbalances are liquidated at the spot price on the date of delivery. By contrast, *non-firm bilateral contracts* are contracts with conditional delivery, meaning that they are exercised only if their holders decide that it is in their best interests to do so [19].

Generally speaking, the network usage resulting from bilateral contracts may need to be approved by the system operator. For firm bilateral contracts, the SO may need to confirm that the full amount of approved power could be scheduled, except in cases of emergency. In order to withdraw from such contracts, or even curtail them, the parties may need not only the consent of each other, but also a permission from the SO. On the other hand, non-firm bilateral contracts are not guaranteed, and would be scheduled only as the operation conditions allow.

Derivatives are contracts whose values depend on (or derive from) the values of other, more basic, underlying variables, typically the prices of traded assets [22]. The term derivative comes from how the price of a contract is derived from the price of some underlying commodity, security or index or the magnitude of some event. It is used to refer to a set of financial instruments that includes forwards, futures, options and swaps [16].

2.2.2.1 Forwards, Futures, Options and Swaps

Forward bilateral contracts are agreements to sell or buy a specific amount of electricity at a certain future time for a specific price [22]. One of the parties assumes a *long position* and agrees to buy the energy on a future date for a predetermined price, and the other assumes a *short position* and agrees to sell the energy on the same date for the same price. The payoff from a long position in a forward contract on one unit of electricity is the difference between the spot price (SPr) at maturity date and the delivery price (DPr).[11] Similarly, the payoff from a short position is the difference: $DPr - SPr$. These payoffs can be positive or negative. If enough sellers and buyers are interested in trading electricity in advance of delivery, a *forward market* for energy will develop. This market is essentially a decentralized market in which electricity is sold using forward bilateral contracts. The delivery time can range from days to years in the future [1].

Often forward bilateral contracts use standard terms making possible to resell them. The establishment of a secondary market where the trading parties can buy and sell standardized forward contracts helps them to manage their exposure to fluctuations in the spot price [19]. The price at which forward contracts are traded—the resale price—may be higher or lower than the price agreed by the originators of the contracts. Interestingly, two parties may also negotiate some terms and conditions of customized (or tailored) forward contracts, particularly long-term contracts, designed to cover the delivery of large amounts of power over longs periods of time. Such contracts are very flexible since the trading parties can specify any terms and conditions they desire.[12]

Clearly, participation in a secondary market does not have to be limited to physical participants—those entities who produce or consume electrical energy. Specifically, parties that cannot take physical delivery may want to buy contracts for delivery at a future date, in the hope of being able to sell them later at a higher price. Similarly, they can sell a contract first, hoping to buy another one later at a lower price. These parties are often referred to as speculators—individuals who take a position in the market. Since these contracts are not backed by physical delivery, they are called future contracts. Like a forward contract, a *future bilateral contract* is an agreement between two parties to buy or sell energy at a certain time in the future for a specific price [22]. A *futures market* is essentially a market for financial instruments conditioned on delivery at a particular time and place. Both futures and forward contracts are firm contracts [19]. Futures differ from forwards, however, in their standardization, exchange trading, margin requirements, and daily settling [1].

---

[11] The maturity date is the end of the life of a contract [22].

[12] Note that customized (or tailored) long-term contracts are typically negotiated in bilateral marketplaces outside organized markets. For the sake of clarity, and also simplicity in exposition, the category "market architecture of the conceptual framework considers organized submarkets (or markets) only (see Table 2.1). Future work aims at extending the framework by considering marketplaces for negotiating the terms and condition of different types of customized (or tailored) long-term bilateral contracts.

Options are contracts enabling their holders to buy or sell energy at a specified price [22]. There are two types of options: calls and puts. A *call option* gives the right to buy a given amount of energy by a certain date for a specific price. Similarly, a *put option* gives the holder the right to sell a given amount of energy at a certain price. The price in an option contract is known as the *exercise price* or *strike price* and the date as the expiration date. Although options give the holder the right to do specific actions, he/she does not have to exercise this right. The holder of a forward or future contract, by contrast, is obligated to buy or sell the underlying quantity of energy. This is a fundamental difference between options and forwards or futures. Also, options offer a way for investors to protect themselves against adverse price movements in the future while still allowing them to benefit from favorable price movements. Additionally, options involve the payment of an up-front fee—there is a cost to acquire an option, whereas it costs nothing to enter into a forward or future contract. As noted, options are non-firm contracts—that is, they are exercised only if their holders decide that it is advantageous to do so. European options can be exercised only on the expiration date, while American options can be exercised at any time up to the expiration date.[13]

Swaps are used to lock in a fixed energy price for a predetermined—though not necessarily constant—energy quantity.[14] Several major types of swaps can be distinguished, notably vanilla, variable volume, differential, participation, double-up, and extendable. A *vanilla swap* is a contract in which the parties exchange a floating energy price for a fixed energy price [27]. The contract defines the fixed quantity of energy over a specific period of time. The buyer or receiver of the swap pays a fixed price and receives the floating price either by receiving the cash value of the spot energy or the spot energy itself. The swap provider receives the fixed price and either supplies the spot energy or its cash equivalent. The contract is settled at predetermined regular intervals over its period—typically monthly, quarterly, semi-annually or annually.[15]

A *variable volume swap* is essentially identical to a vanilla swap except that the underlying energy quantity is not known in advance. Also, a *differential swap* is similar to a vanilla swap except that the parties exchange the difference between two different floating prices—the differential—for a fixed price differential. In a *participation swap*, the fixed price payer is fully protected when prices rise above the agreed fixed price, but they participate in a certain percentage of the savings if prices fall. For a *double-up swap*, the fixed price payer can achieve a better swap price than the market price but in return the swap provider has the option to double the energy quantity before the pricing period starts. An *extendable swap* is a contract similar to a double-up swap except that the swap provider has the option to extend the period of the contract for a specified period of time [27].

---

[13]The terms American and European do not refer to location. This means that some options traded on North America can be European.

[14]Swaps are also known as *contracts for difference* (CFDs) [27].

[15]The predetermined regular intervals over the period of a contract are often referred to as *settlement dates* or *reset dates*.

Caps provide price protection for buyers above a certain level—the cap price. *Floors* guarantee the minimum price that will be paid or received at a certain level (the floor price).[16] A generic cap may be viewed as a portfolio of European call options with strike prices equal to the cap level and maturity dates equal to the settlement dates of the cap. Similarly, a generic floor is essentially a portfolio of European put options with strike prices equal to the floor level and maturity dates equal to the settlement dates of the floor. *Collars* are combinations of long positions in caps and short positions in floors.[17] Specifically, a generic collar is a portfolio of a long position in a cap and a short position in a floor. Standard caps, floors and collars are for specific energy quantities and usually cash settled at regular intervals over the period of the contracts [27].

Buyers and sellers of electrical energy may eventually be obliged to trade solely through centralized markets—that is, they may not be allowed to enter into bilateral agreements to reduce their exposure to price risks. In such situations, market participants may want to resort to contracts for difference that operate in parallel with a centralized market. A *contract for difference* (CFD) is a bilateral contract in which the purchaser pays the seller the difference between the contract price—the *strike price*—and some market price, usually the spot price [1]. The trading parties agree on a strike price and an energy quantity and then take part in a centralized market. They sell and buy their power through the pool, at the pool marginal price, and then in a separate financial transaction compensate each other for the difference between the strike and actual prices. Specifically, a CFD is settled as follows [19]:

- If the strike price is higher than the market price, the buyer pays the seller the difference between these two prices times the agreed amount of energy.
- Conversely, if the strike price is lower than the market price, the seller pays the buyer the difference between these two prices times the amount agreed.

Thus, as noted, a CFD can be described as a combination of a call option and a put option with the same exercise price. Unless the market price is exactly equal to the strike price, one of these options will necessarily be exercised.

2.2.2.2 Derivatives Exchanges and Over-the-Counter Markets

Derivatives are traded in two kinds of markets [16]: exchanges and over-the-counter (OTC) markets. *Derivatives exchanges* exist for a long time and are essentially markets where individuals trade standardized contracts, defined by the exchanges [22].[18] Once two trading parties agree on a trade, it is handled by an exchange clearing house. Specifically, a clearing house takes care of the credit risk by requiring each of the traders to deposit funds (known as the margin).

---

[16] In some markets, caps are sometimes known, together with floors, as *one-way CFDs*.

[17] Collars are also referred to as *two-way CFDs* [27].

[18] A (derivatives) exchange denotes a market for financial instruments, such as forwards and futures. In contrast, a (power) exchange denotes a centralized market in which supply bids and demand offers are aggregated to find a clearing price at which supply and demand are equal [1].

Traditionally, derivatives exchanges used an *open outcry system* involving traders physically meeting on the floor of the exchanges, shouting, and using a complicated set of hand signals to indicate the trades they would like to carry out. Open outcry has its advocates, but, as time passes, has become less and less used. Currently, exchanges have largely replaced this system by *electronic trading platforms* that automatically match the bids and offers from market participants. Futures are normally traded on exchanges and options can be traded both on exchanges and in over-the-counter markets [22]. The trading of these derivatives on exchanges is usually conducted through brokers and not dealers [16].

Electronic trading involves the use of computer programs to initiate trades, often without human intervention, and has become an important feature of derivatives markets. Market participants submit bids to sell energy and offers to buy energy in a computerized marketplace. The prices and quantities submitted may be observed by all participants but they typically do not know the identity of each other. When a participant submits a new bid, a computer program checks for a matching offer for the period of delivery of the bid. In case it finds an offer whose price is greater than or equal to the price of the bid, a deal is struck and the associated price and amount of energy are displayed to all participants. If no match is found, the bid is added to the list of outstanding bids and will remain there until either the submission of a matching offer, or the withdrawal of the bid, or the closing of the market for the delivery period. A similar procedure is used when a new offer to buy energy is submitted to the computerized marketplace. A flurry of trading activity often takes place in the instants before the closing of the market as participants fine-tune their proposals ahead of the delivery period [19].

Over-the-counter markets are markets where buyers and sellers agree to derivatives transactions without involving exchanges [22]. The parties can clear over-the-counter (OTC) trades either bilaterally or by presenting them to central counter-parties (CCPs). In a bilaterally-cleared OTC market, the parties usually sign an agreement covering all transactions.[19] Typically, this agreement covers various issues, such as the circumstances under which outstanding transactions can be terminated, how settlement amounts are determined in the event of a termination, and how the collateral that should be posted by each party is calculated. The standard OTC transactions that are not cleared bilaterally are cleared through CCPs. These are essentially clearing houses that perform much the same role as exchange clearing houses. For instance, in a forward contract where buyer B reaches an agreement to buy a specific amount of energy from seller S in one year for a certain price, the clearing house (CCP) agrees to: (i) buy the energy from S in one year for the agreed price, and (ii) sell the energy to B in one year for the agreed price. Thus, it takes on the credit risk of both B and S. Forward contracts are traded in over-the-counter markets and can be customized if necessary. Swaps are also over-the-counter agreements.

[19]The agreement often includes an annex, referred to as the credit support annex, requiring each of the parties, or both, to provide collateral. Practically speaking, the collateral is similar to the margin required by exchange clearing houses [22].

Over-the-counter derivatives markets are organized along three main lines [16]: "traditional" dealer markets, electronically brokered markets and proprietary trading platform markets.[20] Traditionally, participants in OTC derivatives markets contacted each other by phone and email, or found counter-parties for their trades using inter-dealer brokers [22]. These individuals or institutions acted mainly as market makers by maintaining bid and offer quotes to market participants.[21] The trading process of negotiation was generally conducted over the telephone, although it could be enhanced through the use of electronic bulletin boards.[22]

Electronically brokered markets are characterized by the use of electronic brokering platforms (also known as electronic brokering systems). These platforms are similar to the electronic trading platforms used by exchanges and can automatically match bids and offers to execute trades. Put simply, an electronic brokering platform is essentially a system in which market participants can submit bids and offers, and observe other participants entering bids and offers, and then observe as the quotes are matching according to an algorithm and then executed [16]. Typically, the firm operating the platform acts only as a broker and does not take a position (or act as a counter-party to any trade).

Proprietary trading platform markets are mainly a composite of traditional and electronically brokered markets, where OTC derivatives dealers set up their own proprietary electronic trading platforms. These are dealer platforms—that is, bids and offers are posted exclusively by dealers and other market participants can execute trades by signaling the acceptance of posted bids and offers. Electronic trading, or dealing, platforms are often described as one-way multilateral environments because only dealers' quotes are observable and those of other participants might at best be inferred from changes in execution prices. In these platforms, the dealer is the counter-party to every trade and holds the credit risk in the market [16].

2.3 Market Structure

The concept of market structure appeared as part of the "structure-conduct-performance" paradigm of industrial organizations in the early fifties. More recently, Stoft [1] advocated that market structure refers to market properties closely tied to technology and ownership. Accordingly, the cost structure of a power market describes both the costs of generation and the costs of transmission. Also, two key determinants of the competitiveness of a market's structure are supply concentration and elasticity of demand.

[20] Although it is useful to distinguish among these three key terms, we note that they are not used consistently in the extensive literature on derivatives markets.

[21] Throughout this book, *bid quotes* and *offer quotes* are expression of prices at which agents are willing to sell or buy, respectively.

[22] *Electronic bulletin boards* are systems whereby dealers or all market participants can post bids and offers, but they are not matched or executed [16].

Table 2.2 Conceptual framework: the category "market structure"

Group (or category)	Dimension	Element (or characteristic)
Market structure	Market sectors	• Wholesale
		o Regulated (no competition)
		o Deregulated (full competition)
		• Retail
		o Regulated (no competition)
		o Deregulated (full competition)
		• Central coordination and transmission
	Market entities	• Generating companies
		• Retailers
		• Power marketers
		• Market operator
		• Independent system operator
		• Transmission companies
		• Distribution companies
		• Aggregators
		• Consumers
		o Large consumers
		o Small consumers
		• Other market entities (e.g., regulators)

In this chapter, however, *market structure* encompasses both market sectors and market entities (see Table 2.2). A power market is typically divided into several sectors (or structural components), notably a wholesale sector, a retail sector and a central coordination and transmission sector. *Wholesaling* is a regulatory term of art and essentially means sales to resellers.[23] The wholesale sector may be completely regulated—that is, electricity prices may be set by a regulator or a central government—or fully deregulated—electricity prices may be set in a competitive wholesale market.[24] In other words, there may be either no wholesale competition or full wholesale competition.[25] *Retailing* is basically sales to final customers and may include several commercial functions. Similarly to the wholesale sector, the retail sector may be completely regulated or fully deregulated (i.e., electricity prices may be set in a competitive retail market).[26] See Sect. 2.3.1 for a more in-depth discussion.

[23] Strictly speaking, wholesaling also means sales to large customers, since they are often allowed to purchase electrical energy directly on wholesale markets.

[24] A *wholesale market* is a market where competing generators offer their electricity output to resellers and possibly other market participants.

[25] The level of competition may be viewed as forming a spectrum in itself (i.e., it is not necessarily binary in nature).

[26] A *retail market* exists when end-use customers have the possibility to choose their suppliers from competing electricity retailers and possibly other market participants.

The trading coordination and transmission sector is a central component of new and emerging electricity markets. It may include several market segments represented by different entities, notably transmission owners (TOs), ancillary service providers (ASs), scheduling coordinators (SCs), the power exchange (PX) and the independent system operator (ISO) [28]. Existing markets have adopted different structures for this sector—that is, a variety of structures have been proposed, considered, and experimented with in different countries around the world. Accordingly, the key market entities representing the sector and their responsibilities and activities vary widely. Srivastava et al. [29] present an overview and a comparative study of five existing electricity markets. The authors point out, for instance, that there is no PX in the Electric Reliability Council of Texas (ERCOT), nor SC in the power pool of Alberta. Also, the Pennsylvania-New-Jersey-Maryland Interconnect (PJM) in USA is an example of a merged ISO/PX, whereas the markets of California, Norway and Alberta are examples of separated ISO and PX.[27]

The daily operation of a power market involves many buyers and sellers of electrical energy as well as many other companies and organizations that function independently. *Market entities* are the entities that take part in a market—here, they are categorized into the independent system operator (and the market operator) and the several market participants. Representative examples of *market participants* include generating companies (GenCos), retailers (RetailCos), power marketers, transmission companies (TransCos), distribution companies (DistCos), aggregators, and consumers (see also Sect. 2.3.2, below).

2.3.1 Competition and Market Sectors

The electric power industry throughout the world is physically and operationally similar. The *physical functions* of the industry are generation, transmission, distribution, and system operation [18]. Electricity is most often generated at power stations, transmitted at high-voltages to multiple substations located near populated areas, and distributed at medium and low-voltages to end-use customers. Electric power transmission and distribution are transport functions—electricity transmission involves a power network and the flow can typically reverse, whereas electricity distribution is usually radial and, traditionally, the flow is one-way (to the customers). The transport system requires constant attention and needs to be managed on a continuous, real-time basis. System operation is basically the function that coordinates the generating plants with the instantaneous usage of end-use customers to maintain a stable transmission system.[28]

[27] A detailed comparative analysis of the central sectors of existing electricity markets is well beyond the scope of this chapter (but see, e.g., [29] for more details).

[28] Generally speaking, generation, transmission and distribution account for about 35–50%, 5–15%, and 30–50% of the final cost of electricity, respectively [18]. These percentages are given here as a general indication only, since the cost of the main functions of the electric industry in different countries varies widely.

Opening up electricity production to competition is an important tool to improve the efficiency of the electric power industry. Competitive forces can drive producers to innovate and operate in more efficient and economic ways. Innovation can lead to lower prices and a better use of energy resources. Practically speaking, competition in the electricity industry means competition in the production of electricity and also in the commercial functions of wholesaling and retailing [18]. The transport functions—transmission and distribution—are natural monopolies and need to be regulated. However, market participants have open access to transmission and distribution lines, so as to freely engage in electricity trades between any two points in a power system, with no discrimination in the opportunity to use the lines or in the cost to use them. The separation of transmission ownership from transmission control ensures a fair and nondiscriminatory access to transmission services. Furthermore, system operation is also a monopoly. An independent entity called the system operator maintains the security of the system operation, coordinates maintenance scheduling, and has a role in coordinating long-term planning [17].

Four models of competition are commonly described in the literature [18, 19]. The models chart the evolution of the electricity industry from regulated monopolies to full competition. They are as follows:

- *Monopoly*: corresponds to the traditional monopoly utility integrating either: (i) generation, transmission and distribution of electricity, or (ii) generation and transmission and selling energy to local monopoly distribution companies.
- *Single buyer or purchasing agency*: allows independent power producers to sell to a single utility (purchasing agency). Also, distribution companies can purchase energy for their customers from the purchasing agency.
- *Wholesale competition*: allows distribution companies and large industrial customers to purchase electricity directly from competitive producers.
- *Retail competition*: gives choice of supplier to all customers.

All of these models assume continued monopoly over transmission, distribution, and system operation. However, they consider progressively more choice.

In particular, the wholesale competition model considers the existence of a wholesale electricity market taking the form of a pool or of bilateral transactions. It introduces competition between generating companies because the wholesale price is determined by the interplay of supply and demand. Distribution companies and large consumers can purchase electricity from generating companies. At the wholesale level, the only functions that remain centralized are the spot market and the transmission network. At the retail level, the system remains centralized—each distribution company operates a distribution network and also purchases electricity on behalf of the consumers located in its area. Retail prices remain regulated and consumers cannot choose their supplier from competing electricity suppliers [19]. The main advantage of this model is to consider competition in production, which is where most of the benefits are. However, the model suffers from several problems, notably boundaries, the need for contracts, and the difficulties of making these contracts [18].

The retail competition model considers both a wholesale market and a retail market (see Fig. 2.1). All generation is deregulated and sells into a competitive wholesale

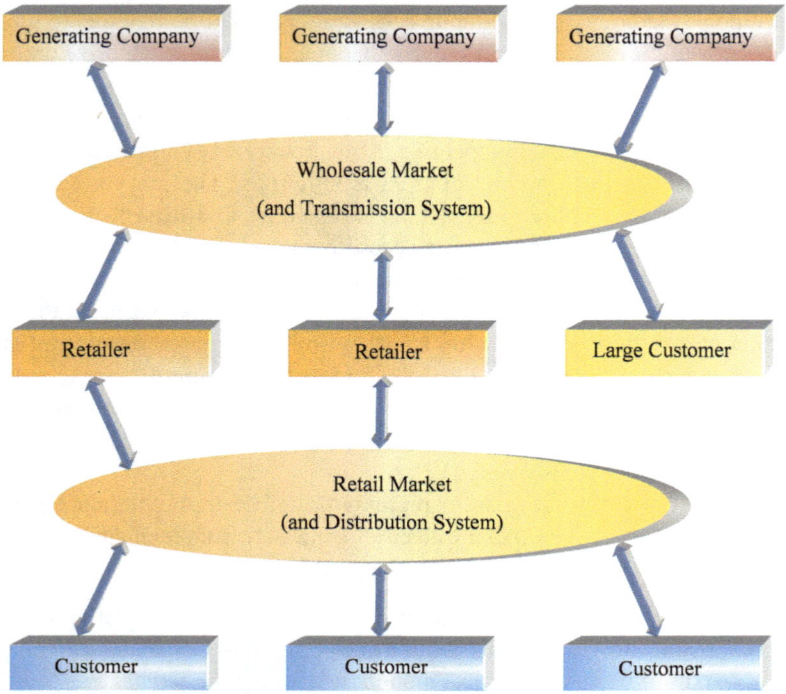

Fig. 2.1 Full wholesale and retail competition (based on [19])

market. Retailers and large customers can purchase competitively in the wholesale market. Medium and small customers normally buy through retailers and aggregators, because of wholesale transaction costs. However, all end-use customers are free to choose their electricity supplier. Retails prices are no longer regulated—all customers can change supplier when they are offered a better price. The "wires business" of distribution companies—that is, the transport of electricity from the transmission system to customers—is usually separated from the retail activities, since each company no longer has a local monopoly for the supply of energy in a specific area. Both at the wholesale and retail levels, the remaining monopoly functions are the provision and operation of the transmission and distribution networks [19]. The costs associated with these two networks are still charged on a regulated basis. Full retail competition is a logical end point and the retail competition model is now in place, or being put in place, in many countries around the world. The big drawback of the model, however, is the cost of the settlement system for all the small consumers and also the need to get them educated [18].

2.3.2 Key Market Entities

The restructuring of the electric industry has changed the role of traditional entities—separated some functions and combined others—and created new entities that can function independently. The independent system operator (ISO) and the market operator (MO) are the leading entities that play a central role in power markets. This section introduces these two entities and also other key market participants, including GenCos, RetailCos, TransCos, DistCos, marketers, aggregators, and consumers (large and small consumers). Power markets have evolved at different rates and in somewhat different directions around the world, and thus not all entities may be found in a specific market. Also, a single entity may perform more than one of the functions presented below. Figure 2.2 illustrates the flow of information among the key entities operating in a competitive market. The figure also illustrates the traditional flow of power from bulk generation resources through transmission and distribution networks to end-use customers.[29]

Generating companies (GenCos) operate and maintain power plants.[30] They may own a single generating plant or a portfolio of plants of different technologies. GenCos may sell electrical energy either to organized markets or directly to retailers and other market participants through bilateral contracts. In addition to real power, they may trade several services—such as reserve, regulation and voltage control—needed by the ISO to maintain quality and security. Their key objective is to maximize profit by selling electricity and eventually these services [31]. To this end, GenCos may choose to take part in whatever markets (energy markets, ancillary-services markets, reserves markets, etc.) and take whatever actions (e.g., gaming).

Retailers (RetailCos) buy electricity in wholesale markets and re-sell it to customers in retail markets (typically, end-use customers that are not allowed, or do not want, to participate in wholesale markets). They do not generally own production units and need to purchase all the electric power and other services needed to provide energy to final customers. Their key objective is to maximize profit by selling energy to customers—profit margins are usually narrow as RetailCos should provide their clients with the lowest possible prices to avoid them to change supplier. RetailCos may deal indirectly with end-use customers through aggregators (see Fig. 2.2).

Marketers are entities that buy and sell power without necessarily owning generating facilities [17]. They may handle both marketing and retailing functions—that is, they may intermediate between producers and retailers and also play the same role as retailers (see Fig. 2.2). Marketers take position and therefore risk [18]. They may sell power to retailers and final consumers.

[29] A cautionary and explanatory note is in order here. Demand-side management and distributed generation may result in a bi-directional flow of power, which in turn calls for a bi-directional flow of information. Thus, both the flow of power and the flow of information may be bi-directional (see, e.g., [30]).

[30] *Non-dispatchable generating companies* are producers with non-dispatchable sources, such as wind or solar-thermal power plants. Also, *independent power producers* are generating companies that coexist with vertically integrated utilities [19].

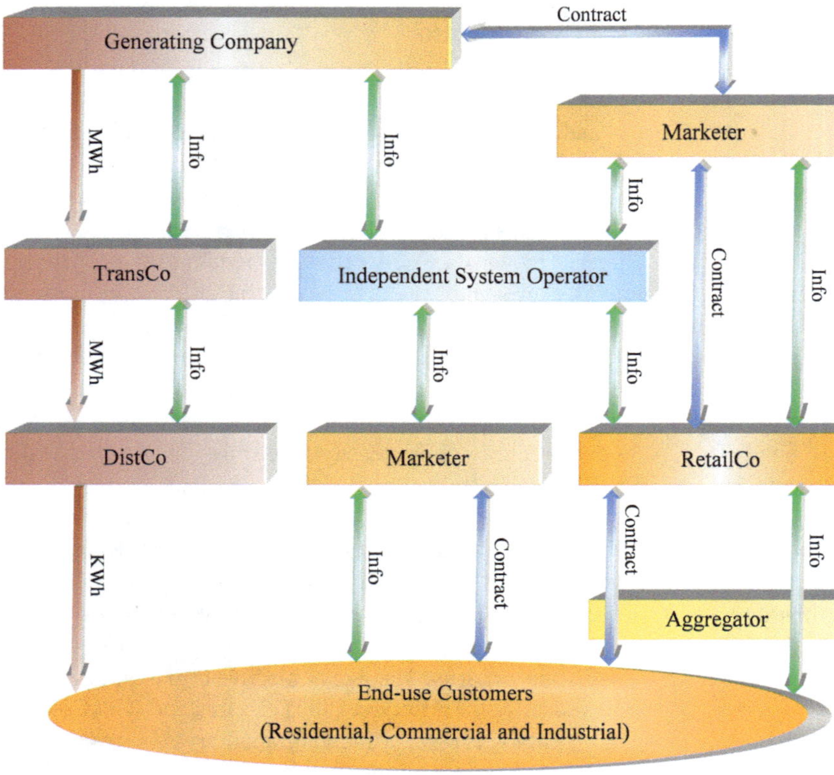

Fig. 2.2 Key market entities and flow of information and power (based on [32])

The *independent system operator* (ISO) is the entity responsible for the reliability and security of the power system. The ISO should not be under the control of any group of market participants, nor favor or penalize one group over another. This neutral, independent, and typically non-profit entity usually combines system operation responsibility with the role of running the market where generation and load are balanced in real time [19]. In particular, the ISO can administer transmission tariffs, coordinate maintenance scheduling, and play a role in coordinating long-term planning. The ISO may also be responsible to commit and dispatch system resources and to curtail loads to maintain system security [17].

As we mentioned earlier, there are two major perspectives regarding the role of the independent system operator. The *MinISO* structure favors decentralization and is mainly concerned with security—the market role of the ISO is very modest. On the contrary, the *MaxISO* structure includes a power exchange running an energy market—the ISO is directly involved in the trading process. Both structures have their merits and shortcomings, and their application has been extensively discussed and analyzed in the literature (see, e.g., [1, 18, 24]).

The *market operator* (MO) is usually responsible for running the markets that close some time ahead of real time (e.g., the day-ahead market). This independent, and typically for-profit, entity ensures a competitive marketplace. Market participants should provide extensive data, such as energy cost for every generator and daily demand for every retailer (or large consumer). The MO then runs a computer program that matches the bids to sell energy with the offers to buy energy and calculates the market-clearing price based on the highest price bid in the market. Also, the MO takes care of the settlement process by forwarding payments from buyers to sellers following the delivery of electricity [19].

Transmission companies (TransCos) are profit-making companies responsible for building, maintaining, and operating the transmission system.[31] They own transmission assets—such as lines, cables, transformers and reactive compensation devices—and transmit electricity using a high-voltage, bulk transport system from generating companies to distribution companies for delivery to final customers. However, they operate these transmission assets according to the instructions of the ISO. Specifically, transmission maintenance and expansion is coordinated between TransCos and the independent system operator. TransCos are regulated to provide non-discriminatory connections and comparable service for cost recovery [17]. They are typically independent.

Distribution companies (DistCos) are entities that own and operate the distribution system—the system of lines, cables, transformers, and other equipment used to transport electricity to end-use customers, usually within a local region. Thus, they receive bulk energy from transmission grids and distribute electricity through their facilities to customers. They are responsible to respond to distribution network outages and power quality concerns. In a competitive market, the operation, maintenance and development of a distribution network is decoupled from the sale of electricity, meaning that several retailers compete to perform this energy sale activity.[32] DistCos are often regulated by state regulatory agencies.

Aggregators are entities that support groups of end-use customers in trading electrical energy [17]. They typically act as agents between retailers and groups of end-use customers (see Fig. 2.2). Each group pursues the objective of buying large blocks of electric power at cheaper prices, when compared to prices for single customers. Aggregators that purchase power for resale to individual or groups of end-use customers act mainly as retailers, and are usually qualified as retailers.

Consumers are the end-users of electricity with facilities connected to the transmission system, in the case of large consumers, or more often connected to the distribution system, in the case of small consumers (see Figs. 2.1 and 2.2). Their key

---

[31] Sometimes, TransCos are subsidiaries of companies that own generating plants. Also, *independent transmission companies* are transmission companies that do not own generating plants and also act as ISOs [19].

[32] A retailer may be a subsidiary of a local distribution company that owns and operates a distribution network.

objective is to minimize the procurement cost or to maximize the utility obtained from electricity usage [31].

Large consumers can take an active role in the market by buying electrical energy in the pool or by signing bilateral contracts (e.g., with producers). They may have direct access to generators or contracts with other providers of power, and choose packages of services with the best overall value that meets their needs. *Small consumers*, on the other hand, buy energy from retailers and possibly other market participants (e.g., marketers). Their participation in the market involves mainly the selection of a specific retailer among several competing retailers.

2.4 Conclusion

This chapter has pointed out that agent-based simulation of EMs is an active area of research and mentioned that a number of prominent models and systems have been proposed in energy-related journals. Since there is a large degree of heterogeneity in describing details of existing models and systems, the chapter has highlighted the need of a generic framework to compare the usage of key concepts in one contribution with usage in other contributions, and also to analyze disparate research efforts. It has also claimed that the development of such a framework can be very important and instructive, providing a coherent set of concepts related to agent-based simulation of EMs, helping to assess progress in this encompassing research area, and facilitating the development of future models and systems.

Accordingly, the chapter has introduced the first part of a generic framework for agent-based simulation of EMs. The complete framework includes three groups (or categories) of dimensions, namely:

1. Market architecture.
2. Market structure.
3. Software agents.

The first group, labeled "market architecture", contains the following three key dimensions:

1. Submarkets (e.g., day-ahead, forward and futures markets).
2. Market types (e.g., energy exchanges and pool markets).
3. Market linkages (e.g., arbitrage, timing and location).

The chapter has introduced the concept of "market architecture", presented various types of markets, and discussed two key market models: pool trading and bilateral trades or contracts. More specifically, it has described the two main variations of marginal pricing (system marginal pricing and locational marginal pricing), highlighted the importance of the bid format, and discussed the concepts of "market" power and "price volatility". It has also explained the distinction between physical and financial bilateral contracts as well as the distinction between firm and non-firm contracts. Furthermore, the chapter has introduced the term "derivative", discussed four key types of contracts (forwards, futures, options and swaps), and described two central markets for trading derivatives: exchanges and over-the-counter markets.

The second group of dimensions, labeled "market structure" is composed by two distinct, yet interrelated dimensions:

1. Market sector (e.g., wholesale and retail).
2. Market participants (e.g., generating companies and retailers).

The chapter has introduced the concept of "market structure", described the main market sectors and discussed several key market entities. In particular, it has presented the four models of competition commonly mentioned in the literature (monopoly, single buyer or purchasing agency, wholesale competition and retail competition). It has also discussed the key market entities that take part in electricity markets, notably generating companies (GenCos), retailers (RetailCos), marketers, independent system operator (ISOs), market operators (MOs), transmission companies (TransCos), distribution companies (DistCos), aggregators, large consumers and small consumers.

The third group of dimensions, labeled "software agents", includes two broad dimensions: agent architectures and agent capabilities. For reasons associated with the nature and specificity of the area of intelligent agents, as well as for the sake of simplicity in exposition, this group will be the subject of the next chapter. Also, the discussion of fruitful areas for future work—such as potential extensions of the complete framework by adding new dimensions—will be deferred to Chap. 3.

Acknowledgements This work was performed under the project MAN-REM: Multi-agent Negotiation and Risk Management in Electricity Markets (FCOMP-01-0124-FEDER-020397), supported by FEDER funds, through the program COMPETE ("Programa Operacional Temático Factores de Competividade"), and also National funds, through FCT ("Fundação para a Ciência e a Tecnologia"). The author wishes to acknowledge the valuable comments and suggestions made by Hannele Holttinen, from the VTT Technical Research Centre of Finland, João Santana and Rui Castro, from INESC-ID and also the Technical University of Lisbon (IST), and João Martins and Anabela Pronto, from the NOVA University of Lisbon.

References

1. Stoft, S.: Power System Economics: Designing Markets for Electricity. IEEE Press and Wiley Interscience, New York (2002)
2. Ventosa, M., Baíllo, A., Ramos, A., Rivier, M.: Electricity market modeling trends. Energy Policy **33**(7), 897–913 (2005)
3. Kahn, E.P.: Numerical techniques for analyzing market power in electricity. Electr. J. **11**(6), 34–43 (1998)
4. Hobbs, B.F.: Linear complementarity models of nash-cournot competition in bilateral and POOLCO power markets. IEEE Trans. Power Syst. **16**(2), 194–202 (2001)
5. Day, C.J., Hobbs, B.F.: Oligopolistic competition in power networks: a conjectured supply function approach. IEEE Trans. Power Syst. **17**(3), 597–607 (2002)
6. Nash, J.F.: The bargaining problem. Econometrica **18**(2), 155–162 (1950)
7. Weidlich, A., Veit, D.: A critical survey of agent-based wholesale electricity market models. Energy Econ. **30**, 1728–1759 (2008)

8. Rothkopf, M.H.: Daily repetition: a neglected factor in the analysis of electricity auctions. Electr. J. **12**(3), 60–70 (1999)
9. Koritarov, V.: Real-world market representation with agents: modeling the electricity market as a complex adaptive system with an agent-based approach. IEEE Power Energy Mag. **2**(4), 39–46 (2004)
10. Wooldridge, M.: An Introduction to Multi-agent Systems. Wiley, Chichester (2009)
11. Jennings, N.R.: On agent-based software engineering. Artif. Intell. **117**, 277–296 (2000)
12. Jennings, N.R., Sycara, K., Wooldridge, M.: A roadmap of agent research and development. Auton. Agents Multi-agent Syst. **1**, 7–38 (1998)
13. Lopes, F., Coelho, H., Santana, J.: A framework for agent-based electricity markets: preliminary report. In: Spies, M., Wagner, R., Tjoa, A. (eds.) 26th Database and Expert Systems Applications (DEXA 2015), pp. 57–61. IEEE (2015)
14. Lopes, F., Coelho, H., Santana, J.: Towards a conceptual framework for agent-based electricity markets. In: Tjoa, A., Vale, Z., Wagner, R., (eds.) 27th Database and Expert Systems Applications (DEXA 2016), pp. 143–147. IEEE (2016)
15. Bodie, Z., Kane, A., Marcus, A.: Investments. McGraw-Hill Education, New York (2014)
16. Dodd, R.: The structure of OTC derivatives markets. The Financier **9**(1–4), 1–5 (2002)
17. Shahidehpour, M., Yamin, H., Li, Z.: Market Operations in Electric Power Systems. Wiley, Chichester (2002)
18. Hunt, S.: Making Competition Work in Electricity. Wiley, Chichester (2002)
19. Kirschen, D., Strbac, G.: Fundamentals of Power System Economics. Wiley, Chichester (2004)
20. Zhang, X.-P. (ed.): Restructured Electric Power Systems. Wiley, Chichester (2010)
21. Wood, A., Wollenberg, B., Sheblé, G.: Power Generation, Operation, and Control. Wiley, Chichester (2014)
22. Hull, J.: Options, Futures, and Other Derivatives. Pearson Education, New York (2015)
23. Schweppe, F., Caramanis, M., Tabors, R., Bohn, R.: Spot Pricing of Electricity. Kluwer Academic Publishers, Boston (1988)
24. Hogan, W.: Competitive Electricity Market Design: A Wholesale Primer. Harvard University, Cambridge (1998)
25. Cain, M., Alvarado, F.: Implications of cost and bid format on electricity market studies: linear versus quadratic costs. In: Large Engineering systems Conference on Power Engineering (LESCOPE-04), pp. 2–6. IEEE (2004)
26. Cardell, J., Hitt, C., Hogan, W.: Market power and strategic interaction in electricity networks. Resour. Energy Econ. **19**, 109–137 (1997)
27. Clewlow, L., Strickland, C.: Energy Derivatives: Pricing and Risk Management. Lacima Publications, London (1998)
28. Rahimi, F., Vojdani, A.: Meet the emerging transmission market segments. IEEE Comput. Appl. Power **12**(1), 26–32 (1999)
29. Srivastava, A., Kamalasadan, S., Patel, D., Sankar, S., Al-Olimat, K.: Electricity markets: an overview and comparative study. Int. J. Energy Sect. Manag. **5**(2), 169–200 (2011)
30. Rahimi, F., Ipakchi, A.: Transactive energy techniques: closing the gap between wholesale and retail markets. Electr. J. **25**(8), 29–35 (2012)
31. Conejo, A., Carrión, M., Morales, J.: Decision Making Under Uncertainty in Electricity Markets. Springer, Heidelberg (2010)
32. Cheung, K., Rosenwald, G., Wang, X., Sun, D.: Restructured electric power systems and electricity markets. In: Zhang, X.-P. (ed.) Restructured Electric Power Systems: Analysis of Electricity Markets with Equilibrium Models, pp. 2–6. Wiley, New Jersey (2010)

Chapter 3
Electricity Markets and Intelligent Agents Part II: Agent Architectures and Capabilities

Fernando Lopes and Helder Coelho

Abstract Agent technology is a relatively new and rapidly expanding area of research and development. The major motivations for the increasing interest in intelligent agents and multi-agent systems include the ability to provide solutions to problems that can naturally be regarded as a society of autonomous interacting components, to solve problems that are too large for a centralized agent to solve, and to provide solutions in situations where expertise is distributed. Electricity markets (EMs) are complex distributed systems, typically involving a variety of transactive techniques (e.g., centralized and bilateral market clearing). The agent-based approach is an ideal fit to the naturally distributed domain of EMs. Accordingly, a number of agent-based models and systems for EMs have been proposed in the technical literature. These models and systems exhibit fairly different features and make use of a diverse range of concepts. At present, there seems to be no agreed framework to analyze and compare disparate research efforts. Chapter 2 and this companion chapter claim that such a framework can be very important and instructive, helping to understand the interrelationships of disparate research efforts. Accordingly, Chap. 2 (Part I) and this chapter (Part II) introduce a generic framework for agent-based simulation of EMs. The complete framework includes three groups (or categories) of dimensions: market architecture, market structure and software agents. The first two groups were the subject of Chap. 2. This chapter discusses in considerable detail the last group of dimensions, labeled "software agents", and composed by two distinct yet interrelated dimensions: agent architectures and agent capabilities.

F. Lopes (✉)
LNEG–National Laboratory of Energy and Geology,
Estrada do Paço do Lumiar 22,
1649-038 Lisbon, Portugal
e-mail: fernando.lopes@lneg.pt

H. Coelho
University of Lisbon,
Bloco C6, Campo Grande,
1749-016 Lisbon, Portugal
e-mail: hcoelho@di.fc.ul.pt

F. Lopes and H. Coelho (eds.), *Electricity Markets with Increasing Levels of Renewable Generation: Structure, Operation, Agent-based Simulation, and Emerging Designs*,
Studies in Systems, Decision and Control 144, https://doi.org/10.1007/978-3-319-74263-2_3

3.1 Introduction

Intelligent agents and multi-agent systems (MAS) represent a relatively new way to conceptualize and implement complex systems [1]. Such systems are composed of autonomous components that interact with one another for their individual and/or common interests—that is, they are characterized in terms of distributed and concurrent problem solving, but also by sophisticated patterns of interaction (e.g., cooperation, coordination and negotiation). The agent-based approach is ideally suited to represent problems that have multiple problem solving entities, multiple perspectives and multiple problem solving methods [2]. This approach offers a powerful repertoire of tools, techniques, and metaphors that have the potential to considerably improve the way in which researchers analyze, design, and implement a diverse range of software solutions.

Agent technology has been used to solve real-world problems in a range of industrial and commercial applications. Broadly speaking, agent-based applications range from personal software assistants—agents play the role of proactive assistants to users—to distributed systems—agents become processing elements in a distributed system [2]. Typical areas in which agent-based approaches were applied include manufacturing, telecommunications, air-traffic control, traffic and transportation management, electronic commerce, business process management, and information retrieval and management (see, e.g., [3, 4]).

Electricity markets (EMs) are open systems that promote competition among suppliers and provide consumers with a choice of services. EMs are distributed in nature, complex (e.g., fluctuations of supply and demand, non-storability of electricity, etc.) and may involve a variety of transactive techniques, including bidding, auctions, centralized and bilateral market clearing, and settlements. In particular, centralized (sub-) markets involve no direct negotiations between the parties—all participants who wish to either sell or buy electricity on a specific delivery day submit their bids and offers [5, 6]. Tailored (or customized) bilateral contracts are essentially agreements between buyers and sellers to trade electricity at specific terms, notably price and quantity (see, e.g. [7]).

Agent-based simulation (ABS) presents itself as a promising approach to accurately model and analyze in detail the behavior of EMs over time. Accordingly, several researchers have paid attention to agent-based simulation of EMs over the last years and a number of prominent models and systems have been proposed in the literature. Ventosa et al. [8] present a survey of the most relevant publications regarding electricity market modeling, identifying three major trends: optimization models, equilibrium models and simulation models (including agent-based models). The authors highlight the need to identify, classify and characterize the somewhat confusing diversity of publications that can be found in the technical literature on EMs. To this end, they introduce a taxonomy of these publications involving the following seven attributes: degree of competition, time scope, uncertainty modeling, inter-period links, transmission constraints, generation system representation and market modeling.

Zhou et al. [9] summarize the main features of several popular agent-based systems for electricity markets, notably SEPIA [10], EMCAS [11], NEMSIM [12], AMES [13], PowerACE [14] and MASCEM [15]. Based on the common features of the analysed systems, the authors propose an ABS framework to facilitate the development of future models for EMs. The main components of the framework are the following: physical system and configuration, agents and their interactions, role of the independent system operator, market model, and decision making and adaptation of each agent. Sensfuß et al. [14, 16] point out that agent-based simulation overcomes several weaknesses of conventional approaches to modeling the electricity sector and acknowledge the lack of a standard piece of work providing an overview and a systematization of the relevant work in the field. Accordingly, the authors present a literature survey structured into three main categories: analysis of market power and design, modeling agent decisions, and coupling of short-term and long-term decisions. Each category includes several major sub-categories (e.g., market models, players and grid constraints for the market category, or agent architecture and learning capabilities for the agent category).

Weidlich and Veit [17, 18] present a critical survey of the most relevant agent-based wholesale electricity market models. The authors categorize the models according to the learning algorithms employed by agents. The major categories are reinforcement learning (involving the Erev–Roth and Q-learning algorithms), evolutionary concepts (especially genetic algorithms and learning classifier systems) and model-based adaptation algorithms (naive or intuitive formulations tailored to specific designs of the simulated markets). Within each category, the authors consider four main sub-categories, namely market mechanism and power, transmission grid constraints, demand side representation, and the learning behavior of agents. They point out that the field of agent-based simulation of EMs has already departed from its infancy, but on its way to adulthood several methodological questions need to be addressed to increase the comparability of different models. These include sound arguments for the choice of learning algorithms, reliable and well documented empirical validation techniques, and a more careful description of the models.

Guerci et al. [19] present a broader review of the literature on agent-based simulation of competitive wholesale electricity markets. Specifically, the authors review 49 articles in detail, identifying three major modeling trends: research issues, behavioral adaptive models, and market models. They start by classifying the publications according to eight major research issues, namely market performance and efficiency, market mechanism comparison, market power, tacit collusion, multi-settlement markets, technological aspects affecting market performance, diversification or specialization, and divestiture or merging of generation assets. Next, they concentrate on market participants (software agents) and analyse the following three major aspects: type of adaptive behavioral model (e.g., reinforcement learning), features of generation companies (e.g., format of selling bids), and demand model adopted (e.g., fixed inelastic demand). Finally, they focus on market models and examine market types and their inter-dependencies (e.g., single or multi-settlement markets), pricing methods (e.g., locational marginal pricing), and incorporation of network constraints.

Overall, agent-based simulation of energy markets is, at the time of writing, an active area of research—a growing number of researchers have simulated (and replicated) several key characteristics of energy markets using software agents and a number of prominent models and systems have been proposed in energy-related journals. These models and systems exhibit fairly different features and make use of a diverse range of concepts (e.g., electricity pricing and agent architecture), so assessing and relating individual research contributions is difficult. At present, there seems to be no agreed framework to compare the usage of such concepts in one scientific contribution with usage in other contributions, nor to compare disparate research efforts. We believe that such a framework can be very important and instructive, providing a coherent set of concepts, helping to understand the interrelationships of disparate research efforts, and providing a foundation for the development of future models and systems.

Against this background, Chap. 2 (Part I) and this chapter (Part II) introduce a generic framework for agent-based simulation of EMs. The complete framework includes the following three groups (or categories) of dimensions:

1. Market architecture.
2. Market structure.
3. Software agents.

Chapter 2 has discussed, in considerable detail, the first and the second groups of dimensions, labeled "market architecture" and "market structure", respectively.

In this companion chapter, we discuss the last group of dimensions, labeled "software agents", and including two broad dimensions: agent architectures (e.g., model-based, goal-based, utility-based and learning) and agent capabilities (e.g., autonomy, reactivity, pro-activeness, social ability and adaptability). Specifically, the purpose of this chapter is threefold:

1. to examine the literature on software agents and to identify the main strands of work in this active area of research;
2. to introduce the last part of a generic framework for agent-based simulation of EMs;
3. to describe in considerable detail the various components of the last group of dimensions and to discuss the key features of autonomous software agents.

This chapter is not meant as a survey of the area of agent-based simulation of electricity markets. Rather, the description of the various components of the conceptual framework is generally undertaken with particular reference to work from multi-agent systems, artificial intelligence, and computer science generally. Thus, this chapter does not present new theorems nor important experimental results, but, instead, aims at providing both a coherent set of concepts and a comprehensive and systematic basis for objectively comparing and contrasting different agent-based models and systems for EMs. Such a basis can be, we believe, an important step for the definition of the core elements and features required by software agents able to simulate EMs in a realistic way, and thus (clearly) helping the area of agent-based simulation of EMs to reach a higher level of stability and maturity.

The remainder of this chapter is structured as follows. Section 3.2 discusses various agent architectures (first subsection) and a core set of agent capabilities (second subsection). Section 3.3 illustrates the applicability of the (complete) framework for agent-based simulation of EMs. Specifically, this section summarizes the work in which the authors are currently involved, realized in an agent-based simulation tool for EMs, called MATREM, and classifies the tool according to the various elements of the framework. Section 3.4 presents some concluding remarks. Finally, the Appendix presents some theoretical notes on agency.

3.2 Software Agents

Market entities in agent-based models and system are represented by software agents (also referred to as computational agents). Although the term "agent" has been widely used, by many researchers working in closely related areas, it defies attempts to produce a single universally accepted definition (see, e.g., [20] for several common definitions). There is much ongoing debate and controversy on this very subject, and indeed no universal definition has been accepted. Nevertheless, some sort of definition is important—otherwise, there is a danger that the term will lose all meaning, becoming a term subject to both abuse and misuse, to the potential confusion of the research community.

Wooldridge and Jennings [21] distinguish two general usages of the term agent: a weak and relatively uncontentious, and a stronger and potentially more contentious. Researchers following the week notion of agency generally mean an agent to be a hardware or (more usually) a software-based computer system that enjoys four key properties: autonomy, social ability, reactivity and pro-activeness. On the other hand, researchers following the stronger notion of agency use the term agent to denote a computer system that, in addition to having the above properties, makes use of concepts frequently applied to humans, notably mentalistic notions such as belief, desire, intention and obligation.

More recently, Wooldridge [2] states that autonomy is central to the notion of agency and proposes the following definition:

> An *agent* is a computer system that is *situated* in some *environment*, and that is capable of *autonomous* action in this environment in order to meet its delegated objectives.

There are several points to note about this definition. First, an agent is situated (embedded) in a particular environment—it receives input related to the state of the environment through sensors and acts on the environment through effectors. Second, the agent is autonomous—it can act without the direct intervention of humans or other systems, having control over its own behavior. Like agency itself, autonomy is a somewhat tricky concept to tie down precisely, and we will elaborate this point below. In particular, we will see that autonomy forms a spectrum in itself and is not binary (yes-no) in nature. Finally, the agent is designed to fulfill a specific purpose—it typically has specific tasks to accomplish.

Table 3.1 Conceptual framework: the category "software agents"

Group (or Category)	Dimension	Element (or Characteristic)
Software agents	Agent architectures (Agent types)	• Purely reactive
		• Model-based
		• Goal-based
		• Utility-based
		∘ Perfectly rational
		∘ Limited rational
		• Learning
		• Other (e.g., hybrid)
	Agent capabilities	• Autonomy
		• Reactivity
		• Pro-activeness
		• Social ability
		• Adaptability
		• Other (e.g., planning horizon)

In this section, software agents are characterized in terms of two interrelated dimensions: agent architectures and agent capabilities (see Table 3.1).[1] The term "agent architecture" has also been the subject of a somewhat heated debate in the agent-based community, and unfortunately has been used in a variety of different ways. Here, we consider what is basically an abstract view of an *agent architecture*: a general methodology for designing software agents [22]. It specifies how the overall problem can be decomposed into subproblems—that is, how the development of an agent can be decomposed into the development of a set of component modules and how these modules should be made to interact. The total set of modules and their interactions should provide an answer to the question of how the sensor data and the current internal state of an agent determine the actions and the future internal state of the agent [23]. Thus, we are writing very much from the point of view of abstract architectures, also referred to as (conceptual) agent models, and the material that follows clearly reflects this bias.

Agent capabilities can reflect a variety of behaviors and lead to strikingly different types of agents. Indeed, there is a host of agent capabilities, and no single set is widely agreed upon as fundamental to characterize software agents [24]. Nevertheless, we refer here to a core set that we find central to the definition and development of software agents to operate in competitive energy markets. These include the four aforementioned key features of agency: autonomy, reactivity, pro-activeness and social ability. To this, we would add a fifth feature: adaptability.

[1]Although it is conceptually useful to distinguish between the two broad dimensions of "agent architectures" and "agent capabilities", the distinction is not absolute (and may at times be somewhat arbitrary). Accordingly, the reader may find some overlap between these dimensions.

3.2.1 Agent Architectures

The various agent architectures described in this section are essentially abstract architectures for decision making systems that are embedded in a particular environment.[2] Agents take sensory input from their environment and act upon it through actuators—they execute actions that affect their environment. The key problem facing agents is that of deciding which of their actions they should perform in specific circumstances. Typically, the complexity of the decision-making process depends on several environmental properties (e.g., a discrete vs. continuous environment). Also, the interaction agent-environment is usually an ongoing, non-terminating one.[3]

Purely reactive agents decide what actions to perform based on the present only, with no reference to the past. They select actions on the basis of the current perceptual input. In many agents, the map of perceptual input to actions is done by simple if-then rules (also called condition-action, situation-action, or production rules). Conceptually, purely reactive agents have the admirable property of being simple. Furthermore, economy, computational tractability, robustness against failure, and elegance all make such agents appealing. However, purely reactive agents are of limited intelligence—they work only if the correct decisions can be made on the basis of only the current perceptual input, i.e., the environment is assumed to be fully observable [25]. There are also several fundamental, unsolved problems, with such agent architectures (see, e.g., [2] for details).

Model-based agents are able to handle partial observability by keeping track of the part of the environment they can't observe at any given instant—that is, they maintain some kind of *internal state* that depends on the perceptual input history and thereby can reflect some of the unobserved aspects of the current environment state. The behavior of a model-based agent is based, at least in part, on the internal state information and can be summarized as follows. The agent starts in some initial internal state, observes the environment, and generates a perceptual input. The internal state of the agent is then updated—the current perceptual input is combined with the initial state to generate a new state description, based on the agent's model of how the environment works. Next, the agent selects an action to perform based on the updated state description. This action is then performed, and the agent enters another cycle, perceiving the world, updating its state, and choosing another action to perform. Typically, updating the internal state information as time goes by requires both information about how the environment evolves independently of the agent and information about how the agent's own actions affect the environment. Information about "how the world works" is called a model of the world [25].

---

[2]The architectures discussed in the first part of this section are based on the five basic types of agents presented in [25, Chap. 2]. The present section, however, is not intended as a summary of the authors' views on agent architectures, and also presents a top-level view of a software agent.

[3]For convenience, throughout this section we use the term "environment" to denote a *generic* agent environment. For software agents that represent market entities operating in a competitive energy market, the term should denote, naturally, this *particular* market environment.

Goal-based agents contain an explicitly represented model of the world as well as information about the goals they try to accomplish. They have some internal data structure to record information about their goals—that is, information about the situations that are desirable. Also, they are able to perform goal-based action selection, i.e., to combine goal information with information about the world to select actions that achieve their goals. The decisions about what actions to perform are based on information represented explicitly. Notice that purely reactive agents do not make use of explicitly represented information, since the built-in rules map directly from perceptual input to actions [25].

Goal-based action selection alone may not be adequate to generate high-quality behavior in complex environments (e.g., when there are various sequences of actions that allow agents to achieve their goals, but some are more reliable and feasible than others). Hence, agents often need some kind of performance measure allowing them to compare different environment states according to exactly how these states are desirable—that is, the performance measure defines the criterion of success. *Perfectly rational agents* are able to select the actions that are expected to maximize their performance measure, given the evidence provided by the perceptual input history and whatever built-in knowledge the agents have [26]. They can generate a sequence of actions that cause the environment to go through a desirable sequence of states. This notion of desirability is captured by the performance measure.

Economists and computer researchers use the term *utility* to mean the quality of being useful and typically formalize this generic performance measure by defining a utility function. *Perfectly rational utility-based agents* make use of utility functions that measure their preferences among states of the environment, *always* choosing the actions that maximize their utility [25]. *Perfect rationality* denotes the capacity to generate maximally successful behavior given the available information [26]. Now, notice that partial observability and decision making under uncertainty are ubiquitous in complex environments. Accordingly, rational agents act at every instant in such a way as either to maximize their utility, or when there is uncertainty, to maximize their *expected utility*—that is, the utility they expect to derive, on average, given the probabilities and utilities of each action outcome.

Clearly, the selection of the utility-maximizing course of action at every instant is a difficult task, requiring sophisticated algorithms and computational tools. Furthermore, even with these algorithms and tools, perfect rationality is usually not considered feasible in complex environments due to the inherent computational complexity—the demands are so high that it is often not considered a realistic objective. Mechanisms take time to process information and select actions, hence the behavior of agents cannot immediately reflect changes in environments and will generally be suboptimal [27]. *Limited rational agents* are able to act appropriately when there is not enough time to do all the computations they might like to do.[4]

[4]Formally speaking, several conceptions of limited rationality for software agents have been proposed in the literature, notably *bounded optimality*—the capacity to generate maximally successful

Learning agents can improve the performance of their component modules in order to select (and perform) better actions. Conceptually, learning agents can be divided into four central components [25]: the performance element, the critic, the learning element and the problem generator. The behavior of a learning agent can be summarized as follows. The *performance element* represents basically what we have considered to be an entire agent of any of the aforementioned four basic agent types (i.e., purely reactive, model-based, goal-based or utility-based agents). Thus, it observes the environment, updates its information, and decides on actions to execute. The *critic* also takes in a perceptual input and passes some information (feedback) to the learning element—it indicates to this conceptual component how well the agent is doing with respect to a fixed performance standard. The observation of successive environment states can allow the agent to learn "how the environment evolves". Also, the observation of the results of its actions can allow the agent to learn "what its actions actually do". Furthermore, the performance standard can understand and analyze part of the incoming perceptual input as a reward (or a penalty) that may provide important feedback on the quality of the agent's behavior.

Both the performance element—responsible for selecting actions—and the learning element—responsible for making improvements—are the central components of the learning agent. As noted, the performance element consists of whatever data and control structures as well as algorithms the agent has for selecting its driving actions. The *learning element* uses feedback from the critic to determine how the performance element should be updated to perform better in the future. This conceptual component may suggest (and make) changes to any of the individual components of the performance element—that is, to any of the components modules of the four agent types previously considered. The *problem generator* is the last component of the learning agent and is responsible for suggesting exploratory actions that may lead to more effective behavior. It can identify aspects needing improvement and suggest new and informative experiments.

At this stage, we hasten to add three explanatory notes. First, learning allows software agents to operate in an initially unknown environment and to become more competent than their initial knowledge alone might allow. The initial configuration of the agents may reflect some prior knowledge of the environment, but as they gain experience this may be modified and augmented. They can learn to compensate for partial or incorrect prior knowledge. Second, based on the type of feedback, learning can be classified into the following three main types [25]: supervised (agents observe input-output pairs and learn functions that map from input to output), reinforcement (agents learn from a series of reinforcements—rewards or punishments), and unsupervised (agents learn patterns in the input even though no explicit feedback is supplied). For each type of learning, powerful techniques have been developed that have proved effective for many problems (see, e.g., [18] and also [28–30]). Finally, some researchers believe that the problem of how to express all the knowledge that a computational system needs—the problem of "knowledge bottleneck"—may be

behavior given the available information and computational resources. Bounded optimal agents behave as well as possible, given their computational resources (see, e.g., [26, 27] for details).

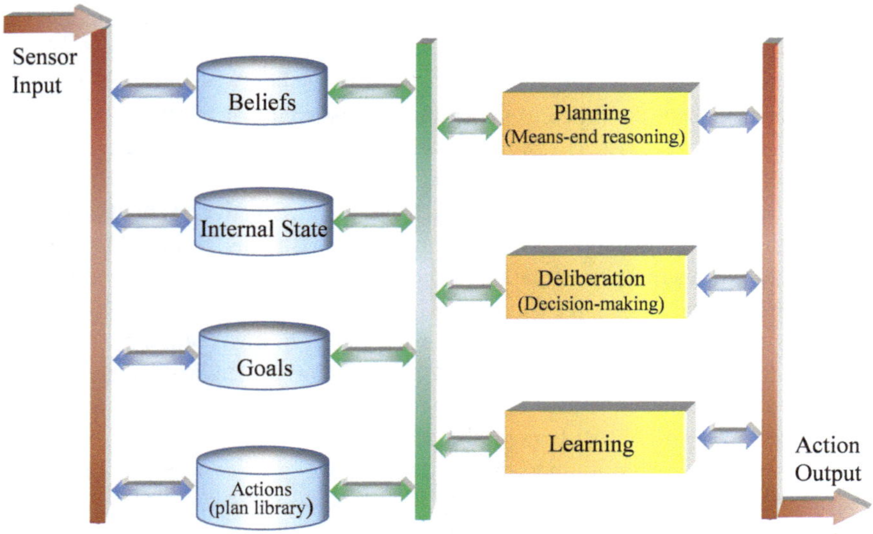

Fig. 3.1 Abstract, top-level view of a software agent (based on [24])

solved by learning methods rather than hand-coded knowledge engineering, provided the learning algorithms have enough data to go on (see, e.g., [31]).

Figure 3.1 presents an abstract architecture for a software agent—it specifies the agent's key information stores and the main processes operating on these information stores.[5] We use cylinders to denote information stores (or data structures) and rectangular prisms to denote processes (or modules). The agent interacts with the environment by receiving sensory input and acts upon that environment by performing actions that affect it.[6] The contents of the *beliefs* data store are essentially facts about the agent itself—that is, it comprises a perspective of the agent's skills and capabilities. The internal data structure records information about the environment state and history (typically, it includes facts about static properties of the environment, current observations about the environment, and conclusions derived from these observations).[7] The *goals* data store may include different types of goals, notably achievement goals—that is, goals having well-defined sets of start and final

[5]Figure 3.1 gives an *abstract* view of an agent. Specific details about each component module and the control flow among modules need further architectural refinement (e.g., details about the decision-making mechanism). See [32, Chap. 2] and [33] for representative surveys of *concrete* agent architectures up to 1998, and [2, Chaps. 3–5] for a description of subsequent work.

[6]The agent commonly operates in an environment populated by other agents and interacts with them to meet its design objectives. In Fig. 3.1, we have not included components that explicitly support such interaction (but see the next subsection).

[7]For the sake of simplicity, we consider fairly direct representations of the agent's beliefs and internal state. However, the internal state may be seen as a subset of the beliefs, namely beliefs about the environment where the agent operates.

states such that, upon arriving in a final state, a particular goal may be considered achieved and thus able to be terminated.

The agent has a set of possible *actions* to execute that represents its effectoric capabilities. Actions may have pre-conditions associated with them, which define the possible situations in which they can be applied. Also, actions can be structured into plans-as-recipes [34], which describe how certain sequences of actions may be performed to achieve particular goals under specific conditions.[8] The three key processes—planning, decision-making and learning—allow the agent to be goal-oriented, to make educated decisions, to behave rationally, and to learn. In short, the agent can generate plans to achieve its goals, reason about alternative plans and decide rationally upon the best one, and act accordingly. The results of its actions can serve as feedback for learning, to improve its performance and to better meet its goals. A more detailed description of each key process follows.

The agent selects actions that further its goals, based on its conception of the environment. Goal-based action selection may be straightforward—for example, when goal satisfaction results immediately from a single action—or it may more complicated—for example, when the agent needs to consider a sequence of actions to achieve a goal. The *planning* process computes sequences of actions whose execution lead to the achievement of particular goals. The generation of a plan can be based either on a formalization of the actions in the plan (planning from first principles) or on a plan library (planning from second principles). Given an initial world state, a goal (final) state and a set of actions, planning from first principles consists basically in searching the space of possible actions to find some sequence of actions that turn the initial state into the goal state. In contrast, planning from second principles is guided by domain knowledge—basically, it consists in adapting plans-as-recipes to specific situations.

The planning process may propose a number of options, all of which are means to a particular end. Given several competing alternative courses of action, the *decision-making* process weighs these alternatives and decides on the preferred one based on a performance measure (typically, the expected utility). Thus, plans that achieve a goal with some effort are preferred to other plans achieving the same goal, but involving a higher effort.

The abstract architecture shown in Fig. 3.1 includes capabilities for means-end reasoning and for the weighting of alternative courses of action. However, a *concrete* agent architecture should also specify how these two capabilities interact. Furthermore, such an architecture should address the problem of limited rationality, since planning and deliberation takes time, and the more time spent on these processes, the more chances there are that the environment will change in an important way. A detailed description of such an architecture is beyond the scope of this

[8]Plans-as-recipes are often stored in an internal data structure called *plan library* (see, e.g., [34]). Also, researchers working in the area of agent architectures use different terms to denote structures similar in function to plans-as-recipes (e.g., knowledge areas [35], or plan templates or schemata [36]).

chapter (but see, e.g., [2, Chap. 4] and the Appendix for a discussion of *belief-desire-intention* (BDI) architectures for *practical reasoning*). Finally, the general design for a *learning* agent described above is classic in the machine learning literature [25].

3.2.2 Agent Capabilities

Autonomy is one of the most important and distinctive agent properties. Broadly speaking, *autonomy* refers to the ability of agents to perform unsupervised actions and to pursue their design objectives without being explicitly programmed or instructed for doing so [24]. As noted above, autonomous agents have some kind of control over their actions and behavior—they can decide for themselves, at least in part, what they need to do in order to satisfy their goals. Autonomy is best represented by a spectrum of values [37]. At one extreme on this spectrum are absolutely autonomous systems that may do anything they please [38]. At the other extreme are fully controlled systems that simply do what they are told. The point between these two extremes that agent designers are largely interested in is a system to which they can delegate goals, and then have this system decide for itself how best to meet its goals. Accordingly, for most agent designers, autonomy means the ability and requirement to decide how to act so as to accomplish the delegated goals [2].

Furthermore, agent designers may want to equip agents with *adjustable autonomy*, which allows them to transfer control for their key decisions to human users (and other agents) whenever certain conditions are met [39]. Thus, agents can dynamically vary (reduce) their own autonomy and let users or other agents make decisions in key situations. Determining whether and when the transfer of decision-making control should occur is a central issue in adjustable autonomy. Practically speaking, agents that always come back to their users for help with decisions will probably be unhelpful, and agents that never seek assistance will probably also be useless. There is a need to strike a difficult balance: agents should transfer control to human users or other agents whenever they provide superior decision making expertise, while the number of such transfers should be minimized (but see [39]).

Reactivity and pro-activeness are two other important properties of agents. *Reactivity* means that agents are able to perceive their environment and respond in a timely fashion to changes that occur in it to satisfy their design objectives. *Pro-activeness* means that agents are able to exhibit goal-directed behavior by taking the initiative—that is, they do not simply act in response to their environment, but can also initiate actions to satisfy their objectives [21]. Although autonomous agents may exhibit both reactive and goal-directed behavior, pro-activeness suggests a high-level of agent autonomy [24].

Now, building a system that exhibits a simple form of goal-directed behavior is often not considered a hard task. Crudely, software developers do it every time they write a method in a programming language (e.g., Java)—the method is simply a plan or recipe for achieving a goal. This model of goal-directed programming makes, however, some important limiting assumptions. In particular, it assumes that both the environment does not change while the method is executing and the goal remains valid at least until the method terminates. These assumptions may not be reasonable in multi-agent environments. In such dynamic environments, agents should be able to exhibit some form of reactive behavior [2].

A purely reactive system—that is, a system that continually responds to its environment—is also not considered difficult to construct. However, building a system that achieves an effective balance between goal-directed and reactive behavior is indeed a hard task. Specifically, the system should be able to pursue its goals in a systematic way, but should not continue blindly performing a task (e.g., executing a complex method) in an attempt to achieve a particular goal either when such goal is for some reason no longer valid or when it is clear that the method will not work. In such circumstances, the system should be able to react to the new situation, in time for the reaction to be of some use. Clearly, this effective integration of goal-directed and reactive behavior is one of the key problems facing agent designers [2].

Several prominent approaches to build agents that can reconcile reaction and deliberation have been proposed in the literature.[9] *Layering* is probably the most popular and powerful approach—it allows agent designers to structure the functionalities of an agent into two or more interacting layers to achieve coherent behavior of the agent as a whole [32]. *Layered architectures* contain at least two layers, to deal with reactive and proactive behaviors, although they may contain many more layers (notably, a social layer to deal with social behavior). Such architectures have several inherent advantages, including the modularization of agents, making their design more compact and increasing their robustness. InteRRaP [32] is a relevant example of a layered architecture. Also, Stanley [40] is a hybrid agent with an architecture containing 30 different independently operating modules, embodied in the Volkswagen Touareg R5 that completed the 132-mile course in the Nevada desert in less than 7 hours, winning the DARPA Grand Challenge.[10]

A fourth relevant agent property is social ability.[11] Practically speaking, *social ability* means that agents are capable of interacting with other agents (and possibly human users) to satisfy their design objectives [21]. To interact and communicate about some domain, agents need to agree on the terminology to use to describe this domain. An ontology is basically a specification of a set of terms intended to provide a common basis of understanding about some domain. More specifically, an *ontology* is an explicit specification of the objects, concepts, and other entities that are presumed to exist in some domain and the relationships that hold among them [41]. The most typical type of ontology used in constructing agents involves a

[9] Software agents that can combine both reactive and deliberative reasoning are commonly refereed to as *hybrid agents*.

[10] At this stage, a natural question to ask is: "Which of the architectures described in the previous subsection should be considered by agent designers?" The answer is: "All of them" [25, Chap. 27].

[11] Although the three agent properties discussed earlier—autonomy, reactivity and pro-activeness—are mainly related to the *micro aspects* of agent technology (the *agent level*), social ability is closely related to the *macro aspects* of agent technology (the *social level*). In other words, we now move from the micro level of individual agents to the macro level of multi-agent systems.

taxonomy of class and subclass relations coupled with their properties and allowed values, and makes use of the inheritance between classes. An ontology together with a set of individual instances of classes constitutes a *knowledge base* [42]. In practice, there is a fine line where an ontology ends and a knowledge base begins.

A general methodology to help agent designers to develop new ontologies is presented in [42]. It involves the following seven main steps: determine the domain and scope, consider reusing, enumerate the important terms, define the classes and the class hierarchy, define the properties of classes, define the properties of properties, and create instances. The web ontology language (OWL) is probably the most important and influential ontology language [43]. OWL is basically a collection of several XML-based ontology frameworks, within which ontologies can be expressed [2]. The extensible markup language (XML) [44] is not an ontology language, although it is usually the language of choice for defining simple ontologies, created for specific purposes.

The notion of *social ability* presented above also means that agents are able to communicate with their peers—typically, by exchanging messages in an expressive *agent communication language*. The knowledge query and manipulation language (KQML) is a standard message-based language for agent communication [45]. KQML provides designers with a standard syntax for messages, defines a collection of performatives for defining the intended interpretation of messages, and does not mandate any specific language for message content. Crudely, each message has a performative (e.g., tell and reply) and a number of parameters (e.g., sender, receiver and content). The inspiration for these message types comes largely from the *speech act theory* [46, 47]. The key idea underlying this theory is to treat communication as a specific type of action, in the sense that they change the state of the world in a way analogous to physical actions. Thus, communicative utterances are modeled as actions that alter the mental state of the communication participants—typically, they are performed by a speaker with the intention to bring about some particular mental state in a listener. Three widely recognized categories of speech acts are representatives (the paradigm example is informing), directives (the paradigm case is requesting), and commissives (the paradigm case is promising).

The success of KQML within the agent research community was significant and several KQML-based implementations were developed (see, e.g., [48]). KQML was, however, criticized on a number of grounds, notably because the lack of a clear semantics, allowing to tell whether two agents claiming to be talking KQML were in fact using the language properly. In other words, the "meaning" of KQML performatives was mainly defined using informal, English language descriptions, open to different interpretations [2]. This and other criticisms led to the development of a new, but rather closely related, language: the FIPA agent communication language [49]. Both the structure of the messages and the message attribute fields are very similar in KQML and FIPA ACL. However, FIPA ACL provides a somewhat different collection of performatives [50]. Also, FIPA ACL has a comprehensive formal semantics based on a theory of speech acts as rational action (see [50, Annex A] and also [51, 52]). As a result, various software tools have been developed that support FIPA messaging. Most tools provide structures corresponding to (simple) agents and facilities for agent interaction and communication. The best known and most widely used is the Java Agent DEvelopment framework (JADE) [53]. It provides software packages to allow agent developers to deploy FIPA agent systems.

Several notes about the degree of interaction inherent in the agent capability *social ability* are in order here. First, *interaction* means some type of collective action in a multi-agent system, wherein one agent takes an action or makes a decision that has been influenced by the presence or knowledge of another agent [54]. Thus, interaction is inherently distributed and dependent upon the action of at least two agents. Second, interaction is a central concept in multi-agent systems (MAS), often associated with a number of different dimensions. For example, Bobrow [55] consider the following three dimensions of interaction: communication, coordination, and integration. Bond and Gasser [54] go far and propose six dimensions of interaction—specifically, among whom the interaction takes place, when the interaction occurs, what is the content of the interaction, how the interaction is accomplished, why the action occurs, and what the basis of commonality is.

Third, the possible types of interaction among autonomous agents operating in a multi-agent system form a spectrum [37]. At one extreme of this spectrum are very simple, as well as uniform types of interaction (e.g., two computers exchanging binary information). At the other end of the spectrum are sophisticated patterns of interaction, typically involving *cooperation* (agents working together towards a common goal), *coordination* (agents organizing their activities so that harmful interactions are avoided or beneficial interactions are exploited), and *negotiation* (agents working towards an agreement acceptable to all the parties involved). It is the flexibility and high-level nature of these patterns of interaction which distinguishes multi-agent systems from other forms of software and which provides the underlying power of the paradigm [3].

Fourth, to consider the range of possible interaction types, the architecture depicted in Fig. 3.1 should include sociality-supporting software constructs. Specifically, it should include a component module for communication and also modules for collaboration, coordination and negotiation. Also, meaningful interaction with other agents in the environment often needs that software agents maintain models of themselves, the other agents and the environment. Furthermore, the three processes shown in Fig. 3.1 should be extended to support multi-agent interaction. Finally, the agent research community has paid a great deal of attention to cooperation, coordination and negotiation over the past two decades, and a number of prominent models and systems have been proposed in the literature. The discussion and analysis of the most important is, however, well beyond the scope of this chapter (but see, e.g., [56–58]).

Before closing this section, we hasten to add a final note about the host of agent capabilities. In addition to the aforementioned four key capabilities—autonomy, reactivity, pro-activeness and social ability—various other attributes are commonly discussed in the context of agency, including planning horizon [59], and soundness [60] (an extensive overview is provided in [38]). Despite the intense interest—and some controversy—that the attributes of agency (and the usage of the term "agent") have evoked, we believe that these four attributes constitute the basic building block of a software agent. Naturally, for certain types of applications, some properties may be more important than others.

Certainly, for some industrial and commercial applications, agent designers may find it helpful to consider additional properties. This book is about software agents for competitive energy markets and, for this particular real-world application, some researchers consider highly desirable that agents exhibit some type of adaptive behavior (see, e.g., [18]). *Adaptability* is indeed considered a dimension of agenthood [24]. *Adaptive agents* can adapt to changes that occur both in the environment and in the behavior of the other agents—typically, they are able to learn which actions to take in order to improve their performance over time. To support this capability, the architecture presented in Fig. 3.1 may be extended with an adaptation module. Alternatively, the three processes—planning, deliberation and learning—may be extended with adaptability-supporting software constructs.

3.3 Practical Application of the Framework

An ongoing study at LNEG is looking at using software agents to help manage both the complexity of wholesale energy markets and the unique challenges of bilateral contracting in retail markets.[12] The main aim is to provide an agent-based simulation tool to analyze the behavior and outcomes of EMs, particularly markets with increasing penetrations of variable generation. This section first gives an overview of the tool (first subsection),[13] and then illustrates the applicability of our generic framework by categorizing the main features of the tool (second subsection).

3.3.1 Agent-Based System for Energy Markets: An Overview

The system, called MATREM (for Multi-Agent TRading in Electricity Markets), allows the user to conduct a wide range of simulations regarding the behavior of energy markets under a variety of conditions. In each simulation, different agents are used to capture the heterogeneity of restructured markets, notably generating companies, retailers, aggregators, large and small consumers, market operators, and system operators. The agents are currently being developed using both JADE [53]—an agent-oriented middleware built on top of the Java programming language, with a flexible infrastructure facilitating the development of agent-based applications—and Jadex [61]—a reasoning engine that runs over JADE, enabling the development of belief-desire-intention (BDI) agents. The target platform for the system is a 32/64-bit computer running Microsoft Windows. A graphical interface allows the user

[12] http://www.lneg.pt/iedt/projectos/473/ (access date: September 2016).

[13] Chapter 8 is entirely devoted to the agent-based system and presents a detailed description of its main features. The reader is therefore referred to it for details.

to specify, monitor and steer all simulations. The interface is fully integrated into the Windows environment and employs the familiar look of other desktop Windows applications.

MATREM agents are essentially computer systems capable of flexible action and able to interact, when appropriate, with other agents to meet their design objectives. Each agent has is its own Java thread, using it to control its life cycle and decide autonomously when to perform which actions. The path of execution of a thread involves three main tasks [53]: (i) agent creation (initialization operations and addition of initial behaviours), (ii) agent "life" (execution of behaviours from the pool of active behaviours), and (iii) agent termination (clean-up operations).

The agents communicate by sending and receiving messages in strict accordance with the FIPA specifications. Each agent can initiate communication with other agent—or group of agents—at any time it wishes and can also be the object of an incoming communication at any time. The communication paradigm is based on asynchronous broadcast message passing and involves four main tasks [53]: (i) a sender agent (S) prepares and sends a message to a receiver agent (R), (ii) the JADE run-time posts the message in the message queue of the agent R, (iii) the agent R is notified about the receipt of the new message, and (iv) the agent R gets the new message from the message queue and processes it.

In the earlier versions of the tool, the conceptual model (or abstract architecture) that underpinned MATREM agents was a "traditional" deliberative model. For convenience, and also in the interests of completeness, a description of the key features of the agents equipped with this model follows. The agents have an internal data store to record the beliefs about themselves, about the environment—that is, a competitive electricity market—and about the other agents operating in the market. They can access the contents of the beliefs data store and also change that contents (i.e., add new beliefs and revise the current beliefs). Although the operations of adding sentences, revising sentences, and querying what is known may involve inference—that is, deriving new sentences from old ones—the agents can exhibit only very restricted inferential capabilities.

Also, the agents have an internal data store to record their (top-level) achievement goals (e.g., "maximize-profit" or "calculate-market-clearing-price"). They are able to perform a simple form of goal-based action selection, namely to combine information about their goals with the contents of their beliefs data store to select actions that further specific goals. The effectoric capabilities of the agents are represented by low-level actions, including numerical computations and communicative actions, performed by executable methods or code fragments.

Although simple and intuitive, the "traditional" deliberative model has been progressively "replaced" by a simplified belief-desire-intention model. Accordingly, each agent has now four major components: beliefs, desires, intentions and plans (or, more precisely, plans-as-recipes). Desires represent preferences over well-defined future states of the environment. Goals are desires that should be pursued by the agent. Plan templates, or recipes, specify how top-level goals can be incrementally refined into sub-goals until a sufficiently fine-grained level of abstraction is reached, which is suitable for execution. Intentions represent committed goals that the agent typically tries to achieve until either it believes they are satisfied or it believes they are no longer achievable (Chap. 8 delves into the technical details of the BDI model, and the interested reader is therefore referred to it).

MATREM supports a power exchange and a derivatives exchange. The power exchange comprises a day-ahead market and a shorter-term market known as intra-day market (see, e.g., [62–64]). Most energy transactions take place in the day-ahead market, i.e., the intra-day market is mainly used to make adjustments in the positions of market participants as delivery time approaches. Both system marginal pricing (SMP) and locational marginal pricing (LMP) are supported. The derivatives exchange comprises a futures market for trading standardized bilateral contracts. This exchange uses an electronic trading system that automatically matches the bids and offers from various market participants.

The tool also supports a bilateral marketplace for negotiating the details of two types of tailored (or customized) long-term bilateral contracts: forward contracts (see, e.g., [65, 66]) and contracts for difference (see, e.g., [67, 68]). Buyers and sellers are equipped with a negotiation model that handles two-party and multi-issue negotiation [69]. The negotiation process involves three main phases or stages: (i) pre-negotiation (focuses on preparation and planning for negotiation and is marked by each party's efforts to formulate an agenda, emphasize points of difference, and posture for positions), (ii) actual negotiation (seeks a solution for a dispute and is characterized by extensive interaction, strategic maneuvers, and movement toward a mutually acceptable agreement), and (iii) post-negotiation (centers on details and implementation of a final agreement).

Furthermore, MATREM supports six different types of market entities: generating companies (GenCos), retailers (RetailCos), aggregators, consumers, market operators (MOs) and system operators (SOs). GenCos may sell electrical energy either to the organized markets or directly to retailers and other market participants through tailored bilateral contracts. RetailCos buy electricity in the organized markets and re-sell it in the retail market. Aggregators support groups of end-use customers in trading electrical energy. Large consumers can take an active role in the market by buying electrical energy directly through the centralized markets (and can also sign tailored bilateral contracts). Small consumers, on the other hand, buy energy from retailers (or deal indirectly with retailers through aggregators). Market operators are responsible for running both the power exchange and the derivatives exchange. A system operator agent is currently being developed and will be responsible for running the market in which load and generation are balanced in real time (this market is currently under development).

The mouse-and-keyboard-based interface handles all interactions with the user and incorporates four key functions: (i) agent management, (ii) scenario construction, (iii) simulation management, and (iv) report analysis. MATREM considers two main types of software agents: market agents and assistant agents. Market agents represent the entities that take part in the different simulated markets. Assistant agents are categorized into interface managers (responsible for managing the interfaces of the simulated markets) and intelligent assistants (provide support to the user in making strategic decisions and can act as "information assistants", "trading assistants", etc.).

Table 3.2 Key features of the agent-based system for electricity markets

Group (or Category)	Dimension	Element (or Characteristic)	MATREM
Market architecture	(Organized) Submarkets	• Energy market	✓
		o Day-ahead market	✓
		o Intra-day market	✓
		o Real-time market	±
		• Ancillary-service market	×
		• Transmission-rights market	×
		• Capacity market	×
		• Forward market	×
		• Futures market	✓
		• Options market	×
		• Swap market	×
	Market types	• Bilateral market	
		o Direct-search market	×
		o Bulletin-board market	✓
		o Brokered market	±
		• Mediated market	
		o Dealer market	×
		o Exchange (or auction) market	✓
		o Pool market	×
	Market linkages	• Implicit	×
		o Arbitrage linkages	
		o Spacial linkages	
		o Temporal linkages	
		• Explicit	×

The system either supports (✓), partially supports (±), or does not support (×) a feature

The human-computer interaction paradigm is based on a creative integration of direct manipulation interface techniques with assistant agents (again, the reader is referred to Chap. 8 for details).

3.3.2 System Classification

The main features of MATREM are summarized in Tables 3.2, 3.3 and 3.4. Both a day-ahead market and an intra-day market sell energy to RetailCos (and possibly other market participants) and buy energy from sellers in advance of time when the energy is produced and consumed.[14] The day-ahead market is cleared during the day before the day of operation (e.g., at 12 noon). The intra-day market sets prices and schedules a few hours ahead to facilitate balancing on advance of real time.

[14]As noted earlier, a market to match the imbalances caused by the variability and uncertainty present in power systems is currently being developed.

Table 3.3 Key features of the agent-based system for electricity markets (continued)

Group (or Category)	Dimension	Element (or Characteristic)	MATREM
Market structure	Market sectors	• Wholesale	✓
		∘ Regulated (no competition)	
		∘ Deregulated (full competition)	✓
		• Retail	✓
		∘ Regulated (no competition)	
		∘ Deregulated (full competition)	✓
		• Central coordination and transmission	±
	Market entities	• Generating companies	✓
		• Retailers	✓
		• Power marketers	×
		• Market operator	✓
		• Independent system operator	±
		• Transmission companies	×
		• Distribution companies	×
		• Aggregators	✓
		• Consumers	
		∘ Large consumers	✓
		∘ Small consumers	✓

The system either supports (✓), partially supports (±), or does not support (×) a feature

This market is cleared several times once the day-ahead market has been cleared. In addition to the day-ahead and intra-day markets, a futures market provides both financial and physical products that span from days to several years.

Tailored (or customized) bilateral trades are defined by privately negotiated bilateral contracts designed to cover the delivery of energy over longs period of time. Their terms and conditions are very flexible and can be negotiated to meet the objectives and needs of the negotiating parties. Negotiation proceeds by an iterative exchange of offers and counter-offers according to the rules of an alternating offers protocol [70]. An offer is essentially a set of issue-value pairs (e.g., "energy price" = 45 €/MWh, "contract duration" = 12 months, and so on). A counter-offer is an offer made in response to a previous offer. The negotiation process may end with either agreement or no agreement.

MATREM supports coalitions of end-user customers—that is, two or more customers can ally into a coalition to strengthen their bargaining positions and pursue a superior negotiation outcome. The customers can come together to pool their efforts in search for a solution that meets common or overlapping goals. Simply put, various customers intentionally form a coalition who interacts and negotiates with a seller agent (e.g., a RetailCo agent) to achieve a desired outcome that meets shared objectives (but see, e.g., [71, 72]).

The system relies on multiple autonomous agents to simulate the central functions of wholesaling and retailing. A competitive wholesale market allows the trading of

Table 3.4 Key features of the agent-based system for electricity markets (continued)

Group (or Category)	Dimension	Element (or Characteristic)	MATREM
Software agents	Agent architectures (Agent types)	• Purely reactive	×
		• Model-based	✓
		• Goal-based	✓
		• Utility-based	✓
		o Perfectly rational	
		o Limited rational	
		• Learning	×
		• Hybrid	×
	Agent capabilities	• Autonomy	✓
		• Reactivity	×
		• Pro-activeness	✓
		• Social ability	✓
		• Adaptability	×

The system either supports (✓), partially supports (±), or does not support (×) a feature

electrical energy between GenCos, RetailCos and large consumers. For reasons of transaction costs, only the largest consumers can purchase energy directly on the wholesale market. Also, a competitive retail market allows the trading of energy between RetailCos, end-use customers and other participants (e.g., aggregators).

As noted, MATREM agents can be broadly classified as market agents (e.g., GenCos and RetailCos) and assistant agents. Assistant agents are further categorized into interface managers and intelligent assistants. Interface manager agents are responsible for managing the interfaces of the various simulated markets. Intelligent assistant agents provide support to the user in making strategic decisions.

Computationally, the agents have their own thread of control, taking sensory input from the environment—that is, a competitive electricity market—and acting upon that environment by executing actions that affect it. They are autonomous and have (some degree of) control both over their own internal state, and over their behavior. Also, they have their own beliefs—that is, information about themselves, the market, and the other agents competing in the market—their own goals—that is, future states to pursue—and their own intentions—that is, chosen or committed goals that they try to achieve.

The agents are able to act pro-actively, i.e., to exhibit goal-directed behaviour by taking the initiative and selecting actions that will (eventually) lead to the achievement of their goals. To this end, they make use of plan templates, or recipes, specifying particular courses of action that should be undertaken in order to achieve their intentions (committed goals). They are equipped with utility functions and may seek to maximize their utility (e.g., by adopting profit maximization models) or, alternatively, may choose actions that will lead to good (rather than optimal) solutions (e.g., by adopting heuristic strategies during bilateral contracting of electricity). In this way, the agents can reason and make decisions to generate maximally successful behaviour, or they can act appropriately given the available information, by selecting utility-acceptable courses of action, rather than utility-maximizing courses of action.

3.4 Conclusion

Chapter 2 and this chapter have highlighted the need of a generic framework for agent-based simulation of electricity markets. Both chapters have claimed that the development of such a framework can be very important and instructive, providing a coherent set of concepts related to agent-based simulation of EMs, helping to understand the interrelationships of disparate research efforts, and providing a foundation for the development of future models and systems. Accordingly, Chap. 2 (Part I) and this companion chapter (Part II) have introduced a generic framework for agent-based simulation of EMs. The complete framework includes the following three groups (or categories) of dimensions:

1. *Market architecture*: includes the dimensions submarkets, market types and market linkages.
2. *Market structure*: composed by the dimensions market sector and market participants.
3. *Software agents*: includes the dimensions agent architectures and agent capabilities.

The first and the second groups of dimensions, labeled "market architecture" and "market structure", respectively, were the subject of Chap. 2. The last group of dimensions, labeled "software agents", was the subject of this chapter. More specifically, this chapter has described in considerable detail several key features of autonomous software agents. It has introduced the concept of "agent architecture" and discussed six key types of software agents:

1. *Purely reactive agents*: decide what actions to perform based on the present only (with no reference to the past).
2. *Model-based agents*: maintain some kind of internal state that depends on the perceptual input history.
3. *Goal-based agents*: maintain information about the goals they try to accomplish and choose actions that will (eventually) lead to the achievement of such goals.
4. *Utility-based agents*: select actions based on a specific performance measure (defined by an utility function).
5. *Learning agents*: learn which actions to take in order to improve their performance over time.

Also, this chapter has discussed the concept of "agent capability" and introduced a core set of capabilities, which we believe are central to the definition and development of software agents operating in competitive energy markets. These include the following key features of agency: *autonomy*, *reactivity*, *pro-activeness*, *social ability* and *adaptability*. To illustrate the applicability of the complete framework, the chapter has also given an overview of our agent-based simulation tool for electricity markets, called MATREM (for Multi-Agent TRading in Electricity Markets), currently under development, and then described the main features of the tool according to the various components of the framework.

The complete framework can be taken as a starting point from which to develop finer-grained frameworks. Accordingly, two fruitful areas for future work will be to demonstrate the efficacy of the framework and to extend it by incorporating new dimensions. Specifically, future work will focus on the following:

1. Effectiveness of the framework: to select a representative sample of the most prominent models and systems that exist in the technical literature on power markets and software agents and to classify them according to the various elements of the conceptual framework. The aim will not be on providing an exhaustive classification of all the extant models and systems, but rather to choose a number of exemplar models and systems to show how the framework can both help to understand the interrelationships of disparate research efforts (because they are similar on various dimensions) and differentiate seemingly research efforts (because they differ on one or more dimensions).
2. Extension of the framework: to add new dimensions to the three aforementioned groups (or categories) of dimensions and to consider new groups of dimensions to achieve more flexibility and adaptability.

Acknowledgements The work described in this chapter was performed under the project MAN-REM: Multi-agent Negotiation and Risk Management in Electricity Markets (FCOMP-01-0124-FEDER-020397), supported by FEDER Funds, through the program COMPETE ("Programa Operacional Temático Factores de Competividade"), and also National Funds, through FCT ("Fundação para a Ciência e a Tecnologia"). The authors also wish to acknowledge the valuable comments and suggestions made by Hannele Holttinen, from the VTT Technical Research Centre of Finland.

Appendix: Notes on Agents and Agent-based Modeling

Software Agents. Agents are computer systems that perceive the environment with sensors and are able to react over it through actuators. They have several important capabilities, notably autonomy (they decide for themselves which actions to perform in order to satisfy their design objectives) and social ability (they interact with other agents, either to achieve their objectives or to manage the dependencies that ensue from being situated in a common environment). The interactions can vary from simple communication of information to cooperation, collaboration, coordination and negotiation.

An important question is whether the (abstract) architecture or (conceptual) model that underpins software agents should be relatively simple or more sophisticated in

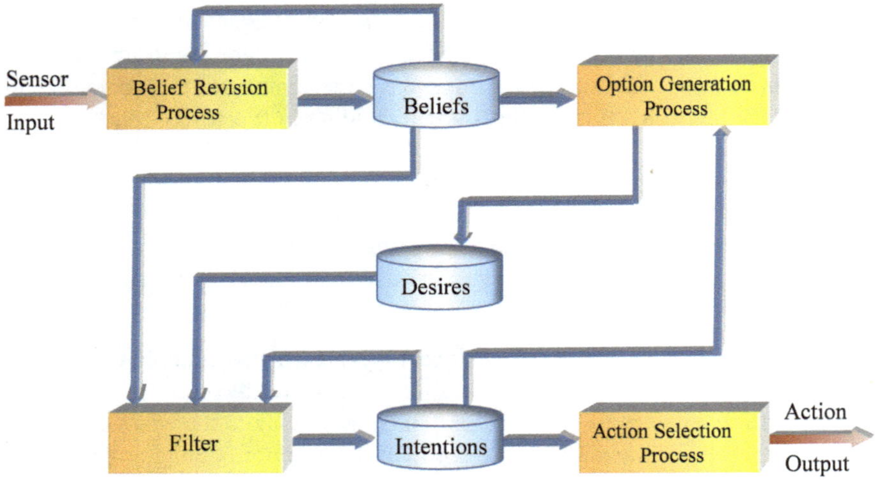

Fig. 3.2 Generic belief-desire-intention (BDI) architecture

nature. A simple and abstract view of an architecture considers sensors and inputs, some sort of internal state, actions and outputs. A more complex and concrete view, based on cognition, considers deliberation (generation of goals or plans, reconsideration of goals, etc.), decision making (choice of options, commitment, etc.) and execution of actions (rules of action, movement, etc.).

An even more sophisticated view, based on the folk-psychology concepts by which human behavior is normally predicted and explained, is shown in Fig. 3.2. Put simply, software agents reason and act according to three key mental attitudes: belief, desire, and intention (and, mainly for this reason, they are called BDI agents). They perceive the world, acquire and update information (beliefs), reason about the objectives to achieve (desires), and deliberate (choose based on preferences) to find the objectives to commit to (intentions).

The belief-desire-intention (BDI) model of practical reasoning is arguably the dominant force in the theoretical foundations of rational agency (see, e.g., [73–76]). However, the question of exactly which combination of mental attitudes is most appropriate to characterize software agents has been the subject of some debate. As a result, several different models that predict and explain agent behavior according to combinations of mental attitudes different from beliefs, desires and intentions, yet often interrelated, have been proposed in the technical literature. Put simply, there are alternatives to the popular use of beliefs, desires and intentions. For example, Shoham [77] suggests that the notion of choice is fundamental. Broersen et al. [78, 79] propose the beliefs-obligations-intentions-desires (BOID) architecture. Schut et al. [80] discuss the integration of the BDI model with partially observable Markov decision processes (POMDP). Simari and Parsons [81] analyze, in detail, several key relationships between the BDI model and (fully observable) Markov Decision Processes. Nair and Tambe [82] present the BDI-POMDP framework for multi-agent teaming. Dimuro et al. [83] extend the BDI-POMDP framework with a module based on the hidden Markov model (HMM). Despite these and other relevant efforts, however, comparatively little work has yet been done on comparing the suitability of different combinations (of mental attitudes) to characterize agents.

Agent-based Modeling versus Equation-based Modeling. Two different types of approaches are face-to-face in competition: system level (equation-based), the traditional type, and individual level (agent-based), the "modern" type. They differ in what is the model and the execution. Equation-based modeling (EBM) operates with variables, and evaluates or integrates sets of equations relating such variables. The model is a set of equations and the execution is supported by evaluating them. Agent-based modeling (ABM) is based on a multitude of agents that encapsulate the behaviors of the diverse individuals that compose a system. The execution consists of emulating such behaviors.

EBM and ABM have common objectives, but differ in both the essential relationships among the entities they model and the level at which they focus attention. Both approaches identify two entities, with a temporal feature: the individuals and the observables. Individuals are characterized by observables and affect their values by specific actions. Observables are related to one another by equations. Individuals interact with one another through their behaviors.

It is worth to highlight a key feature of the models underlying the two approaches. EBM has the equation as the basic unit whereas ABM represents the internal behavior of each individual. This diversity in model structure gives to ABM a significant advantage in most commercial and industrial applications, because the natural unit of system decomposition is the individual rather than the equation, and the physical distribution of computation across multiple processors is naturally desirable.

Agent-based systems are often easier to construct and facilitates the distinction between the physical and the interaction space. Also, they offer an additional level of validation, support direct experimentation, and are easier to translate back into practice. Typically, ABM gives more realistic results than EBM.

References

1. Macal, C., North, M.: Tutorial on agent-based modelling and simulation. J. Simul. **4**, 151–162 (2010)
2. Wooldridge, M.: An Introduction to Multi-agent Systems. Wiley, Chichester (2009)
3. Jennings, N.R., Sycara, K., Wooldridge, M.: A roadmap of agent research and development. Auton. Agents Multi Agent Syst. **1**, 7–38 (1998)
4. Pěchouček, M., Mařík, V.: Industrial deployment of multi-agent technologies: review and selected case studies. Auton. Agents Multi Agent Syst. **17**, 397–431 (2008)
5. Stoft, S.: Power System Economics: Designing Markets for Electricity. IEEE Press and Wiley Interscience, New York (2002)
6. Wood, A., Wollenberg, B., Sheblé, G.: Power Generation, Operation, and Control. Wiley, Chichester (2014)
7. Kirschen, D., Strbac, G.: Fundamentals of Power System Economics. Wiley, Chichester (2004)

8. Ventosa, M., Baíllo, A., Ramos, A., Rivier, M.: Electricity market modeling trends. Energy Policy **33**(7), 897–913 (2005)
9. Zhou, Z.Z., Chan, W.K., Chow, J.H.: Agent-based simulation of electricity markets: a survey of tools. Artif. Intell. Rev. **28**, 305–342 (2007)
10. Harp, S.A., Brignone, S., Wollenberg, B.F., Samad, T.: SEPIA: a simulator for the electric power industry agents. IEEE Control Syst. Mag. **20**(4), 53–69 (2000)
11. Koritarov, V.: Real-world market representation with agents: modeling the electricity market as a complex adaptive system with an agent-based approach. IEEE Power Energy Mag. **2**(4), 39–46 (2004)
12. Batten, D., George Grozev, G.: NEMSIM: finding ways to reduce greenhouse gas emissions using multi-agent electricity modelling. In: Perez, P., Batten, D. (eds.) Complex Science for a Complex World Exploring Human Ecosystems with Agents, pp. 227–252. Australian National University Press, Canberra (2006)
13. Sun, J., Tesfatsion, L.: Dynamic testing of wholesale power market designs: an open-source agent-based framework. Comput. Econ. **30**, 291–327 (2007)
14. Sensfuß, F.: Assessment of the impact of renewable electricity generation on the German electricity sector: an agent-based simulation approach. Ph.D. Dissertation, Karlsruhe University (2007)
15. Vale, Z., Pinto, T., Praça, I., Morais, H.: MASCEM - electricity markets simulation with strategically acting players. IEEE Intell. Syst. **26**(2), 9–17 (2011)
16. Sensfuß, F., Genoese, M., Ragwitz, M., Möst, D.: Agent-based simulation of electricity markets—a literature review. Energy Stud. Rev. **15**(2), 1–29 (2007)
17. Weidlich, A., Veit, D.: A critical survey of agent-based wholesale electricity market models. Energy Econ. **30**, 1728–1759 (2008)
18. Weidlich, A.: Engineering Interrelated Electricity Markets. Physica-Verlag, Heidelberg (2008)
19. Guerci, E., Rastegar, M., Cincotti, S.: Agent-based modeling and simulation of competitive wholesale electricity markets. In: Rebennack, S., Pardalos, P., Pereira, M., Iliadis, N. (eds.) Handbook of Power Systems II, pp. 241–286. Springer, Heidelberg (2010)
20. Franklin, S., Graesser, A.: Is it an agent, or just a program?: a taxonomy for autonomous agents. In: Müller, J., Wooldridge, M., Jennings, N. (eds.) Intelligent Agents III: Agent Theories, Architectures, and Languages, pp. 21–35. Springer, Heidelberg (1997)
21. Wooldridge, J., Jennings, N.: Intelligent agents: theory and practice. Knowl. Eng. Rev. **10**(2), 115–152 (1995)
22. Kaelbling, P.: A situated-automata approach to the design of embedded agents. SIGART Bull. **2**(4), 85–88 (1991)
23. Maes, P.: The agent network architecture (ANA). SIGART Bull. **2**(4), 115–120 (1991)
24. Shehory, O., Sturm, A.: A brief introduction to agents. In: Shehory, O., Sturm, A. (eds.) Agent-Oriented Software Engineering, pp. 3–11. Springer, Heidelberg (2014)
25. Russell, S., Norvig, P.: Artificial Intelligence: A Modern Approach. Pearson Education Inc., New Jersey (2010)
26. Russell, S.: Rationality and intelligence. Artif. Intell. **94**, 57–77 (1997)
27. Russell, S.: Rationality and intelligence. In: Elio, R. (ed.) Common Sense, Reasoning, and Rationality, pp. 37–57. Oxford University Press, New York (2002)
28. Kaelbling, L., Littman, M., Moore, A.: Reinforcement learning: a survey. J. Artif. Intell. Res. **4**, 237–285 (1996)
29. Sen, S., Weiss, G.: Learning in multi-agent systems. In: Weiss, G. (ed.) Multi-Agent Systems: A Modern Approach to Distributed Artificial Intelligence, pp. 259–298. MIT Press, USA (1999)
30. Bishop, C.: Pattern Recognition and Machine Learning. Springer, Heidelberg (2006)
31. Halevy, A., Norvig, P., Pereira, F.: The unreasonable effectiveness of data. IEEE Intell. Syst. **24**(2), 8–12 (2009)
32. Müller, J.: The Design of Intelligent Agents: A Layered Approach. Springer, Heidelberg (1996)
33. Müller, J.: Architectures and applications of intelligent agents: a survey. Knowl. Eng. Rev. **13**(4), 353–380 (1998)

34. Bratman, M., Israel, D., Pollack, M.: Plans and resource-bounded practical reasoning. Comput. Intell. **4**, 349–355 (1988)
35. Georgeff, M., Ingrand, F.: Decision-making in an embedded reasoning system. In: Proceedings of the Eleventh International Joint Conference on Artificial Intelligence (IJCAI-89), Detroit, Michigan, pp. 972–978 (1989)
36. Ferguson, I.: Touring machines: autonomous agents with attitudes. IEEE Comput. **25**(5), 51–55 (1992)
37. Sridharan, N.: 1986 workshop on distributed AI. AI Mag. **8**(3), 75–85 (1987)
38. Huhns, M., Singh, M.: Agents and multiagent systems: themes, approaches, and challenges. In: Huhns, M., Singh, M. (eds.) Readings in Agents, pp. 1–23. Morgan Kaufmann, San Francisco (1998)
39. Scerri, P., Pynadath, D., Tambe, M.: Adjustable autonomy for the real world. In: Hexmoor, H., Castelfranci, C., Falcone, R. (eds.) Agent Autonomy, pp. 211–241. Springer Science+Business Media, New York (2003)
40. Thrun, S., Montemerlo, M., Dahlkamp, H., Stavens, D., Aron, A., Diebel, J., Fong, P., Gale, J., Halpenny, M., Hoffmann, G., Lau, K., Oakley, C., Palatucci, M., Pratt, V., Stang, P., Strohband, S., Dupont, C., Jendrossek, L.-E., Koelen, C., Markey, C., Rummel, C., van Niekerk, J., Jensen, E., Alessandrini, P., Bradski, G., Davies, B., Ettinger, S., Kaehler, A., Nefian, A., Mahoney, P.: Stanley: the robot that won the DARPA grand challenge. In: Buehler, M., Iagnemma, K., Singh, S. (eds.) The 2005 DARPA Grand Challenge, pp. 1–43. Springer, Berlin (2007)
41. Gruber, T.: A translation approach to portable ontology specifications. Knowl. Acquis. **5**(2), 199–220 (1993)
42. Noy, N., McGuinness, D.: Ontology development 101: a guide to creating your first ontology. Technical report KSL-01-05, Knowledge Systems Laboratory, Stanford University, USA (2001)
43. Bechhofer, S., van Harmelen, F., Horrocks, I., McGuinness, D., Patel-Schneider, P., Stein, L.: OWL Web Ontology Language Reference (2004). http://www.w3.org/TR/2004/REC-owl-ref-20040210/. Accessed 12 Jan 2017
44. W3C Recommendation: The Extensible Markup Language (XML) 1.0 (2008). http://www.w3.org/TR/REC-xml/. Accessed 12 Jan 2017
45. Finin, T., Fritzson, R., McKay, D., McEntire, R.: KQML – a language and protocol for knowledge and information exchange. In: Proceedings of the 13th International Distributed Artificial Intelligence Workshop, pp. 93–103. AAAI Press, Menlo Park, California (1994)
46. Austin, J.: How to Do Things With Words. Oxford University Press, London (1962)
47. Searle, J.: Speech Acts: An Essay in the Philosophy of Language. Cambridge University Press, London (1969)
48. Jeon, H., Petrie, C., Cutkosky, M.: JATLite: a java agent infrastructure with message routing. IEEE Internet Comput. **4**(2), 87–96 (2000)
49. FIPA: ACL Message Structure Specification. Foundation for Intelligent Physical Agents, Document Number SC00061G (2002). http://www.fipa.org/specs/fipa00061/. Accessed 12 Jan 2017
50. FIPA: Communicative Act Library Specification. Foundation for Intelligent Physical Agents, Document Number SC00037J (2002). http://www.fipa.org/specs/fipa00037/. Accessed 12 Jan 2017
51. Cohen, P., Levesque, H.: Rational interaction as the basis for communication. In: Cohen, P., Morgan, J., Pollack, M. (eds.) Intentions in Communication, pp. 221–256. The MIT Press, Cambridge (1990)
52. Bretier, P., Sadek, D.: A rational agent as the kernel of a cooperative spoken dialogue system: implementing a logical theory of interaction. In: Müller, J., Wooldridge, M., Jennings, N. (eds.) Intelligent Agents III (LNAI 1193), pp. 189–203. Springer, Heidelberg (1997)
53. Bellifemine, F., Caire, G., Greenwood, D.: Developing Multi-agent Systems with JADE. Wiley, Chichester (2007)
54. Bond, A., Gasser, L.: An analysis of problems and research in DAI. In: Bond, A., Gasser, L. (eds.) Readings in Distributed Artificial Intelligence, pp. 3–35. Morgan Kaufmann Publishers, San Mateo (1988)

55. Bobrow, D.: Dimensions of interaction. AI Mag. **12**(3), 64–80 (1991)
56. Lopes, F., Wooldridge, M., Novais, A.Q.: Negotiation among autonomous computational agents: principles, analysis and challenges. Artif. Intell. Rev. **29**, 1–44 (2008)
57. Weiss, G. (ed.): Multiagent Systems. The MIT Press, Cambridge (2013)
58. Lopes, F., Coelho, H. (eds.): Negotiation and Argumentation in Multi-agent Systems. Bentham Science, The Netherlands (2014)
59. Poole, D., Mackworth, A.: Artificial Intelligence: Foundations of Computational Agents. Cambridge University Press, Cambridge (2010)
60. Goodwin, R.: Formalizing properties of agents. Technical report CMUCS93159, School of Computer Science, Carnegie-Mellon University, Pittsburgh (1993)
61. Braubach, L., Pokahr, A., Lamersdorf, W.: Jadex: a BDI-agent system combining middleware and reasoning. In: Unland, R., Klusch, M., Calisti, M. (eds.) Software Agent-Based Applications, Platforms and Development Kits, pp. 143–168. Birkhäuser Verlag, part of Springer Science+Business Media, Basel, Switzerland (2005)
62. Vidigal, D., Lopes, F., Pronto, A., Santana, J.: Agent-based simulation of wholesale energy markets: a case study on renewable generation. In: Spies, M., Wagner, R., Tjoa, A. (eds.) 26th Database and Expert Systems Applications (DEXA 2015), pp. 81–85. IEEE (2015)
63. Algarvio, H., Couto, A., Lopes, F., Estanqueiro, A., Santana, J.: Multi-agent energy markets with high levels of renewable generation: a case-study on forecast uncertainty and market closing time. In: Omatu, S. et al. (eds.) 13th International Conference on Distributed Computing and Artificial Intelligence, pp. 339–347. Springer International Publishing (2016)
64. Algarvio, H., Couto, A., Lopes, F., Estanqueiro, A., Holttinen, H., Santana, J.: Agent-based simulation of day-ahead energy markets: impact of forecast uncertainty and market closing time on energy prices. In: Tjoa, A., Vale, Z., Wagner, R. (eds.) 27th Database and Expert Systems Applications (DEXA 2016), pp. 166–70. IEEE (2016)
65. Lopes, F., Rodrigues, T., Sousa, J.: Negotiating bilateral contracts in a multi-agent electricity market: a case study. In: Hameurlain, A., Tjoa, A., Wagner, R. (eds.) 23rd Database and Expert Systems Applications (DEXA 2012), pp. 326–330. IEEE (2012)
66. Algarvio, H., Lopes, F., Santana, J.: Bilateral contracting in multi-agent energy markets: forward contracts and risk management. In: Bajo, J. et al. (eds.) Highlights of Practical Applications of Agents, Multi-Agent Systems, and Sustainability: The PAAMS Collection (PAAMS 2015), pp. 260–269. Springer International Publishing (2015)
67. Sousa, F., Lopes, F., Santana, J.: Contracts for difference and risk management in multi-agent energy markets. In: Demazeau, Y., Decker, K., Pérez, J., De la Prieta, F. (eds.) Advances in Practical Applications of Agents, Multi-Agent Systems, and Sustainability: The PAAMS Collection (PAAMS 2015), pp. 339–347. Springer International Publishing (2015)
68. Sousa, F., Lopes, F., Santana, J.: Multi-agent electricity markets: a case study on contracts for difference. In: Spies, M., Wagner, R., Tjoa, A. (eds.) 26th Database and Expert Systems Applications (DEXA 2015), pp. 88–90. IEEE (2015)
69. Lopes, F., Coelho, H.: Strategic and tactical behaviour in automated negotiation. Int. J. Artif. Intell. **4**(S10), 35–63 (2010)
70. Osborne, M., Rubinstein, A.: Bargaining and Markets. Academic Press, London (1990)
71. Algarvio, H., Lopes, F., Santana, J.: Multi-agent retail energy markets: bilateral contracting and coalitions of end-use customers. In: 12th International Conference on the European Energy Market (EEM 2015), pp. 1–5. IEEE (2015)
72. Algarvio, H., Lopes, F., Santana, J.: Multi-agent retail energy markets: contract negotiation, customer coalitions and a real-world case study. In: Demazeau, I., Takayuki, I., Javier, B., Escalona, M. (eds.) Advances in Practical Applications of Scalable Multi-agent Systems (The PAAMS Collection), LNAI 9662, pp. 13–23. Springer International Publishing (2016)
73. Bratman, M.: Intentions, Plans, and Practical Reason. Harvard University Press, Cambridge (1987)
74. Haddadi, A., Sundermeyer, K.: Belief-desire-intention agent architectures. In: O'Hare, G., Jennings, N. (eds.) Foundations of Distributed Artificial Intelligence, pp. 169–185. Wiley, New York (1996)

75. D'Inverno, M., Luck, M., Georgeff, M., Kinny, D., Wooldridge, M.: The dMARS architecture: a specification of the distributed multi-agent reasoning system. Auton. Agents Multi Agent Syst. **9**, 5–53 (2004)
76. Burmeister, B., Arnold, M., Copaciu, F., Rimassa, G.: BDI-agents for agile goal-oriented business processes. In: Berger, M., Burg, B., Nishiyama, S. (eds.) International Conference on Autonomus Agents and Multi-agent Systems (Industry and Applications Track) IFAAMAS, pp. 37–44 (2008)
77. Shoham, Y.: Agent-oriented programming. Artif. Intell. **60**, 51–92 (1993)
78. Broersen, J., Dastani, M., Hulstijn, J., Huang, Z., van der Torre, L.: The BOID Architecture. In: André, E., Sen, S., Frasson, C., Müller, J. (eds.) International Conference on Autonomous Agents (AGENTS-01), pp. 9–16. ACM Press, New York (2001)
79. Broersen, J., Dastani, M., Hulstijn, J., van der Torre, L.: Goal generation in the BOID architecture. Cogn. Sci. Q. **2**(3–4), 428–447 (2002)
80. Schut, M., Wooldridge, M., Parsons, S.: On partially observable MDPs and BDI models. In: dInverno, M., Luck, M., Fisher, M., Preist, C. (eds.) Foundations and Applications of Multi-Agent Systems (UKMAS Workshops 1996-2000, Selected Papers), LNAI 2403, pp. 243–259. Springer, Berlin (2002)
81. Simari, G., Parsons, S.: Markov Decision Processes and the Belief-Desire-Intention Model: Bridging the Gap for Autonomous Agents. Springer, Heidelberg (2011)
82. Nair, R., Tambe, M.: Hybrid BDI-POMDP framework for multiagent teaming. J. Artif. Intell. Res. **23**, 367–420 (2005)
83. Dimuro, G., Costa, A., Gonçalves, L., Pereira, D.: Recognizing and learning models of social exchange strategies for the regulation of social interactions in open agent societies. J. Braz. Comput. Soc. **17**, 143–161 (2011)

Part II
Electricity Markets with Large Penetrations of Variable Generation: Current and Emerging Designs

Chapter 4
Market Prices in a Power Market with More Than 50% Wind Power

Klaus Skytte and Poul Erik Grohnheit

Abstract Denmark has the highest proportion of wind power in the world. Wind power provided a world record of 39.1% of the total annual Danish electricity consumption in 2014 with as much as 51.7% in Western Denmark. Many would argue that the present power markets are not designed for such high shares of wind power production and that it would be hard to get good and stable prices. However, analyses in this chapter show that the Nordic power market works, extreme events have been few, and the current infrastructure and market organization has been able to handle the amount of wind power installed so far. It is found that geographical bidding areas for the wholesale electricity market reflect external transmission constraints caused by wind power. The analyses in this chapter use hourly data from West Denmark—which has the highest share of wind energy in Denmark and which is a separate price area at the Nordic power exchange. Data have been collected from the last ten years and periods with extreme wind conditions are used as case studies to illustrate the robustness of our findings.

4.1 Introduction

Denmark is one of the Nordic countries that have set up ambitious long-term targets of carbon neutral energy systems with a fossile free electricity sector. To reach the ambitious energy and climate goals of carbon neutral energy systems in the Nordic countries, a large share of variable renewable energy sources will be deployed, especially wind power, in addition to other traditional storable renewable energy sources such as biomass and hydropower. In 2020, it is assumed that 50% of the yearly electricity generation will come from wind energy (ENS [1]). And in 2035,

K. Skytte (✉) · P. E. Grohnheit
DTU Management Engineering, Technical University of Denmark, Building 426, Produktionstorvet, 2800 Kongens Lyngby, Denmark
e-mail: klsk@dtu.dk

P. E. Grohnheit
e-mail: pogr@dtu.dk

F. Lopes and H. Coelho (eds.), *Electricity Markets with Increasing Levels of Renewable Generation: Structure, Operation, Agent-based Simulation, and Emerging Designs*, Studies in Systems, Decision and Control 144, https://doi.org/10.1007/978-3-319-74263-2_4

the political goal is to have 100% renewable energy based electricity generation in Denmark, which includes a high share of wind energy. By nature, the temporal supply of wind power is highly variable because it is determined by weather conditions, it is uncertain due to forecasting errors, and it is location specific as the primary energy carrier cannot be transported like coal or biomass (Borenstein [2], Hirth et al. [3]).

Such properties imply major integration and interfacing challenges of wind power with the energy system. Previously, the challenges related to variability has been studied by Lamont [4], Borenstein [5], Joskow [6], Nicolosi [7], Holttinen et al. [8], and Hirth [9]. Wind power will affect balancing costs as a consequence of forecasting errors (Smith et al. [10], Holttinen et al. [8], and Hirth [9]), and increase the costs of distribution and transmission networks (Brown and Rowlands [11], Lewis [12], and Hamidi et al. [13]).

In addition, wind power may influence the costs of firm reserve capacity and lower utilization rates and cause more cycling and ramping of traditional plants (Ueckerdt and Hirth [14]). At high wind power penetration rates, the overall integration costs could be substantial (Ueckerdt and Hirth [14], Hirth et al. [3], and IEA [15]). Consequently, cost effective integration of wind power has become a pressing challenge in the energy sector.

In liberalized markets with perfect competition (Olsen and Skytte [16], Skytte [17]), it is assumed that power supplier submit their bids at the power markets at marginal cost. However, wind power generation, as well as hydro power and photovoltaic, have no fuel cost and can be assumed to have almost zero marginal costs; all current costs are fixed costs that depends on non-negative generation. Whereas most thermal power generation (e.g., biomass or natural gas fired power plants) have fuel costs and can thereby submit bids at the power market at marginal cost that are higher than zero.

The zero marginal cost in addition to the variable supply of wind power may imply that negative prices occur at the power market. Negative prices are a signal from the market that there is excess production. Production in these hours means that you have to pay to "sell" your production. For example in hours with lower electricity demand than wind power generation (Nicolosi [7]).

Therefore, many would argue that the present power markets are not designed for high shares of wind power production and that it would be hard to get good and stable electricity prices (Skytte and Ropenus [18–20]). However, so far the Nordic power market has shown relatively stable prices, extreme events have been few, and the current infrastructure and market organization has been able to handle the amount of wind power installed. The reasoning behind this may be found in the technology mix and in the design of the Nordic power market. Jacobsen and Zvingilaite [21] and Grohnheit et al. [22] discuss this for the years 2006–2008 and for 2004–2010.

The share of wind has increased a lot during the last years. In this chapter, we analyze the market data up to 2014 from the Nordic power exchange Nord Pool with respect to wind power data, in order to see which influence wind power has. We take departure in the Western part of Denmark, which is the area with the highest share of wind power. In Sect. 4.2, we describe the design of the Nordic power market.

The prices observed in Western Denmark are discussed and analyzed in Sect. 4.3. Section 4.4 looks at extreme hours, and finally Sects. 4.5 and 4.6 discuss perspectives and findings.

4.1.1 Wind Energy in Denmark

Denmark as a wind energy country is a nation that many others are looking to in order to discover sustainable energy solutions for the future. Denmark is a pioneer within wind energy. In the 1950s Johannes Juul, a Danish engineer, made a number of experimental turbines. It was Juul who was the first to connect a wind turbine with an (asynchronous) AC generator to the electrical grid, and around 1956 Juul built the stable three-bladed wind turbine, the Gedser wind turbine, which today underlies the Danish wind turbine design and has become a global standard. The Gedser wind turbine was in operation for many years and gave confidence in the technology.

After the energy crises in the 1970s wind turbines became popular. The first two wind turbines were connected to the power grid in 1976, and from 1978 the sales of serial produced wind turbines started. In the 1980s and 1990s many single turbines were deployed in Denmark. This resulted in many small turbines but with a low total share of the total domestic power consumption. From 1996 the number and installed capacity increased a lot.

Electricity in Denmark is divided into two geographical markets (east and west of the Great Belt), each with strong connection to the neighbor markets. The Western part of Denmark has had the largest deployment and has therefore the largest share of wind power.

In 2004, wind power provided 23.4% of the electricity consumption in Western Denmark (on a national level the share was 18.8% of the total Danish electricity consumption). Ten years later in 2014, the figure had risen to 51.7% in Western Denmark and to 39.1% on a national level of the domestic electricity consumption [23]. This is a new world record.

On a monthly basis the share was much greater in some months. Only in January 2014 the proportion was 61.4 percent at the national level. On an hourly basis wind power provided more than 100% of the Danish power consumption in several hours of the year—leading to export to the surrounding countries.

Today, total wind energy capacity in Denmark is almost 5,000 MW with nearly 1,300 MW (2014) located offshore.

Thus, wind power has a significant impact not only on the hourly price on the day-ahead spot market, but also the intraday or real-time markets. The impact of the volatility of wind power is reduced by market-driven trade.

Wind energy is traded at the Nordic power market in addition to receiving either feed-in premium (Denmark) or green certificates (Norway-Sweden). Jensen and Skytte [24, 25], Morthorst et al. [26], and Skytte [27] discuss the interactions between the power and green certificate markets.

4.2 The Nordic Power Market

Denmark is part of the Nordic power market. The common Nordic power market dates back to 1971 and was originally designed to balance out variations in precipitation and water inflow to hydro power stations. From 1993 it became a market place open for all generators and consumers of electricity in Norway, and expanded to the other Nordic countries in the following years (Amundsen and Bergman [28]). It was developed to exploit beneficial interaction between hydro power and large thermal plants, conventional coal power and nuclear power plants. This showed to be very effective. In wet precipitation years, the flexible conventional power plants in Denmark and Finland were dispatched and the excess generation from the Norwegian and Swedish hydro power plants were exported to Denmark and Finland (and later on also to other countries in Northern Europe). In dry precipitation years, Norway and Sweden imported power.

Simultaneously with the liberalization of the energy markets, a market based system (Nord Pool) was introduced and even small power suppliers were allowed to trade, making a very liquid market with reliable prices. The Nordic power exchange, Nord Pool (from March 2010 within Nasdaq OMX Commodities), is now covering Denmark, Norway, Sweden and Finland and parts of Germany and the Baltics (Skytte [17]). Since 1999 and 2000 the two parts of Denmark (West and East) have been bidding areas with separate prices at Nord Pool.

Nord Pool operates a day-ahead spot market with regional hourly prices (Elspot), an intraday market with continuous power trading up to one hour prior to delivery (Elbas – Electricity Balance Adjustment Service), a Regulating power market and a financial market for the following days, weeks, months and annual contracts up to five years. The participants in the markets are power producers, distributors, industries and brokers. Nord Pool Spot AS acts as counterpart in all contracts and all trades are physically settled with respective Transmission System Operators (TSO)s.[1]

Deviations between planned supply and demand in real time must then be covered by balancing power at the regulating power market. Thus, the fundamental reason for having a balancing market is uncertainty about supply and demand. Regulating power is production capacity or consumption offered by the market players to the balance responsible TSO during the actual day of operation. Skytte [29] revealed the pattern of the prices on the Nordic regulating power market. The level of the regulating power price depends on the level of the corresponding spot price and the amount of regulation needed. Compared to the spot price there is a premium of readiness which is independent of the amount of regulating power but depended of the corresponding spot price.

On the day-ahead market a "system price" is calculated covering the whole area of Nord Pool assuming no network constraints. In hours when congestion occurs on interconnections between bidding areas, Finland, Sweden, Norway (divided in two or more areas), and Denmark (east and west) separate day-ahead area prices are calculated on the basis of the bids from each area. It means that congestion is

[1] http://www.nordpoolspot.com/ (accessed on December 2017).

Table 4.1 Electricity flows and capacities in Western Denmark

Western Denmark	2004	2005	2006	2007	2008	2009	2010	2011	2012	2013	2014
Consumption (TWh)	20.9	21.0	21.4	21.6	21.6	20.6	21.1	20.7	20.4	20.1	20.2
Wind production (TWh)	4.9	5.0	4.6	5.6	5.2	5.1	5.9	7.1	7.6	8.7	10.3
Net import, North (TWh)	−1.6	6.2	−2.4	3.5	5.6	1.1	−3.7	2.0	6.9	−1.0	1.7
Net import, South (TWh)	−1.8	−6.7	−2.1	−5.2	−6.6	−3.2	1.6	−1.5	−4.6	1.1	−0.7
Net import, total (TWh)	−3.4	−0.6	−4.5	−1.8	−1.0	−2.2	−2.0	0.5	2.3	0.0	1.0
Share of wind (%)	22	24	22	26	24	25	28	34	37	43	51
Max. load (GW)	3.6	3.7	3.8	3.7	3.7	3.7	3.7	3.7	3.7	3.6	3.5
Min. load (GW)	1.3	1.3	1.4	1.4	1.3	1.3	1.3	1.3	1.2	1.4	1.2
Max. wind (GW)	2.2	2.2	2.2	2.2	3.6	2.9	2.5	2.7	2.9	3.9	3.5

managed by price differences resulting from these implicit auctions (market splitting) on the interconnectors to Norway, Sweden and between eastern Denmark and Germany (KONTEK). Explicit auctions are used on the interconnector between Western Denmark and Germany. The spot market bids are stated before noon for next day's operation (12–36 h before delivery).

4.3 Area Prices in Western Denmark 2004–2014

Detailed market data from all the markets and price areas per hour are available since 2000.[2] For Western Denmark the minimum hourly demand in all the years between 2004 and 2014 was about 1.3 GW, and the maximum wind production varied between 2.2 and 3.9 GW (see Table 4.1). The installed capacities and transmission capacities were nearly the same in all the years from 2004 to 2009, although a large number of small wind turbines were replaced by larger units in on-shore wind parks. Wind production in 2007 was higher than the previous and following years. New off-shore windparks in 2009 and 2010 have lead to higher wind production in 2010 than in 2007.

The years 2004, 2006 and 2010 were dry years in Norway and Sweden, leading to import from Denmark, while 2005 and 2007–2009 have been more wet years with export to Denmark and further to Germany. In all the years except 2010 there were exports to Germany, largest in wet years with large import from the north.

Strong interconnections between Western Denmark and other regions (up to 1.7 GW for export to Northern Germany with very similar conditions for wind power, and 1.7 GW transmission capacity to Norway and Sweden with little wind capacity and large hydro storage capability) will reduce the number of events with consecutive hours with high prices due to lack of generation from wind.

[2]http://energinet.dk/EN/Sider/default.aspx (accessed on December 2017).

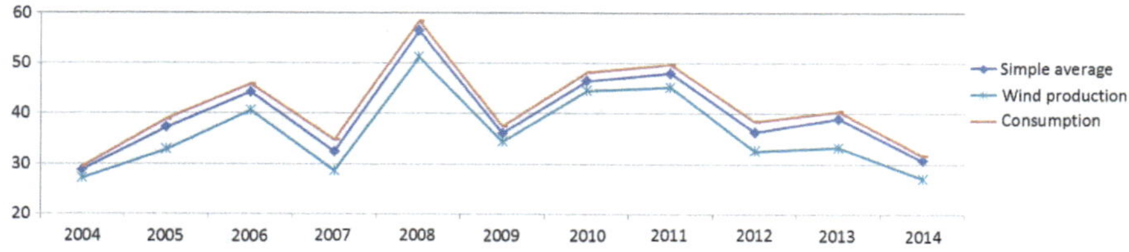

Fig. 4.1 Average Nord Pool area prices for Western Denmark with different weights

Until August 2010, there was no connection between Western and Eastern Denmark, because the thermal capacity of the two systems is very similar, with only small potential gains from trade. This is changed by the larger wind capacity. The capacity on the new HVDC link is 590 MW. After the start of permanent operation more than 90% of the transfer went from west to east during the rest of 2010.

4.3.1 Prices in Western Denmark 2004–2014

Both electricity consumption and wind power supply affect the electricity prices. This can be illustrated by looking at the simple average of hourly prices per year. The simple average of hourly prices—with equal weight to all hours during the year—is slightly lower than the average weighted by total production or consumption, because the larger consumption leads to higher prices on an hourly basis, and total production follows the demand. The average price weighted by wind production is lower than other prices (see Fig. 4.1).

The lower average price weighted by wind production is due to the fact that wind power supply does not follow the hourly demand for electricity. However, the wind generation per week or month is higher in the cold months where the demand is also high. This implies that, on an hourly basis day-ahead prices are negatively correlated with wind production to the extend that the weather forecast for strong wind sometimes can be read from the Nord Pool area prices, whereas there is little correlation between wind production and electricity demand.

4.3.2 Seasonal Variation of Demand and Wind Production

On a seasonal basis—measured on average monthly prices for the years 2004 to 2014—there is no systematic seasonal variation in electricity prices, but Fig. 4.2 clearly shows that both electricity demand and wind power generation are larger during winter months than during summer months, although the wind power production in each month is very different from year to year, while the annual variation in monthly consumption is much smaller. In particular, the monthly wind power

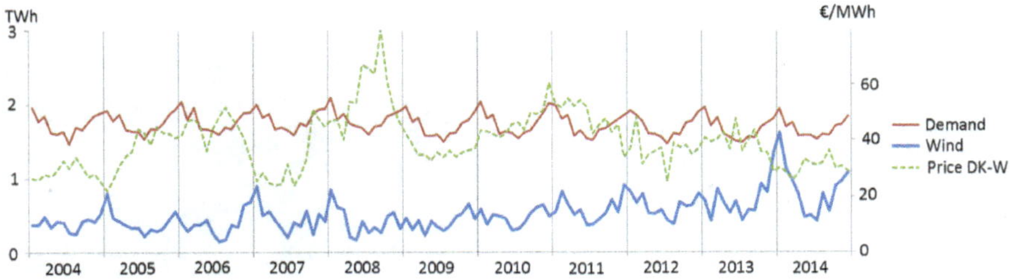

Fig. 4.2 Monthly wind production and electricity demand, and average electricity prices in Western Denmark, 2004–2014

generation was large in January 2005, 2007, 2008, and especially in January 2014 compared to the other monthly data.

2014 was a "normal" average wind year. However it was very windy during the first few months and with very little wind during the summer. For the month of January 2014, the share of the total electricity demand was over 83% for Western Denmark. The month of lowest wind power share was July at 28%.

With the Danish climate the demand for air conditioning during the summer is small. In neighboring Norway the seasonal correlation between wind and demand is even greater due to colder weather and widespread use of electric heating. Thus the increasing amount of wind power will add to the capacity of the hydro reservoirs as seasonal storages for electricity, depending on the management of the existing reservoirs. There may even be a much larger potential for short-term storages by pumping from lower reservoirs to higher. This process of using this potential has not yet started.

The possibility of hydro storage in neighboring countries reduces the variation in hourly prices, but the effect varies among the years, depending on the precipitation and inflow to the storages, and variation in the production from wind increases the variation in prices. In wet years the average price in the Danish area is low and the effect of varying production from wind is limited, while in dry years both the average price and the variation in hourly prices due to wind are high.

4.3.3 International Trade

The area price of Western Denmark is normally between the Nord Pool System price and the price on EEX (European Electricity Exchange, EEX—now EPEX Spot—covering Germany), except for the dry years 2006 and 2010, where prices on all these markets were relatively high. The much higher prices in 2008 are reflecting the much higher EUA (CO2 allowances) prices in 2008 than the almost zero price level for 2007. Figure 4.3 compares the average hourly prices in the five years from north (Nord Pool system price) to south (EEX).

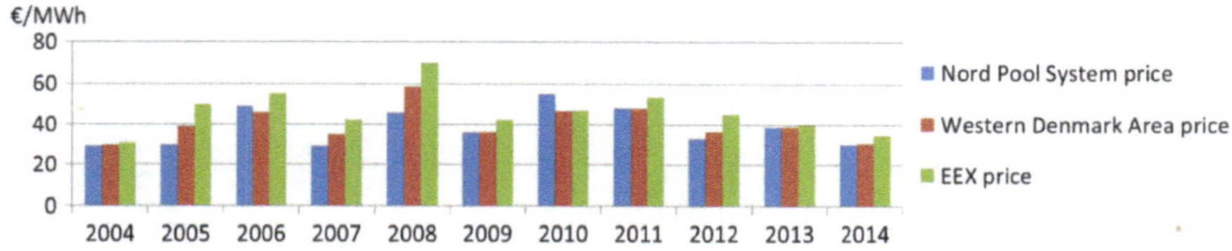

Fig. 4.3 Hourly prices weighted by consumption in Western Denmark

Figure 4.4 shows the number of hours with differences between the area price for Western Denmark and the Nord Pool system price. In all years the area price in Western Denmark was within the interval of ±10 €/MWh from the Nord Pool system price in most of the time (between 73 and 97% in most years, but with only 52% in 2008 and 56% in 2005). Of these hours the area price was identical to the Nord Pool system price in 29% of the hours in 2005 and only 4% in 2008 and 2014. In the remaining hours prices in Western Denmark were different due to transmission constraints. The number of hours with extreme price below 5 €/MWh were negligible with around 1% of the hours, except for 2007 and 2014 which had prices below 5 €/MWh in 2% of the hours. Likewise for prices above 100 €/MWh, except for 2008, which was found in 2% of the hours.

The year 2009 was a "normal" hydrological year in Norway and Sweden. The average Nord Pool system price and the area price of Western Denmark were the same, although the hourly prices were identical in only 14% of the hours. In the wet years 2005, 2007 and 2008 the annual average prices on the EEX were between 13 and 24 €/MWh higher than the Nord Pool prices, while in the remaining years the average EEX prices were between 6 €/MWh above and 8 €/MWh below the Nord Pool System prices.

The regional price differences are reflected in the trade pattern between Western Denmark and the neighboring regions. In all the years, Western Denmark was a net exporter of electricity, ranging from 0.6 TWh in the wet year 2005 to 4.5 TWh in the dry year 2006. The annual variations in wind power are reflected in thermal generation rather than traded volumes.

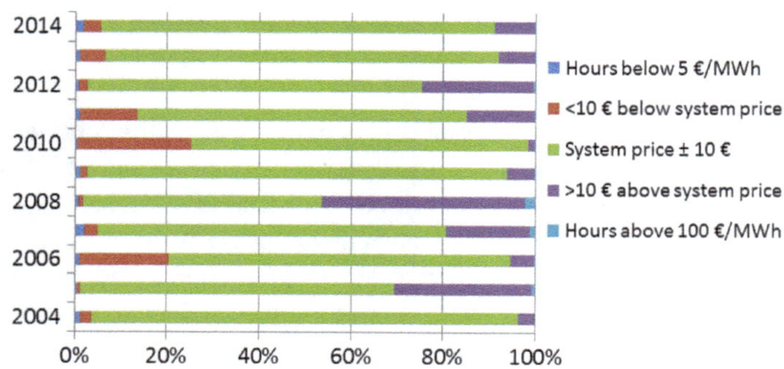

Fig. 4.4 Western Denmark. Differences between area prices and Nord Pool system price, 2004–2014

4.4 Extreme Hours

As noted in Table 4.1 there is a large difference between the minimum load and the maximum wind power generation on an hourly basis. Therefore, one could expect extreme prices (low or negative) in the hours where both the demand load is low and the wind generation is high.

As noted in Fig. 4.2, the winter 2013–2014 was very windy. For 21 December 2013, the wind share of the electricity consumption was 138% in Western Denmark, and for 2 h the share was 178%.

Table 4.2 shows that the number of hours with area prices, below 5 €/MWh or above 100 €/MWh, is quite small, and the number of such extreme hours have decreased since 2007 or 2008. These criteria were used to identify extreme prices. The maximum number of occurrences in a single year is just below 200. Other extremes are chosen for high and low wind. These are less than 1% of the hourly demand (low wind) and more than 100% of the demand (high wind) in the current hour. The number of annual occurrences of "low wind" (<10%) was between 260 and 381, and between 20 and 1225 for "high wind" (>100%). The selected "extreme" hours were used for an analysis focusing on consecutive hours and prices for up-regulation and down-regulation on the balancing market. The tool for this analysis is the filter feature in an Excel database with the hourly observations of market data.

Observations of hourly wind production are not stochastically independent. They follow the pattern of meteorological data. It means that extreme events occur over several hours or days. They are predictable some days before. This is both reflected in the spot and in the regulating power price at Nord Pool. Table 4.2 shows that there are a number of hours with negative down-regulation prices. These negative prices are due to high start and stop costs of decentralized CHP generation in Denmark. The annual number of these events range from 34 in 2005 to 267 in 2010. Also down-regulation by using electricity in electric boilers in district heating systems may lead to negative prices for down-regulation. The price for up-regulation is always equal to or higher than the day-ahead price (Skytte [29]). In one year, 2008, up-regulation prices were more than 100 €/MWh higher than the day-ahead price in nearly 600 h and above 200 MWh higher in 120 h. In all other years the number of these events were much lower.

In the short term negative prices on the spot market are considered as the most important additional measure to address the challenge of the large amount of intermittent generation. Negative price bids were introduced on the German EEX spot market in 2008, and from October 2009 a negative price floor at −200 €/MWh was introduced by Nord Pool. This is significant mainly for Denmark. In the remaining 2009, there were 27 h in Western Denmark with zero prices and 8 consecutive hours with negative prices between 33 and 120 €/MWh. Starting from October 2008, there were 15 h with negative prices in Germany in 2008 and 71 h in 2009. The average of negative prices was −41 €/MWh. In 2010 there were 12 h with small negative values in both Denmark and Germany.

Table 4.2 Number of hours with extreme values in Nord Pool bidding area Western Denmark

	2004	2005	2006	2007	2008	2009	2010	2011	2012	2013	2014
Spot price <0€/MWh, hours						9	11	15	33	39	46
Spot price <5€/MWh, hours	100	55	80	185	62	102	43	93	72	95	164
Spot price >100€/MWh	0	61	11	105	193	4	8	2	46	7	4
Spot price >200€/MWh	0	7	0	26	0	2	0	0	1	5	0
Spot price >400€/MWh	0	0	0	5	0	0	0	0	0	5	0
Wind >100% of demand, hours	20	42	26	50	43	33	71	233	342	851	1225
Wind <10% of demand, hours	3009	2887	3199	2946	3125	2778	2405	1907	1907	1907	1907
Wind <1% of demand, hours	323	298	381	371	352	373	260	231	231	231	231
Price >100% and Wind <1% dem.	0	9	0	13	14	0	0	0	3	0	0
Down-reg. negative price, hours	92	34	201	137	46	127	267	107	59	96	106
Up-reg. 100€/MWh, hours	0	168	68	204	585	77	163	108	213	101	69
Up-reg. 200€/MWh, hours	0	24	1	65	120	23	32	50	55	31	19

4.5 Perspective for Other Regions

Geographically Western Denmark is a small area, some 300 km north-south and less east-west. Congestion in the transmission or distribution network within this area is usually unimportant. On the other hand, the impact of capacity constraints to neighboring bidding areas can be analyzed using available data. Some of the results found in this analysis may be valid for other regions.

In most regions of Europe, existing bidding areas follow national boundaries. This may not be efficient in large countries with a large penetration of wind power in windy areas and bottlenecks in the transmission system. Splitting national markets into bidding areas that reflect these constraints have been practiced in the Nordic region for more than two decades. This leads to prices that reflect the expected amount of supply of wind power in each area, among other variables influencing the energy dispatch. To get the right price signals for generators and consumers, it is becoming increasingly important that the geographical bidding areas for the day-ahead market reflect the pattern of wind variations and transmission constraints.

Market splitting into areas prices will also lead to more transparency concerning the need for new transmission capacity. Large and frequent price differences between neighboring price areas clearly indicate the need for new transmission lines.

The Netherlands is similar to Western Denmark in area, climate, international connections, wind power and cogeneration volumes, but the electricity system is six times larger. This indicates that this type of market splitting is very unlikely to be useful within the Netherlands. In the UK the situation is quite different. There is already an imbalance between the location of generating capacity in the north and population centres in the south. So a large amount of wind power mainly in the north will add to the imbalance.

Some regions in Germany and Spain may have larger shares of wind power than Western Denmark, but so far, these regions are not identified as bidding areas in the electricity markets.

4.6 Findings and Final Comments

The current infrastructure and market organization in Western Denmark and within the Nordic power market is able to handle the current amount of wind power at more than 50% of the regional consumption in Western Denmark. The most important features to handle the variability and unpredictability of wind power in this region are the international transmission lines and the large amount of Hydropower and of CHP systems with heat storages within the region.

The analysis emphasizes the role of the market design and the accumulation of experience to be gained over many years. The long tradition and the design of the Nordic power market (Nord Pool) that was original designed to balance out variations in precipitation and water inflow to the hydro power stations, have implied a very flexible power system with a high share of *energy flexibility* on a yearly basis offered by thermal plants.

The last decades' increasing deployment of variable renewable energy, especially wind power calls for another kind of flexibility in the system, namely *power (effect) flexibility* on the short term basis. The observations for Western Denmark in this chapter have shown that the Nordic power market is able to provide this kind of flexibility.

It was shown that the variations in hydro power (between dry, normal and wet precipitation years) are affecting the price level on a monthly basis. Whereas there is only a very small correlation between the amounts of wind power generated and the monthly prices. In addition, the water inflow to the hydro reservoirs is largest in the summer where the electricity demand is low. Whereas, it is normally more windy in the cold months than in the summer which imply that on average wind power generation follows the electricity demand.

On an hourly basis, the prices do correlate with the wind generation. However, the analysis for 2004–2014 shows that there have been relative few hours with extreme prices or consecutive hours with no wind or maximum wind. Even during the worst storm in ten years, when most of the wind capacity was cut off due to high wind speed, prices on the balancing market were not abnormal.

This illustrates how well wind and hydro power play together in the Nordic area. Hydro power generation is very variable on a yearly level. Whereas, the yearly wind power generation is much more predictable. The opposite is true at the short term (e.g. at the day-ahead market) where wind power is the variable generation and hydro power can be controlled according to the filling of the water reservoirs.

This synergy implies that on the one hand, hydropower can supply short-term power flexibility that facilitates the integration of wind power in the system. On the other hand, wind power can substitute hydro power in windy periods and thereby release more hydro power capacity that can be used to supply short-term power flexibility or be transmitted to neighboring countries. This is in benefits for all. Hydro power helps lowering the system integration costs of wind which increases the value of wind. Wind helps increasing the value of hydro power bye releasing capacity that can be used in high price periods.

Much further penetration of wind power will require additional measures. With the planed expansion of wind power at the Northern European power market in the future, it is doubtful if the electricity market by itself can generate enough flexibility. The Nordic system is able to handle the present amount of wind energy. But the hydro power capacity is limited and with the continuous growth of wind power deployment, it may not be enough in the future.

Therefore, for the future development we will have to look for the potentials of getting flexibility from other energy sectors at low costs. In particular demand response, stronger coupling to other energy markets (heat, gas and transport), and use of new technologies, e.g. electricity storages or electric vehicles, (Lund et al. [30]). The need for more flexibility to counteract variations from wind integration may also require improved system and market operation (IEA [15, 31]).

Finally, the most general recommendation is that bidding areas for the wholesale market should reflect external congestions caused by wind. This will lead to price differences between neighboring areas for a significant number of hours during the year as an incentive for trade that will benefit both. Western Denmark happens to be "born" as such bidding area.

References

1. Energistyrelsen, ENS 2012, Energiaftalen 22. marts 2012. https://ens.dk/. Accessed 26 Feb 2017
2. Borenstein, S.: The private and public economics of renewable electricity generation. J. Econ. Perspect. **26**(1), 67–92 (2012)
3. Hirth, L., Ueckerdt, F., Edenhofer, O.: Integration costs revisited - an economic framework for wind and solar variability. Renew. Energy **24**, 925–939 (2015)
4. Lamont, A.: Assessing the long-term system value of intermittent electric generation technologies. Energy Econ. **30**(3), 1208–1231 (2008)
5. Borenstein, S.: The market value and cost of solar photovoltaic electricity production. CSEM Working Paper 176, University of California Energy Institute, Berkeley, California (2008)

6. Joskow, P.: Comparing the costs of intermittent and dispatchable electricity generation technologies. Am. Econ. Rev.: Pap. Proc. **100**(3), 238–241 (2011)
7. Nicolosi, M.: The economics of renewable electricity market integration. An empirical and model-based analysis of regulatory frameworks and their impacts on the power market. Ph.D. dissertation, University of Cologne (2012)
8. Holttinen, H., Meibom, P., Orths, A., Lange, B., O'Malley, M., Tande, J.O., Estanqueiro, A., Gomez, E., Sder, L., Strbac, G., Charles Smith, J., van Hulle, F.: Impacts of large amounts of wind power on design and operation of power systems. Wind Energy **14**(2), 179–192 (2011)
9. Hirth, L.: The market value of variable renewables: the effect of solar wind power variability on their relative price. Energy Econ. **38**, 218–236 (2013)
10. Smith, C., Milligan, M., DeMeo, E., Parsons, B.: Utility wind integration and operating impact state of the art. IEEE Trans. Power Syst. **22**(3), 900–908 (2007)
11. Brown, S., Rowlands, I.: Nodal pricing in Ontario, Canada: implications for solar PV electricity. Renew. Energy **34**(1), 170–178 (2009)
12. Lewis, G.: Estimating the value of wind energy using electricity locational marginal pricing. Energy Policy **38**(7), 3221–3231 (2010)
13. Hamidi, V., Li, F., Yao, L.: Value of wind power at different locations in the grid. IEEE Trans. Power Deliv. **26**(2), 526–537 (2011)
14. Ueckerdt, F., Hirth, L., Luderera, G., Edenhofera, O.: System LCOE: what are the costs of variable renewables? Energy **63**, 61–75 (2013)
15. International Energy Agency (IEA): The power of transformation: wind, sun and the economics of flexible power systems. OECD/IEA, Paris, France (2014)
16. Olsen, O.J., Skytte, K.: Competition and market power in Northern Europe. In: Glachant, J.-M., Finon, D. (eds.) Competition in European Electricity Markets, pp. 169–193. Edward Elgar Publishing Inc., Cheltenham (2003)
17. Skytte, K.: Fluctuating renewable energy on the power exchange. In: MacKerron, G., Pearson, P. (eds.) The International Energy Experience. Markets, Regulation and the Environment, pp. 219–231. Imperial College Press, London (2000)
18. Skytte, K., Ropenus, S.: Assessment and recommendations. Overcoming in short-term grid system regulatory and other barriers to distributed generation. Deliverable report D2, DG-GRID Project, Contract no. EIE/04/015/S07.38553 (2006)
19. Donkelaar, M., Maly, M., Skytte, K., Ropenus, S., Frias, P., Gomez, T.: Economic, policy and regulatory barriers and solutions for integrating more DER in electricity supply. Annual report, SOLID-DER Project, European Commission (2007)
20. Ropenus, S., Skytte, K.: Regulatory review and barriers for the electricity supply system for distributed generation in the EU-15. Int. J. Distrib. Energy Resour. **3**(3), 243–257 (2007)
21. Jacobsen, H., Zvingilaite, E.: Reducing the market impact of large shares of intermittent energy in Denmark. Energy Policy **38**(7), 3403–3413 (2010)
22. Grohnheit, P.E., Andersen, F., Larsen, H.: Area price and demand response in a market with 25% wind power. Energy Policy **39**(12), 8051–8061 (2011)
23. Energinet.dk. https://en.energinet.dk. Accessed 26 Feb 2017
24. Jensen, S.G., Skytte, K.: Interactions between the power and green certificate markets. Energy Policy **30**(5), 425–435 (2002)
25. Jensen, S.G., Skytte, K.: Simultaneous attainment of energy goals by means of green certificates and emission permits. Energy Policy **31**(1), 63–71 (2003)
26. Morthorst, P.E.: Green certificates and emission trading. Energy Policy **31**(1), 1–2 (2003)
27. Skytte, K.: Interplay between environmental regulation and power markets. EUI-RSCAS Working Papers, RSCAS no. 2006/04, European University Institute (EUI), Robert Schuman Centre for Advanced Studies (RSCAS) (2006)
28. Amundsen, E.S., Bergman, L.: Provision of operating reserve capacity: principles and practices on the Nordic electricity market. In: Competition and Regulation in Network Industries, vol. 2(1), pp. 73–98. Intersentia (2007)
29. Skytte, K.: The regulating power market on the Nordic power exchange Nord Pool: an econometric analysis. Energy Econ. **21**, 295–308 (1999)

30. Lund, P., Lindgren, J., Mikkola, J., Salpakari, J.: Review of energy system flexibility measures to enable high levels of variable renewable energy. Renew. Sustain. Energy Rev. **45**, 785–807 (2015)
31. International Energy Agency (IEA) and Nordic Energy Reserach (NORDEN): Nordic Energy Technology Perspectives. IEA/NORDEN, Oslo, Sweeden (2013)

Chapter 5
Incentivizing Flexibility in System Operations

Erik Ela, Michael Milligan, Aaron Bloom, Audun Botterud, Aaron Townsend and Todd Levin

Abstract Defining flexibility has been a challenge that a number of industry members and researchers have attempted to address in recent years. With increased variability and uncertainty of variable generation (VG), the resources on the system will have to be more flexible to adjust output, so that power output ranges, power ramp rates, and energy duration sustainability are sufficient to meet the needs of balancing supply with demand at various operational timescales. This chapter discusses whether existing market designs provide adequate incentives for resources to offer their flexibility into the market to meet the increased levels of variability and uncertainty introduced by VG in the short-term operational time frame. It presents a definition of flexibility and discusses how increased levels of VG require increased needs for flexibility on power systems. Following this introductory material, the chapter examines how existing market designs ensure that resources have the right incentives to provide increased flexibility, and then discusses a number of emerging market design elements that impact flexibility incentives.

This chapter is based on the detailed discussion of current and emerging market designs to incentivize flexibility in short-term system operations to meet the increased needs from variable generation [1, Sect. 4].

E. Ela (✉)
Electric Power Research Institute, Palo Alto, CA 94304, USA
e-mail: eela@epri.com

M. Milligan · A. Bloom · A. Townsend
National Renewable Energy Laboratory, Golden, CO 80401, USA
e-mail: Michael.Milligan@nrel.gov

A. Bloom
e-mail: Aaron.Bloom@nrel.gov

A. Botterud · T. Levin
Argonne National Laboratory, Argonne, IL 60439, USA
e-mail: abotterud@anl.gov

T. Levin
e-mail: tlevin@anl.gov

F. Lopes and H. Coelho (eds.), *Electricity Markets with Increasing Levels of Renewable Generation: Structure, Operation, Agent-based Simulation, and Emerging Designs*,
Studies in Systems, Decision and Control 144, https://doi.org/10.1007/978-3-319-74263-2_5

5.1 Introduction

The existing wholesale electricity market designs are unique in their complex relationships between economics and the physics of electricity. These markets aim to incentivize resources to provide a variety of services, including energy and various ancillary services. However, it is unclear as to how much the existing markets may be incentivizing flexibility from the market participants in an efficient manner. Questions that should be asked are (1) whether the market designs are incentivizing new resources entering the market to have the needed flexibility capabilities, (2) whether the market designs are incentivizing existing resources to upgrade their technology to offer additional flexibility capabilities if more flexibility capabilities are needed, and (3) whether the market designs are incentivizing resources that have flexible capabilities to offer those capabilities to the short-term energy and/or ancillary services market when flexibility is needed most.

This chapter focuses on the third question and how power system flexibility is incentivized during short-term system operations. First, we describe some definitions and examples of flexibility. Then we discuss how the introduction of variable generation (VG) might be making the topic of flexibility incentives more of a pressing issue. Next, we cover a number of historical, recent, and then proposed market design elements that may affect how flexibility is incentivized in electricity markets. Many of the existing market design elements provided incentives whether or not a unit could be flexible and offer its flexibility to the market operator. However, some of the more recent changes are being designed to more explicitly incentivize an increasing quantity of flexibility, in many ways because to the increasing variability and uncertainty that is brought to the system by VG. Finally, we discuss ongoing issues and remaining questions and provide a summary of this complex topic.

5.2 Defining Flexibility

Flexibility as a term is gaining in popularity in the electric power and energy industry. Although the general meaning is understood, some clarification of the definition is important. In Lannoye et al. [2], flexibility is defined as the ability of a system to deploy its resources to respond to changes in net load. In Ma et al. [3], flexibility describes the ability of a power system to cope with variability and uncertainty in both generation and demand while maintaining a satisfactory level of reliability at a reasonable cost over different time horizons. Both of these definitions focus on the system, but they can be easily disaggregated to individual resources by replacing "ability of a system" with "ability of a resource". A more general definition can be defined as *"the ability of a resource, whether any component or collection of components of the power system, to respond to the known and unknown of power system conditions at various operational timescales"*. For the purposes of this chapter, the changes are those that occur with the active power of individual or aggregate generation, demand, or network elements at multiple timescales. System and market operators should have an objective to utilize the system flexibility to meet the reliability requirements in the most cost-efficient manner possible. A lack of flexibility can lead to imbalance of generation and load, overloading of transmission elements, and other potential reliability issues. Improper utilization of existing flexibility, or unwillingness of resources to provide existing flexibility, can also lead to the same reliability issues, or to higher costs when more expensive flexibility is required when economic flexibility is unwilling to be offered.

A number of characteristics have been emphasized as qualities that are needed for increased flexibility. We focus on active power flexibility of generating units, but this can be applicable to other types of resources that can control active power as well as reactive power. The main characteristics of flexibility will generally fall into three categories: absolute capacity range, speed of power output change, and duration of energy levels. Resources that have a large range of absolute output levels between their minimum and maximum capacity levels can be classified as more flexible because they have greater ability to adjust to changing power system conditions. Resources with greater ramp rates can also be classified as more flexible because they are able to adjust faster to changes in power system conditions of varying speeds. Last, resources that are able to hold energy levels for longer periods of time can be classified as more flexible because they are able to better meet power system conditions that sustain for significant periods.

Many of the existing generating technologies have numerous limitations that may affect absolute power range, speed of power output change, and energy level durations in different ways. For example, a number of constraints will limit thermal plants on how and when they can be committed on or off. This includes minimum on times, minimum off times, maximum starts per day limits, and other commitment constraints. Hydro plants may have rough zones at certain power limits when they cannot provide power without incurring damage [4]. Combined-cycle plants also have constraints on how they can be configured and how configurations can be transitioned [5]. These constraints can affect the absolute power range. Resources that can be easily turned on or off provide increased flexibility because the absolute power range can be taken between zero and maximum capacity. Similar constraints can limit the speed of power output change. Start-up times and shut-down times can limit how fast resources can provide power when needed and provide nothing when not needed. All of these types of "discrete" constraints limit the flexibility and add further complexity when evaluating the flexibility of individual resources and of the system. Further constraints can limit how long resources can provide energy.

It is possible that some resources can provide different levels of flexibility, and that increased flexibility may increase resource costs. For example, a resource may be able to reduce its minimum generation level, but that may in fact cause an inefficient heat rate. Similarly, a resource may have emergency ramp rates it can provide at faster rates, but it may cause more wear and tear to the unit, leading to higher operations and maintenance costs. In these instances, it is important that the extra flexibility that has increased costs is requested only if necessary, and that incentives are present for those periods so that resources can at least recover costs. These topics are crucial and the focus of this chapter.

5.3 The Impact of Increasing Penetrations of VG on the Need for Flexibility

System flexibility is required to meet the known and unknown changes in power system conditions, referred to as variability and uncertainty, respectively. In power systems, the load has variability characteristics, both diurnally, and at shorter timescales. The load is also uncertain, because it cannot be perfectly predicted at all times and over all time horizons. Conventional generation and transmission elements also have uncertainty, because they can fail without any certainty of when and if that may happen. Over time, the industry identified various procedures to accommodate this variability and uncertainty, including operating reserve and security-constrained scheduling models (e.g., n–1 preventative procedures for transmission outages).

VG adds variability and uncertainty to the existing amounts at multiple timescales [6]. The maximum available power changes based on the changing weather driver, such as wind speed and solar irradiance, and cause variability. A number of studies have quantified this variability (see [7–9]). Although the variability of a single turbine or photovoltaic (PV) cell may be quite high, the variability of an entire wind power plant or large PV array is relatively reduced. Further reduction comes from multiple plants in a balancing authority area because of geographic diversity.

Correspondingly, the maximum available power cannot be predicted perfectly, causing uncertainty. These characteristics have been quantified in numerous studies [10, 11]. Intuitively, the farther ahead the horizon of the forecast, the more difficult it is to predict. Wind and solar power uncertainty have very different characteristics because a general pattern is much easier to predict for solar power than it is for wind power. Some studies have now also shown how variability and uncertainty at these different timescales can affect both reliability and efficiency in different ways (see [6, 12]).

The variability and uncertainty of VG can be met by different operational strategies depending on the timescale and time horizon. For instance, short-term (e.g., minute-to-minute) variability might be met by regulation reserve, longer-term (e.g., tens of minutes to hours) variability might be met by flexibility reserve, and long-horizon uncertainty might be met by improved forecasting. It is possible that some of the existing procedures for accommodating the variability and uncertainty of the past can be used to meet the increasing variability and uncertainty of VG; however, it may be that new procedures and tools can do so in a more reliable and efficient manner. Some examples of these evolving strategies include shorter scheduling intervals [13] to meet increased variability and stochastic or robust unit commitment and dispatch solutions for addressing uncertainty [14–17]. In addition, incorporating increased look-ahead horizons in the scheduling model and intelligent operating reserve requirements can help meet increased variability and uncertainty if done efficiently [18]. Finally, and for all the above strategies, increased variability and uncertainty, with all else unchanged, will require increased flexibility requested of the resources.

Besides variability and uncertainty, a few other traits that VG carry may increase the need for flexibility on power systems. VG is not a synchronous machine, nor is it inherently responsive to the frequency of the grid. Although this should not explicitly increase the need for frequency response in itself, increasing penetrations of VG without these characteristics can displace resources that do. This may produce a need for incentives to ensure that this capability is available. For example, primary frequency response and synchronous inertia currently are not part of the ancillary services market, but they may be necessary in the future as penetrations of non-synchronous VG increase [19]. It is also possible that VG can create the controls to provide these services itself [20], but that may also depend on the incentives present and whether any revenues from providing this service will justify VG installing these capabilities.

Other traits include location constraints because VG will be located in areas where there is the highest power production potential. These areas may be located far from load centers. This may increase the need for localized flexibility that has to be able to accommodate the variability and uncertainty of VG without overloading the elements of the transmission network.

All of the above characteristics of VG may require increased needs for flexibility on the power system: most of these characteristics and the need for flexibility is not necessarily new. The past needs of flexibility on the power system may have been met by the current system operating procedures and wholesale market designs. It may be possible, however, that increasing penetrations of VG may push the needs to the point at which the issue of specific flexibility incentives can no longer be ignored. This is a difficult discussion with varying opinions throughout the industry.

Throughout this chapter, we will describe some of the existing wholesale market designs and some recent and proposed changes from the US market operators that may have been implemented and proposed due to increased VG penetration that all attempt to address incentivizing flexibility in system operations. It is important to understand these characteristics, especially those of variability and uncertainty at all operational timescales, as well as asynchronism and non-proximity to load, to understand how different market designs may incentivize the need for increased flexibility in these different manners.

5.4 Traditional Market Design Elements that Impact Flexibility Incentives

A few mechanisms that provide some incentives for market participants to provide flexibility have been in existence since the inception of the US wholesale energy markets or shortly thereafter. The extent to which they provide sufficient flexibility is an ongoing debate. It is not obvious whether some of these elements incentivize flexibility or not. The first step in obtaining flexibility from market participants is to have a mechanism that allows the market operator to commit the resource and dispatch the resource's output when it is needed. A portion of the generating fleet—sometimes as much as 50–70% of the energy—operates through bilateral contracts outside of the pool-based electricity markets. From the market operator perspective, these suppliers, along with others who are not responding to price signals, are self-scheduled resources offering absolutely no flexibility for the market operator to utilize. Although the market operators still receive the energy from the resources to meet the expected demand, they do not have any flexibility from these self-scheduled units to meet ancillary service demands or changing energy demands. The more self-scheduled resources exist in the system, the less flexibility the system operator has access to, holding all else constant. Therefore, in addition to having the ability to quantify needed levels of flexibility, it is important to incentivize suppliers to offer their flexibility into the market. Without this feature, significant levels of physical flexibility may be unavailable to the market operator. Suppliers should be incentivized to allow the market operator to dispatch its output to meet the changing energy and ancillary service demand, while still operating within design parameters.

We focus on five specific examples of traditional market mechanisms in place that in some manner could incentivize some form of flexibility, or at least provide an incentive so that suppliers may offer their flexibility to the market operator. These include centralized scheduling and efficient dispatch, frequent scheduling and frequent settlement intervals, existing ancillary service markets, make-whole payment guarantees, and day-ahead profit guarantees.

5.4.1 Centralized Scheduling and Efficient Dispatch by the Market Operator

Current markets have mechanisms that incentivize resources to offer their operating capabilities to be dispatched by the RTO/ISO. This allows for the market operator to ensure that the resource is flexible, but it does not guarantee how flexible the resource is. The first mechanism is somewhat obvious but often overlooked. When suppliers participate in the pool market, the market operator will operate them at their most efficient operating point based on their offered bid-cost curve. The market operator minimizes the bid-production costs from all these bids to meet the energy and ancillary service demands subject to power system security and unit constraints. The cost of supplying energy for a unit participating in the market and allowing for the market operator to commit and dispatch the supplier's output should theoretically not be greater than the resultant price (reasons that costs can be higher than the price are discussed later in Sect. 5.4.4). When the price increases, the market operator gives the supplier a position that reflects that it is efficient to increase its output, and this allows the supplier to earn more revenue. When the price decreases, the market operator gives the supplier a position to reduce output, because it may be that the current output is no longer efficient when receiving the reduced price.

Under good electricity market design, the supplier output level should always reflect the changing prices and should avoid operating at levels that cost more to produce energy than the price they receive. Self-scheduled resources provide the

Table 5.1 Hypothetical thermal plant, piecewise linear cost curve

Incremental cost ($/MWh)	Energy/Capacity segment (MWh)
35.00	Up to 286.00
47.25	286.1–295.0
47.60	295.1–304.0
47.95	304.1–313.0
48.30	313.1–322.0
48.65	322.1–331.0
49.00	331.1–340.0
49.35	340.1–349.0
49.70	349.1–358.0
50.75	358.1–376.0
52.50	376.1–377.0

market operator with the scheduled output before the market clears, and this schedule is fixed regardless of the price. During periods of high prices, the self-scheduled resource could miss out on additional profit. During low prices, the self-scheduled resource could lose money when the cost to supply energy is greater than the energy payments they receive. When substantial bilateral contracts are self-scheduled into the market, there may come a point at which the flexibility that is available to the market operator is insufficient, inducing a need for other mechanisms to obtain this flexibility. For example, a very high proportion of self-scheduled resources may drive the need for more expensive sources of flexibility, such as additional flexible generating capacity or storage. In cases such as this, the system may possess more flexibility than is needed; however, much of this flexibility may be stranded. It is important to note that the levels of *physical* flexibility may be sufficient; however, some of this may not be contractually available.

To illustrate the potential impacts of self-scheduling, we show a simple example. Table 5.1 shows a bid-in cost curve for a thermal generating unit which is taken from real bid-cost data. This bid-in cost curve reflects representative costs of thermal plants based on a convex, monotonically increasing incremental heat rate. We ignore no-load costs in this example for simplicity. The incremental cost in column 1 is the cost bid for the specific capacity represented in column 2. Therefore, the first 286 MW in this example will always cost $35 \times 286 = \$10\ 010$.

The cost data in this table forms the basis of how this resource would bid into the market. We next turn to the relationship between this cost data and locational marginal prices (LMPs) and an examination of how various self-scheduling strategies compare to how the unit would be dispatched in the absence of self-scheduling.

Table 5.2, column 2, shows a 12 h period of LMPs. Scenario 1 (Market) allows the market operator to efficiently dispatch the resource every hour. For simplicity, ramp rates and other constraints that may cause inefficiencies are ignored. In nearly every time period, the unit's output changes as a function of the LMP. This is in contrast

Table 5.2 Twelve-hour example for allowing the market operator to efficiently dispatch the output of a resource (scenario 1) versus various self-scheduling techniques (scenarios 2 to 5)

Hour	LMP ($/MWh)	Scenario 1 (Market)	Scenario 2 (Min self)	Scenario 3 (Max self)	Scenario 4 (Mid self)	Scenario 5 (Lag LMP)
1	45.41	286	286	377	300	286
2	49.65	349	286	377	300	286
3	52.27	377	286	377	300	349
4	51.37	376	286	377	300	377
5	48.32	322	286	377	300	376
6	46.45	286	286	377	300	322
7	46.35	286	286	377	300	286
8	50.97	376	286	377	300	286
9	49.44	349	286	377	300	376
10	44.70	286	286	377	300	349
11	48.51	322	286	377	300	286
12	51.13	376	286	377	300	322

to each of the self-scheduling scenarios shown in Scenarios 2 through 5. In Scenario 2 (Min Self), the supplier simply schedules itself at its minimum capacity level for all hours. In Scenario 3 (Max Self), the supplier schedules itself at its maximum capacity. In Scenario 4 (Mid Self), the supplier schedules itself at a level in between its minimum and maximum capacity. Finally, in Scenario 5 (Lag LMP), the supplier uses the LMP from the previous hour to predict where it should schedule itself for the following hour.

Table 5.3 shows the revenue, cost, and profit results for all five scenarios. The profits from each of the self-scheduling cases are compared to Scenario 1: Market. Thus, the right-most column shows how much profit the supplier loses by not offering its flexibility into the market. Note that each of the self-scheduling cases results in

Table 5.3 Revenue, cost, profit, and profit lost for various self-scheduling techniques

Scenario	Total revenue ($)	Total cost($)	Total profit ($) (Total revenue–Total cost)	Profit lost($)
Scenario 1 (Market)	195668	147313	48355	N/A
Scenario 2 (Min self)	167326	120120	47206	1148
Scenario 3 (Max self)	220567	173594	46972	1382
Scenario 4 (Mid self)	175517	128079	47438	916
Scenario 5 (Lag LMP)	190452	142909	47542	812

lost profits compared to the market case. Even with intelligence in the self-scheduling strategy (Scenario 5), it would still lose out on $812 during a 12 h period. Although the lost profit is small relative to the total profit, it will be highly dependent on the cost curve and prices during different time periods. For example, if the cost for the first segment of Table 5.1 (up to 286 MW) were $47 rather than $35, the profits lost would be the same as Table 5.3, but the total profits would be an order of magnitude less, making the relative profit loss much more significant.

5.4.2 Five-Minute Scheduling and Five-Minute Settlements

The real-time market scheduling intervals of all the regions that operate wholesale markets in the United States are shorter in time resolution than scheduling intervals of utilities prior to restructuring. In fact, all market regions in the United States now schedule the real-time market and real-time output of resources that offer their flexibility at a 5-min interval, updated every five minutes. This allows for better pricing of actual conditions on a more granular scale and provides incentives for resources that can follow the prices. Because ramp constraints are used in the market-clearing model to constrain the ability of a supplier to sell energy into the market when they do not have the flexibility to follow prices, the selection of supply into the energy market should be based on actual capability when ramp constraints, provided by the resources are based on physical ramp rates. For example, in the hypothetical example in Table 5.2, hour 7 to hour 8, the supplier changes its output from 286 MW to 376 MW. Although it may be possible that this change could be made during 60 min, it may not be possible for most thermal plants to execute this ramp during a 5-min period. If the supplier had a ramp rate of 5-MW/min, it could reach only 311 MW in the next 5-min interval, resulting in $99 of lost profit (although ignoring the fact that it is now MW in 5 min rather than MWh). In this way, the 5-min dispatch provides an incentive for flexibility in response to quickly changing prices.

Most of the 5-min energy markets that are currently in place in the United States do a good job of extracting flexibility without resorting to a separate market for a specific product for ramping capability, as illustrated by the example above. Units that bid into the 5-min energy market are obligated to ramp to their set point by the time the market period begins. Because these set points are calculated so frequently, many units ramp a substantial portion of the time. This allows for the most economic provision of energy given the constraints on the transmission system, and units can take advantage of price volatility when they can ramp faster. However, in some instances, it may be that ramp constraints can give the opposite effect. A resource that is ramp constrained will not set the LMP, because a faster, more expensive unit would have to be used to make up for the slower unit's ramp limitation. Thus, the more expensive unit will set the price, while at the same time giving a higher revenue opportunity for the slower unit.

Milligan and Kirby [21] provide a simple illustration of the issue depicting the real-time market. In this example, a single time period market assumption is used without

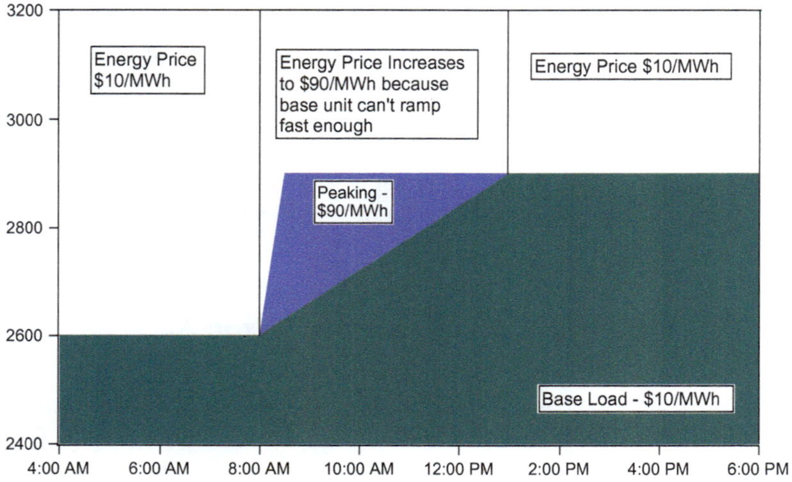

Fig. 5.1 Example of ramp-limited resources and resulting prices

any look-ahead function in the dispatch interval, so that the future expectations cannot affect the current energy pricing. The example is a simplistic power system with only two generators: a base load unit that has a marginal cost of $10/MWh and a peaking unit that has a marginal cost of $90/MWh, as illustrated in Fig. 5.1. During a steep ramp that is beyond the speed that the baseload unit can respond, but within its capacity range, the peaking unit is dispatched to cover the ramp and meet the load. After the baseload unit catches up, the peaking unit is shut off. However, the peaking unit sets the energy price at $90/MWh during the time it is used to meet the load during the ramp period. Although this is not a problem per se, the baseload unit also collects $90/MWh during this period, which does not provide the baseload unit any incentive to become more flexible, and in fact it may provide a disincentive. This topic should be studied further to see what consequences may occur. For example, a multi-period dispatch looks ahead to ensure that units can meet upcoming ramping requirements. This can change the market prices and revenues for market participants, even if pricing is set only for the current interval. In fact, many US market operators have or are developing proposals to move toward multi-period dispatch when solving the real-time market [22].

Although all of the US markets have 5-min real-time energy markets that dispatch and price energy at 5-min intervals, not all of these markets settle at this granularity [23]. Many settle the real-time markets based on the average hourly price of all intervals within that hour. Some areas, including NYISO and SPP, however, do settle at 5-min intervals. SPP states that 5-min settlement incents the submission of ramp capability by resources precisely because the capability to move quickly is rewarded by an LMP commensurate with the 5-min instructions. SPP further explains that without this settlement feature, resources may be disinclined to offer all of their

Table 5.4 Incentive differences when settlements are done on an hourly average versus 5 min

Interval	LMP ($/MWh)	Current hourly average ($/MWh)	Scenario 1 (Market)	Scenario 2 (Moving hourly average)	Scenario 3 (Perfect knowledge of hourly average)
0:05	73.68	73.68	377	377	340
0:10	41.87	57.78	286	377	340
0:15	43.48	53.01	286	377	340
0:20	44.17	50.80	286	376	340
0:25	45.75	49.79	286	358	340
0:30	46.69	49.27	286	340	340
0:35	46.73	48.91	286	331	340
0:40	45.91	48.54	286	322	340
0:45	61.25	49.95	377	358	340
0:50	47.88	49.74	304	358	340
0:55	47.88	49.57	304	349	340
1:00	43.85	49.10	286	340	340

ramp capability, perceiving that they are not being fully compensated for the actions required.

To provide an illustration of how the settlement period can have an impact on incentives for flexible operations, we develop a simple example. Table 5.4 shows 12 5-min LMPs, from real LMP data. Column 3 shows the average LMP for the hour, which is calculated based on the cumulative average LMP from the beginning of the hour to the current time period. We use the same incremental costs for the supplier as shown in the earlier example in Table 5.1. In Scenario 1 (Market), the supplier follows a schedule, as in Sect. 5.4.1, based on the most efficient output level that the market operator computes and directs each 5-min period. In Scenario 2 (Moving Hourly Average), the supplier follows an output that is based on the current hourly average LMP (from column 3), because it gets updated throughout the hour. We ignore any impacts from uninstructed deviation penalties throughout this example. Finally, Scenario 3 (Perfect Knowledge) shows a hypothetical example of what the most efficient output would be from the supplier if it had perfect knowledge of the final average hourly price ($49.10).

An examination of the different dispatches in the table shows that the maximum flexibility is achieved in Scenario 1 (Market). When maximizing profit based on the anticipated or predicted hourly settlement (Scenario 2 and Scenario 3), the unit provides the incorrect level of or no flexibility. It is a simple extrapolation of this scenario that would illustrate similar behavior if the assumption of perfect foresight is relaxed and the unit bids to another hourly average price level. Because price changes in each of the 5-min periods, this is an indication that the system needs a varying level of output; otherwise the price would have remained constant throughout the hour. Next, we turn to an examination of the profits earned in these scenarios.

Table 5.5 Revenue, costs, and profits of different scenarios with 5-min settlements versus average hourly settlements

Scenario	Total revenue ($)		Total cost ($)	Total profit ($) (Total revenue–Total cost)	
	5-min settlements	Hourly average settlements		5-min settlements	Hourly average settlements
Scenario 1 (Market)	15208	13338	10895	4313	2443
Scenario 2 (Moving hourly average)	17479	17441	13373	4106	4068
Scenario 3 (Perfect knowledge of hourly average)	17134	17134	13053	4081	4081

Table 5.5 presents the total revenue, total cost, and total profit for each of these scenarios with both 5-min settlement and hourly average settlement procedures. If settlements are based on the hourly average, as they are in many markets today, the supplier will earn more profit by producing output differently than the dispatch schedule that was given by the market operator. This would result in output levels that are not the most efficient and could potentially result in reliability issues. In the hourly settlement case, the profit almost doubles when the supplier follows the hourly average price compared to the market schedules ($4 068 compared to $2 443). This shows that even though 5-min scheduling is present in almost every US energy market, it is important that the settlement interval length follow the same interval length as the scheduling to incentivize suppliers to provide the flexibility that is needed by the market operator. On the other hand, numerous uninstructed deviation penalties and ex-post pricing rules may also incentivize the supplier to follow the efficient schedules when hourly average prices are used for settlements. There could also be other reasons that require hourly settlements, such as data retention and storage, as well as a desire to limit market complexity. However, from this simplified example, it appears likely that settlements that match the interval length are the most efficient for extracting the desired flexibility needed from market participants.

5.4.3 *Existing Ancillary Service Markets*

A number of ancillary service markets exist in both day-ahead and real-time electricity markets. These services are generally active power capacity that is held as operating reserve and used for various reasons and at various timescales. These markets are usually co-optimized with the energy market so that the market operator is able to efficiently schedule suppliers for both energy and ancillary services and price both services accordingly. In co-optimized energy and ancillary service markets, the ancillary service prices will be set based on the bid-in cost to provide the ancillary service as well as the lost opportunity cost to provide energy or a separate ancillary service.

Ancillary service markets were discussed in *Chap.* 1. Regulating reserve (or regulation), secondary reserve—contingency (or spinning reserve), and tertiary reserve—contingency (nonspinning, 30-min reserve, or replacement reserve) are the three ancillary services that are common among US markets. An important factor is that all of these ancillary services limit the amount of capacity that can be sold to the market based on the response speed of the supplier. For example, regulating reserve typically limits the capacity by how much the resource can provide in 5-min. Secondary contingency reserve is typically limited by 10 min worth of ramp response. The faster the resource can adjust its output, the more it can sell into these markets. For example, if two resources have 50 MW of capacity available, with one having a 1 MW/min ramp rate and another having a 2 MW/min ramp rate, they would be able to provide 5 and 10 MW of regulation reserve, respectively. If the price of regulation were 10$/MW-h, the second resource would receive $50 more revenue than the first, even though they had the same capacity available. This is an obvious incentive for resources to improve flexibility by way of faster response rates.

5.4.4 Make-Whole Payment Guarantee

The mechanism of scheduling and pricing suppliers offering into the market described in Sect. 5.4.1 will theoretically place each market participant in a position to maximize profit, subject to various market and technical constraints. However, because of issues such as non-convex costs, commitment constraints, and out-of-market reliability rules, it is possible for the market operator to direct the flexible supplier to provide an energy and ancillary service quantity and for the market participant to lose money when following this direction. For this reason, additional business rules have been designed as part of the US electricity market design to further incentivize suppliers to offer into the market and allow the market operator to commit and dispatch the supplier's output when market prices alone may not provide sufficient revenue to cover all operating costs. One of these is the make-whole payment, also called bid production cost guarantee (NYISO), revenue sufficiency guarantee (MISO), and operating reserve credit (PJM). This payment ensures that suppliers that offer flexibility into the market are guaranteed to be made whole to their offer cost when that bid clears the market. If the revenue that the supplier makes based on the market prices (LMPs and ancillary service clearing prices) is less than the supplier's bid cost, the supplier is made whole, with the market operator paying the supplier the difference as a side payment. The offer cost will typically include a three-part offer, including no-load (or minimum generation) costs, start-up costs, and incremental energy costs. It may also include the costs that the supplier has bid in to supply ancillary services. The make-whole payment will make it so the total profit is at least zero. A simplified form of the make-whole payment is shown in the equations below. This is typically netted for all hours of a single day, and it is typically performed separately for both day-ahead and real-time markets to incentivize participating as a flexible resource in both markets.

$$\begin{aligned} TotalCost \; = \; & NoLoadCost + StartupCost + \\ & IncrementalCost \times EnergySchedule + \\ & AncillaryServiceCost \times AncillaryServiceSchedule \end{aligned} \tag{5.1}$$

$$TotalRevenue = EnergySchedule \times LMP + AncillaryServiceSchedule \times ASCP \tag{5.2}$$

$$\begin{aligned} & If \;\; TotalRevenue < TotalCost \\ & \qquad MakeWholePayment = TotalCost - TotalRevenue \\ & Else \\ & \qquad MakeWholePayment = 0 \end{aligned} \tag{5.3}$$

If a supplier is the marginal resource and sets the LMP, it will earn enough revenue to recover its incremental energy cost. However, assuming it is marginal for its entire period being online, it will not earn enough revenue to recover its no-load or start-up cost because these costs are generally not part of the LMP (see Sect. 5.5.4 for an exception). This would cause a disincentive for offering flexibility into the market. Similarly, a unit may be needed for voltage or stability constraints, which are typically not part of the market model constraints. To offer the flexibility to maintain reliability, the market operator will guarantee that a flexible supplier recovers all operating costs associated with supplying energy and ancillary services.

Self-scheduled resources would not receive this guarantee, because they are not giving a bid-cost to the market nor are they offering the flexibility for the market operator to commit and dispatch the supplier's output. Therefore, self-scheduled resources will have no guarantee that they will be made whole to their costs, and, depending on pricing outcomes, they can make less revenue than it costs them to be committed and supply energy, leading to negative profits.

5.4.5 Day-Ahead Profit Assurance

Another settlement mechanism in place today to incentivize suppliers to participate in the market and allow for the market operator to commit and dispatch the supplier's output is the day-ahead margin assurance payment (DAMAP). This mechanism ensures that when energy schedules are reduced in the real-time market (RTM) from their day-ahead energy schedules, this will not adversely affect the profit margin the suppliers made in the day-ahead market (DAM). The purpose of the DAMAP is to provide an incentive for the market participants to be flexible in offering into the RTM and to be used by the market operator when conditions in the RTM have changed without being negatively affected. If the real-time market adjusted the supplier output such that it would receive more profit by not operating as the market

Table 5.6 Example of unit receiving day-ahead margin assurance payment (DAMAP) after being reduced in the real-time market

	Day-ahead	Real-time (Combined)
Cost	$50/MWh	$50/MWh
LMP	$60/MWh	$55/MWh
Schedule	200 MWh	–50 MWh (150 MWh)
Revenue	$12 000	$–2 750 ($9 250)
Cost	$10 000	$7 500
Profit	$2 000	($1 750)
DAMAP	$250	

operator suggests and operating as it was scheduled in the DAM, reliability could be adversely affected. The DAMAP will incentivize resources to offer their flexibility in the RTM by guaranteeing the profit it received regardless of real-time outcomes.

$$DAMAP = Max\,\{0,\ (DayAheadEnergySchedule - RealTimeEnergyOutput)\times (DayAheadPrice - RealTimePrice)\} \tag{5.4}$$

An example of a unit receiving a DAMAP payment after providing flexibility in both the day-ahead and real-time markets is shown in Table 5.6. The right-most column also shows the combined effect from day-ahead and real-time in brackets.

The supplier was asked to reduce output in the real-time market. In doing so, it would lose the profit made initially in the day-ahead market because its total profit goes from $2 000 to $1 750 after the real-time market. In this case, the market operator wants the resource to reduce output to maintain reliability and increase efficiency. The DAMAP of $250 ($1 750 + $250 equaling the initial $2 000) is paid to make up for the lost profit and ensure that the supplier will have an incentive to continue to provide its flexibility to the market.

This calculation will ensure that suppliers that offer the flexibility to be adjusted in real time when conditions require a reduction in output are not financially harmed from the position the market operator scheduled them at in the day-ahead market. The DAMAP applies to day-ahead energy and ancillary service markets. Self-scheduled resources that do not offer flexibility are not guaranteed this payment when output changes in real time. This results in a further incentive to offer flexibility to the market rather than self-scheduling.

5.5 Emerging Market Design Elements that Impact Flexibility Incentives

Efficient operation through centralized dispatch, frequent scheduling and short settlement intervals, ancillary service markets, make-whole payments, and day-ahead profit assurance payments are traditional ways that suppliers have been incentivized to offer flexibility to the market operator and allow it to commit and schedule the supplier's output. However, it is unclear whether these existing design elements provide sufficient incentives to ensure an adequate level of system flexibility and that the available flexibility is accessible to the system operator when needed. To some extent, new market designs have recently been implemented to incentivize increased flexibility on the system when that flexibility is needed. We discuss a number of more recent market design changes that may have some influence on incentivizing further flexibility from suppliers.

5.5.1 Flexibility from Nontraditional Resources

The suppliers that have traditionally provided flexibility in the energy and ancillary service markets are thermal and hydro power plants. These resources are able to adjust output at various response speeds with absolute power ranges typically in the range of 50% of total capacity. The wholesale electricity markets were initially designed with these technologies, thermal plants in particular, and their characteristics in mind. Given new characteristics of emerging technologies, adjusting market rules that were designed with other technologies in mind may be required. Some recent changes in the wholesale markets have been made to accommodate such resources as demand response (DR), energy storage, and VG itself.

One of the most significant market rule changes has been made for further adoption of demand-response resources as suppliers of energy and ancillary services. In 2011, the Federal Energy Regulatory Commission (FERC) issued Order 745 [24]. The order directed the wholesale market operators to pay demand-side resources that curtailed their load when directed by the energy LMP, as long as a net-benefits test was used that showed providing the demand response reduced costs per unit to consumers. In addition, many of the market operators have also implemented ways in which demand can participate in ancillary service markets. In ERCOT, nearly half of the contingency reserve needed is supplied by demand-response resources that curtail when system frequency reaches some level below nominal frequency [25]. Other markets have limitations on how much ancillary services can be provided by DR. These limitations do have justification in terms of both reliability and economics, but at the same time the ability to utilize DR introduces a great new source of flexibility that was not historically available. The participation of demand response on wholesale markets is however, subject to significant uncertainty. In 2014, the US Court of Appeals vacated FERC Order No. 745, finding that the FERC overstepped its authority by encroaching on states' jurisdiction of the retail electricity market. The court also noted substantive errors with the FERC's compensation rules. The impacts of this ruling have yet to be realized and may result in substantial changes in how this form of flexibility can participate in wholesale energy markets.

Energy storage is another resource that has tremendous flexibility. Energy storage can effectively double its absolute power range because it can act as a supplier as well as a demand. Most energy storage resources also have superior response rates and much less limiting commitment constraints than thermal units. However, they do have limitations on the amount of time they are able to sustain energy levels. This energy limitation varies depending on the type of energy storage, with some, such as large pumped storage hydro, able to sustain for multiple hours to days; whereas other storage devices, such as flywheels, although extremely fast-responding, can sustain maximum power for only 10–30 min. Some of the market operators have been adjusting market rules to allow for extraction of the tremendous flexibility from energy storage. For example, PJM now has full optimization of pumped storage resources in day-ahead markets [22]. Other markets have adjusted automatic generation control algorithms to allow for extracting the extremely fast speed and flexibility of limited energy storage resources while keeping track of its energy discharge level [26]. Some issues still exist in the markets on furthering the ability to treat storage in a way that incents storage to provide its full flexibility potential [20].

Finally, recent changes have enabled market operators to extract flexibility from VG. Previously, all VG were treated very similar to negative load, in that their outputs were considered fixed and other resources needed to adjust output for the variability and uncertainty that occurred. At first, this practice seemed intuitive. VG has zero or very low variable costs, meaning that it is most cost efficient to use as much power from VG as available. However, due to transmission constraints or minimum generation constraints of thermal units, it was found that there are some instances in which VG could curtail its output, do so quickly, and help balance the system and maintain reliability and security limits. The NYISO proposed this with its wind resource management program. Subsequently, PJM adjusted market rules to allow for negative prices and allow for wind curtailment for economics and reliability. MISO then implemented its Dispatchable Intermittent Resources program, which allows for the economic dispatch of wind energy. Although the rules governing each implementation are slightly different, they all allow for wind to be dispatchable in real-time markets to manage transmission congestion and meet load efficiently. This provided a great new source of flexibility.

To explain why these programs are beneficial, we show an example from [27]. Figure 5.2 shows a simplistic three-bus system, with a cheap generator at Bus 1 (G1), an expensive generator at Bus 3 (G3), and a wind power plant at Bus 2. The 250-MW load (L3) is located at Bus 3, the reactances of all lines are equal (X12 = X13 = X23), and there is a transmission limit on the branch from Bus 2 to Bus 3 of 100 MW (L23). We perform two market solutions, first with wind as a non-dispatchable price taker. The second—similar to the programs in NYISO, PJM, and MISO—allows for wind to be dispatchable in the market at an offer price of $0/MWh.

Table 5.7 shows the production, production costs, and LMPs with wind power fixed, and Table 5.8 shows the same results with wind as a flexible producer. Rows 2 through 4 add 1 MW of load to each bus as a way to approximate what the LMP is (because LMP is the marginal cost of meeting an increment of load at each location). With wind fixed, G3 is required because of the transmission constraint that limits G1's output to 100 MW. The wind power plant receives a negative LMP, and the load must pay the expensive LMP based on G3's marginal cost. A different solution results with a market that allows for wind to be flexible in the market (see Table 5.8). The

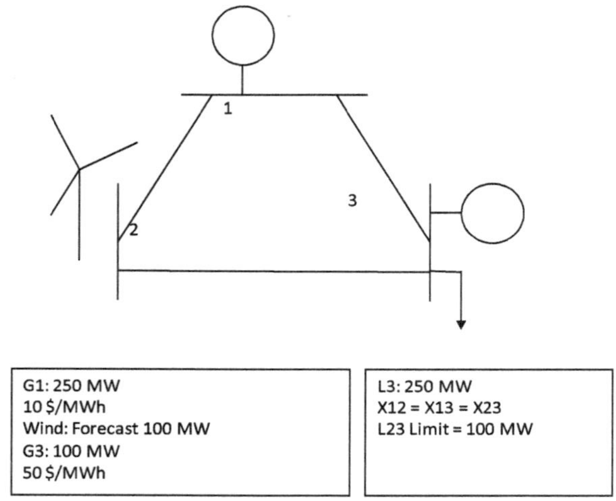

Fig. 5.2 Three-bus example to explain the benefits of wind on dispatch

wind generator reduces its output and no longer has a negative price, the production costs are reduced by more than 40%, and the price the load pays is cut by 60%. This shows that enabling VG to be a flexible resource has great benefits for improving efficiency and increases the flexibility pool.

In addition to VG providing flexibility in the energy market and assisting in congestion management, recent discussions have also explored the ability of VG to offer its flexibility in the ancillary service markets. Wind power can provide various forms of active power control using a combination of mechanical pitch and torque control and power electronics control [28]. In many ways, it can provide a desired response faster than thermal plants are able to. Research has looked at the ability of wind power to participate in regulating reserve markets [29–31]. These works have shown that in certain instances wind power can earn revenue and reduce total costs to consumers by providing regulating reserve.

Table 5.7 Production, production costs, and LMPs with wind as a price taker

	Wind (MW)	G1 (MW)	G1 Cost	G3 (MW)	G3 Cost	Total cost ($)	LMP ($)
Base case	100	100	× $10/MWh +	50	× $50/MWh =	3500	
Add 1 MW to Bus 1	100	101	× $10/MWh +	50	× $50/MWh =	3510	10
Add 1 MW to Bus 2	100	102	× $10/MWh +	49	× $50/MWh =	3470	–30
Add 1 MW to Bus 3	100	100	× $10/MWh +	51	× $50/MWh =	3550	50

Table 5.8 Production, production costs, and LMPs with wind on dispatch

	Wind (MW)	G1 (MW)	G1 Cost	G3 (MW)	G3 Cost	Total cost ($)	LMP at Bus ($)
Base Case	50	200	× $10/MWh +	0	× $50/MWh =	2 000	
Add 1 MW to Bus 1	50	201	× $10/MWh +	0	× $50/MWh =	2 010	10
Add 1 MW to Bus 2	51	200	× $10/MWh +	0	× $50/MWh =	2 000	0
Add 1 MW to Bus 3	49	202	× $10/MWh +	0	× $50/MWh =	2 020	20

5.5.2 Evolving Regulating Reserve Markets

Some recent changes have been made to the ancillary service markets to change the ways in which resources are incentivized. The most significant changes have been made to the regulating reserve markets. In late 2011, Order 755 was issued by FERC on Frequency Regulation Compensation in the Organized Wholesale Power Markets [32]. The order directed market operators that are part of the organized wholesale markets to include market-based payments for regulating reserve performance, lost opportunity costs for all regulating reserve capacity prices, and incentives and rules for accuracy. The order did not require any standardization between markets and also made no changes to the net energy payments that came as a result of the energy from regulating. At present, all markets except ISO-NE have implemented the changes for Order 755. In addition, ERCOT, though not FERC jurisdictional, has initiated a pilot program on fast regulation response service, which is in many ways analogous to the implementations made to meet Order 755 in the other markets.

Although many markets already included lost opportunity costs in regulating reserve markets, the order enforced this. Historically, a few areas had paid only the lost opportunity cost to the suppliers that incurred these costs. With the order, it was decided that the lost opportunity cost is part of the marginal cost of providing regulating reserve capacity and should therefore be a part of the price paid to all regulating reserve suppliers. The order also stated that the market operators are responsible for assigning the lost opportunity costs, but that intertemporal opportunity costs (i.e., by providing regulating reserve in the current hour, a supplier may lose out on energy profits in future hours) must be verifiable and can be included in a supplier's regulating reserve offer.

Historically, ancillary service markets are paid only for the capacity that suppliers held to provide the ancillary service and not the actual utilization of the capacity for the ancillary service [33]. This order adjusts the payments for regulating reserve so that the resource is paid based on how much it was asked to control during each market interval as well as how accurate it followed its automatic generation control signal. The performance price must be market based rather than administratively set, and the performance is based on the absolute amount of movement that a supplier performs in a market interval. Suppliers that are asked to move up and down at a higher frequency would therefore be paid more for performance than those being asked to move more slowly. In addition, the closer in accuracy the supplier followed the automatic generation control signal, the more value it would receive as well. Exactly how the accuracy was measured would vary in each market, but the order required that the accuracy is based on how well a resource follows the control signal and not how well it follows area control error, and that all resources' accuracy is measured by the same means. This design would then incentivize resources that are more flexible and can provide regulating reserve faster and more accurate by providing greater payment than that made to slower and less accurate resources.

Although the order had its objectives toward incentivizing suppliers that provided regulating reserve to be faster and more accurate, there was not a consensus on the benefits of the order. Many commenters on the order believed that the faster response would not have any significant reliability benefit and would only raise costs to consumers. Proponents of the order suggested that the introduction of performance payment would reduce the regulating reserve capacity prices. Other proponents also argued that the use of faster ramping resources would improve efficiency of meeting regulating reserve requirements and thereby reduce the capacity requirement of regulating reserve. This was also shown in other studies, that analyzed the impact of faster responding resources, such as [34]. Table 5.9 shows ancillary service prices for a time period when Order 755 was implemented in NYISO and then the prices for the same time period during the previous year without Order 755. Although there could be many other reasons this occurs rather than Order 755, prices for all ancillary services increased with the new design, not necessarily supporting the efficiency improvement. Although the argument of whether the implementation of Order 755 improves efficiency as well as the argument of how much it improves reliability should continue to be evaluated, it is clear that it does make the regulating reserve market better suited to incentivize response speed as well as response accuracy, giving a great push toward improved flexibility incentives.

Table 5.9 Ancillary service prices of the NYISO during a period with and without regulation performance payment

Before 755 (June 26 to Oct. 22, 2012)	Spin ($/MWh)	Nonspin ($/MWh)	30 min. ($/MWh)	Regulation capacity	Regulation mileage ($/MW)
Average price	$4.00	$1.80	$0.08	$6.44	N/A
Average intervals at 0 price	84.5%	98.0%	99.9%	0.40%	N/A
After 755 (June 26 to Oct. 22, 2013)	Spin ($/MWh)	Nonspin ($/MWh)	30-min.($/MWh)	Regulation capacity	Regulation mileage ($/MW)
Average price	$5.82	$3.26	$1.70	$10.59	$0.23 ($2.30)
Average intervals at 0 price	86.4%	98.1%	99.4%	1.40%	10.2%

5.5.3 *Ancillary Service Markets for Primary Frequency Control*

A common argument to Order 755 discussed in the previous subsection was that it considered only secondary frequency control and not primary frequency control (PFC). The arguments stated that FERC cannot look at only secondary frequency control because of the interrelationship between primary and secondary frequency control. In some ways, the argument was that the faster a supplier can follow the automatic generation control signal, the more it could earn, until the automatic generation control signal is too fast and the supplier follows frequency, in which case the supplier gets paid nothing.

PFC is the response typically from synchronous generator turbine governors that responds proportionally to frequency deviations. The aggregate PFC will arrest frequency decline and bring it to a new steady-state level. Synchronous inertia service, which may or may not be included in the definition of primary frequency control, is typically defined as the immediate injection of active power through the stored kinetic energy of the rotating mass of synchronous machines. This response will slow down the rate of change of the frequency decline. Both primary frequency control and synchronous inertia are crucial services needed to maintain a reliable and secure system and avoid under-frequency load-shedding, machine damage, and potential blackouts.

In the United States, there is no reliability requirement for a balancing authority area or market area to have sufficient synchronous inertia or PFC. A recent draft standard, BAL-003-1, would require a minimum amount of PFC that balancing authority areas must have available at all times. Also, currently there are no incentives in place for individual resources to provide either service. Some studies have shown that the frequency response in the United States, especially in the Eastern Interconnection, has been declining during the past 20 years or more [35]. Some reasons for this include high governor deadbands, generators operating in modes that do not offer frequency-responsive reserve, governors that are not enabled, a reduced percentage of direct drive motor load, and others [36, 37]. Without any controls or changes to meet the needs, increased penetrations of nonsynchronous VG could further degrade frequency response. However, some have claimed that the wholesale electricity market design, lack of incentives, and even the presence of disincentives to provide the service are among the major causes of the decline [38].

A potential disincentive was discovered in Ela et al. [39]. Many market operators have financial penalties in place when suppliers produce outputs different than those that were scheduled. Suppliers providing PFC would automatically adjust output when frequency deviates from its nominal level (60 Hz in the United States) without any control room operator intervention. Few of these markets would use system frequency in their market settlements rules. Therefore, as in the example in Ela et al. [39], in a market with a 3% tolerance band, a supplier with a 5% droop curve would be automatically penalized when frequency deviates more than 90 mHz. Meanwhile, this resource is doing exactly what is required to maintain a reliable and secure power system.

Numerous ancillary service market designs for PFC and synchronous inertia have been proposed in the literature, including that from [28]. As VG increases, displacing resources that typically offer PFC and synchronous inertia, the need for incentivizing this service can become more apparent. It can also incentivize resources that would not typically provide these services, such as VG, to install the capability and offer into that market. Recently, in its ancillary service market redesign initiative, ERCOT was the first market to mention its intentions to implement a PFC ancillary service market [40]. After BAL-003-1 was passed, and with the increasing need to incentivize resources to be more flexible and provide these services, it is likely that this trend will continue in the future.

5.5.4 Convex Hull Pricing

Some differences do occur in the way that each ISO prices energy and ancillary services. The marginal pricing theory for energy and ancillary service markets is based on continuous, convex, monotonically increasing variable costs. As a result of primarily no-load costs and start-up costs, actual costs are not convex, and the lumpiness creates additional requirements to ensure efficient market design [41, 42]. This creates the need for uplift payments so that resources that do not recover their no-load or start-up costs from the LMP, which is typically based solely on incremental energy costs, will get side payments (see Sect. 5.4.4).

For energy markets, pricing in some market regions is not exactly based on the pure marginal cost. MISO has discussed the extended locational marginal pricing (ELMP), which is based on the convex-hull pricing concept [43, 44]. This is similar to the hybrid-dual approach at the NYISO. A question arose in the past on the correct price given when peaking gas turbines were turned on to meet high energy demands. When 1 MW of additional load must be served, and a peaking plant with a minimum capacity of 20 MW is turned on to provide its needed energy, the next cheaper unit would be backed down by 19 MW [45]. In this situation, the marginal cost of energy would be the bid-based cost of the cheaper unit. This means that even though the more expensive unit was needed, the marginal cost of energy (and price) was not increased. The peaking unit would get paid less than its bid-based costs, requiring a make-whole payment, and the rest of the generation fleet would earn lower revenues because the marginal-cost-based price was suppressed. The extended LMP and hybrid-dual pricing concepts consider the non-convex aspects of the resource costs and constraints as part of their pricing rules. For certain resources it will include the no-load and/or start-up cost of the resource as part of its total bid cost, meaning these non-convex costs can influence the price. The benefit to this approach is more transparency in pricing to the more expensive resource by having prices better reflect actual costs. The convex-hull pricing approach is currently an ongoing debate.

Because of the increased variability and uncertainty of VG on the system, additional resources may be committed without being economic according to their energy costs. This could lead to increased times where resources online would not recover their costs because energy prices are based on the marginal cost of energy. These resources would receive uplift payments, reducing the transparency of prices. Future pricing mechanisms, similar to ELMP or hybrid-dual pricing that incorporate these non-convexities into energy prices should be evaluated with increased penetrations of VG.

5.5.5 Flexible Ramping Products

Finally, it is important that the electricity market designs are incentivizing increased flexibility to provide energy when that flexibility is needed. It is debatable whether incentivizing flexibility is being done efficiently in all United States markets today. A few areas have been introducing and proposing changes to their electricity market designs to ensure that energy markets are incentivizing the greater flexibility needed from increased penetrations of variable generation. Other areas are not presently making significant changes to their designs, perhaps believing that the mechanisms described above in Sect. 5.4 are enough to incentivize flexibility in the energy markets. Some market areas, namely CAISO and MISO, have begun to introduce explicit markets for energy flexibility as a new ancillary service. We discuss these next.

CAISO has performed a number of studies to analyze the impacts of integrating significant levels of VG on its system. Two of the more recent studies analyzed the impacts that VG has on the capacity and ramping needs for its energy markets. The first study determined that the amount of ramping needs would increase by up to 30 MW/min to 40 MW/min with 20% renewables [46]. A later study found similar ramp rate increases and determined that the amount of incremental load-following capacity that would be needed as a result of the variability and uncertainty of VG was 845 MW and 930 MW for upward and downward load-following capacity, respectively [47]. Figures 5.3 and 5.4 show the total capability of load-following up and load-following down, respectively, from the generating fleet. Figure 5.5 shows the same information as Fig. 5.4, except that the total load-following down capability is not limited by resources that are self-scheduled and not offering their flexibility to the market. As shown, some of the early morning hours would not be able to meet the total load-following down capability of 930 MW (Fig. 5.4). However, if more resources changed from being self-scheduled resources to flexible resources, the requirement could be met easily (Fig. 5.5). This can support the idea that further incentives are needed for the self-scheduled resources to offer their flexibility to meet the increased flexibility needs resulting from increased VG.

In August 2011, the CAISO board of governors approved a flexible ramping constraint mechanism in the ISO energy market design [48, 49]. This is an additional constraint added to the market-clearing engine that ensures that sufficient ramping capacity is committed and available in the real-time commitment and real-time dispatch process. The use of this constraint reduces infeasibilities in the dispatch procedure compared to when ramp capability is not committed, reduces the need

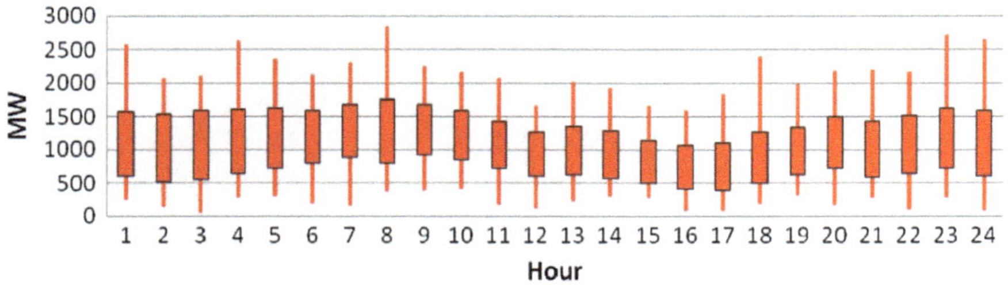

Fig. 5.3 Load-following up capability in generating fleet, summer period 2009–2010 (Source: from [47])

for reliance on regulation reserve and relying on neighboring balancing errors, and eliminates the need to biased hour-ahead forecasts to prepare for potential variations in real time. At present, the constraint was only for upward ramp capability needs. The amount of ramp capability that is required in the constraint is determined by the CAISO operators based on the following:

1. expected level of variability for the interval;
2. potential uncertainty as a result of load and VG forecast error;
3. differences between the hourly, 15-min average net load levels and the actual 5-min net load levels.

These levels are determined from historical data, and the total requirements are published for the various market processes.

Similar to other ancillary service products, there is a potential for a lost opportunity cost for resources that are withholding their capacity to meet this ramping constraint. If a resource foregoes profit in the energy market or other ancillary service market to reserve capacity for the ramping constraint, it has a lost opportunity cost for serving the ramp constraint. In the current market design for this flexible ramping constraint, all resources that meet this ramping constraint with capacity that is not being used for other ancillary service products are paid the marginal resource's lost opportunity cost. The value is based on the incremental cost that would be incurred by the system if increasing the ramping need by one unit. Currently, there is no allowance for other costs associated with ramping to be added—i.e., no separate bid for this product—

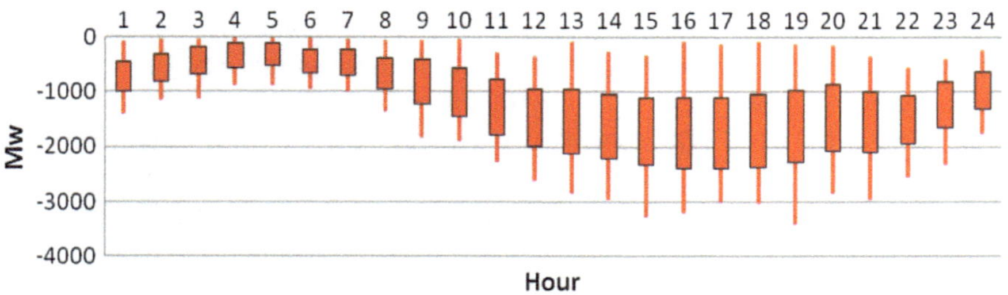

Fig. 5.4 Load-following down capability in generating fleet, summer period 2009–2010 (Source: from [47])

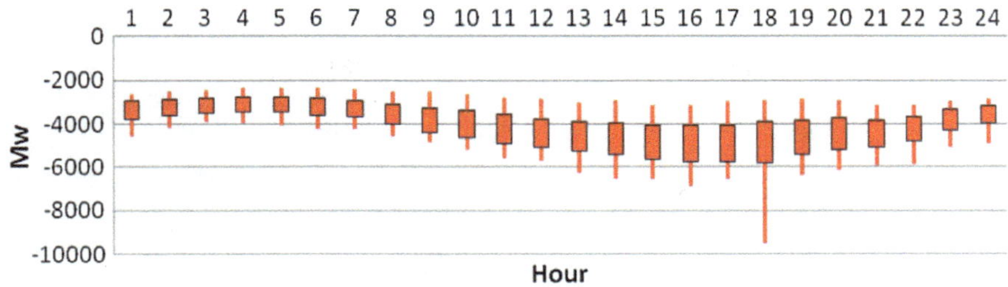

Fig. 5.5 Load-following down capability in generating fleet if self-scheduled resources are included, summer 2009–2010 (Source: from [47])

therefore, the only way for a resource to get paid to provide ramping capability under this ramping constraint would be if a lost opportunity cost were incurred. All of the costs associated with the price paid to suppliers selected for the flexible ramping constraint are currently paid by demand.

During the first few years of the flexible ramping constraint, some observations can be made. In 2012, the total cost to this constraint was approximately $20 million, compared to $35 million for the spinning reserve market [50]. It was found that during this time period much of the flexible ramping capacity was in the northern part of the system and often unavailable to provide assistance in relieving congestion in the southern part of the system. Table 5.10 shows additional statistics from the first year the constraint was enacted. The table shows total payments, the percentage of intervals in which the constraint was binding (i.e., nonzero, when the constraint required change in dispatch), the percentage of intervals in which the flexibility constraint requirement could not be met (i.e., had a procurement shortfall), and the average price of the constraint when it was binding. The spring time period had the highest prices, as well as the most binding and short intervals. Payments seemed to be greatly reduced by the end of the year. However, the first quarter of the following year 2013, the costs of flexible ramping constraint were $10M, half that of the entire previous year, whereas spinning reserve costs were approximately $6M [51]. This was mostly because the ISO increased the requirement more consistently than it did in 2012. The costs reduced by the end of the 2013.

After the flexible ramping constraint was approved, the ISO and its stakeholders proposed a full flexible ramping product similar to other ancillary service products. The ISO, along with stakeholders and its board of governors, agreed that greater market effectiveness could be gained by developing market-based products that can better identify, commoditize, and compensate for this flexibility. The main differences in the flexible ramping product from the constraint described above are the inclusion of downward ramping, the change in using the 5-min real-time dispatch interval rather than the 15-min real-time pre-dispatch model, the inclusion of the product in the day-ahead market, and a flexibility demand curve. Because CAISO uses a multi-period market-clearing engine, ramping requirements are already within the model based on the expected change in net load from one interval to the next. This is the minimum ramping requirement that must be met. The ISO then will require

Table 5.10 Statistics for the first year of flexible ramping constraint in CAISO (Source: from [50])

Month	Total payments to generators ($M)	Intervals constraint was binding (%)	Intervals with procurement shortfall (%)	Average shadow price when binding ($/MWh)
January	2.45	17	1.0	38.44
February	1.46	8	1.3	77.37
March	1.90	12	1.0	42.75
April	3.37	22	1.5	39.86
May	4.11	23	6.0	79.48
June	1.49	13	2.3	77.37
July	1.01	8	1.4	42.75
August	0.77	7	1.2	39.86
September	1.03	13	0.8	79.48
October	0.9	9	1.0	39.19
November	0.23	4	0.5	53.34
December	1.09	9	1.6	61.84

additional ramping capability requirements above the expected ramping requirement to meet the unexpected ramping capability requirement, which can be as high as the 97.5th percentile change in net demand (or 2.5th percentile for the downward-ramp capability). Between the minimum and maximum ramp need, there is a stepped demand curve. This will ensure that the ISO will procure a certain amount of ramping capability based on both the need and the additional cost. The penalty costs that are a part of this flexible ramping product demand curve are based on the probability of power balance violations as a result of not having ramping capability and the penalty cost of those violations. The maximum price of ramping capability of 250 $/MWh is set when there is not enough ramping capability to meet the minimum ramping need, i.e., the expected ramping need.

Other market and settlement rules accompany the new flexible ramping product. Bids for ramping capability are only allowed in the day-ahead market. The prices that occur in real time are based on only the lost opportunity cost incurred by ramping units not able to fully participate in the energy or ancillary service markets, or from the penalty prices that are part of the flexible ramping product demand curve. The settlement between the day-ahead and real-time markets is performed similarly to other products, including energy. The quantity of flexible ramping capability available in the day-ahead market is sold at day-ahead flexible ramping prices, and the difference in real time is paid (or bought back) at the real-time flexible ramping capability price. Note that the difference takes into account the interval resolution of the different markets (i.e., because the day-ahead market is hourly, it is divided by 12 to calculate the difference from the real-time market ramping capability, because the real-time market is in a 5-min resolution).

Table 5.11 Generator properties for flexible ramp example

Generator	Bid cost ($/MWh)	Ramp rate (MW/min)	Pmax
Gen 1	25	100	500
Gen 2	30	10	500

The importance of the flexible ramping product can be illustrated with some short examples from [52]. This importance can be shown simply with the expected ramp capability need before discussing the need from unexpected ramp capability. As discussed, CAISO, like many other ISOs, solves the real-time market using a multi-period market-clearing engine (e.g., multi-period security-constrained economic dispatch). The first example is a two-period dispatch solution with a load of 420 MW in Interval 1 and 590 MW in Interval 2. The second example is the same scenario with a flexible ramping product requirement set to require slightly more ramping capability than the inherent need from the first scenario (170.01 MW in 5-min). Both scenarios use a two-generator set with characteristics shown in Table 5.11. Both generators have zero-cost bids for flexible ramping and minimum generation levels at 0 MW.

Both scenarios have essentially identical operational results (see Tables 5.12 and 5.13). However, the prices that result are quite different in each scenario because of the flexible ramp requirement. The prices are based on the marginal cost of the entire period for providing the service in the associated time interval. Finally, with the different distribution of pricing, it turns out that the revenues for both units will be equal in both scenarios when both intervals are settled. The main issue that CAISO describes in the draft summary is that the current market design will consider only the first interval during multi-period dispatch as the binding settlement interval. Because the price of the second interval will change when it is the binding interval, the overall result is that the revenues of the units will not be the same, even though the operational results will still remain identical. The first scenario would not have the same incentive for providing flexibility as would the second scenario.

The example above shows some benefit of the flexible ramp product when expected variability is present. The product will also increase the ramping need above the expected ramp to be able to meet the unexpected ramp need (ramp needs that are not forecasted). This allows for resources to have capacity and ramp available in case a ramping event occurs. Pricing based on the lost opportunity cost allows all resources to be indifferent whether providing flexible ramping or energy either for certain or uncertain ramping events.

Table 5.12 Scenario 1, multi-interval dispatch

Generator	Interval 1 (LMP = 25 $/MWh)		Interval 2 (LMP = 35 $/MWh)	
	Energy	Flex	Energy	Flex
G1	380	0	500	0
G2	40	0	90	0

Table 5.13 Scenario 2, flex reserve product

Generator	Interval 1 (LMP = 30 \$/MWh, FRP = 5 \$/MWh)		Interval 2 (LMP = 30 \$/MWh)	
	Energy	Flex	Energy	Flex
G1	379.99	120.01	500	0
G2	40.01	50	90	0

MISO has proposed a similar product to CAISO, the up-ramp capability and down-ramp capability [53]. MISO had claimed that the most common reason for scarcity pricing conditions in its area was not caused by limited capacity but by insufficient ramp capability. These scarcity conditions were causing large price spikes in the energy and ancillary service markets. The product would be introduced in both day-ahead and real-time markets to ensure enough ramping capability would be available in the future. The product has similar concepts to CAISO, including a demand curve for insufficient ramp capability, pricing based solely on the lost opportunity costs from other products, and a requirement based on historical information to meet both expected and unexpected ramp requirements. MISO has also included deliverability requirements within the ramp capability conceptual design. The schedules for up-ramp capability and down-ramp capability would be made such that, when combined with energy, the full deployment of that capacity will not violate any transmission constraints assuming pre-defined locations of the variability and uncertainty that caused the ramp capability need. This ensures that the locations of reserving the ramp capability will be able to be deployed without transmission constraints when needed. The product is being filed with FERC, and it is not likely to be introduced into the market until 2015.

One key difference between the CAISO and MISO approaches is how the payments for the ramping capability will be allocated. In the current CAISO proposal, the allocation will be based on a cost causation principle while the MISO proposal will be based on primary beneficiaries of the product. In CAISO, the allocation is proposed to be distributed among loads, suppliers, and fixed ramp resources (e.g., external transactions and internal self-scheduled resources). Loads causing need for ramping will be allocated based on their 10-min movement. Suppliers will be allocated based on their 10-min uninstructed deviations from real-time schedules. Fixed ramp resources will be allocated similarly to load based on 10-min net movement. This allocation proposal from CAISO will be the first which an ancillary service costs are reimbursed through cost causation principles as opposed to it being reimbursed fully by load-serving entities.

Other market operators are not seeing the need for this new ramping product to incentivize flexibility. For example, NYISO has suggested that the DAMAP, make-whole payments, and optimal operating resources at their most efficient production level are sufficient and fully incorporate load-following (flexibility) services into pricing and scheduling outcomes. An additional payment for this flexibility would be unnecessary. Although ERCOT has a supplemental reserve service in its new ancillary service market redesign, similar to those products in CAISO and MISO, it also notes that such a service will not ultimately be required in the long term [40]. This is contrary to the direction of CAISO and MISO. The differences between these market designs should be further studied. For example, the need in CAISO may be because of the higher amount of self-scheduled bilateral agreements that have been in place since the energy crisis in 2000 and 2001, limiting the amount of flexibility available to the system. MISO has a very large coal fleet, which has limited flexibility. NYISO has a significant amount of flexible natural gas on their system, which could be why they have not seen a need yet for incentivizing further flexibility. One size does not fit all when it comes to electricity market designs, and each area has solved its historical issues through specific designs involving the stakeholders and market participants in each area. The topic of flexibility incentives should take a more holistic view to see what the reasons are for further market design changes, how these changes should occur, and how increasing amounts of variable energy resources may affect these market design changes.

5.6 Conclusion

This chapter discussed the importance of short-term flexibility in system operations and how that flexibility is needed to better accommodate the increased variability and uncertainty of VG. Flexibility, though often hard to define, is an important characteristic that should be incentivized so that it can be called upon when it is needed. There are many different forms of flexibility and different ways that market operators can extract that flexibility. Market designers have established numerous traditional mechanisms to incentivize resources to offer their flexibility to the market and new, recent mechanisms that attempt to further incentivize increasing levels of flexibility when that flexibility is needed. Things like make-whole payments and evolving ancillary service markets can incentivize flexibility while self-scheduling and hourly settlement intervals can inhibit needed flexibility.

The various regional markets have not converged on approaches to incentivize flexibility in short-term markets. For example, a number of market regions have proposed new flexible ramping ancillary service products to incentivize further flexibility, whereas others have decided this is unnecessary and current mechanisms may already get the needed flexibility. Whether these differences are because of regional differences in the system characteristics, generating portfolios, or existing market rules or procedures is still undetermined. Further research should study these different designs with varying system characteristics to make conclusions on whether certain market mechanisms do in fact fit different characteristics more than others.

A number of new scheduling software programs have been developed that show ways of providing flexibility at low cost and improved reliability with increased levels of VG. The ways in which the pricing is determined in these new scheduling models is not always as straightforward. When new methods appear to reduce costs or improve reliability, it is important that the resulting prices are analyzed to determine whether the resources providing the flexibility to improve system operations are actually being incentivized to do so and that unintended consequences are avoided. Otherwise, although it looks like costs are reduced or reliability is improved in the short term, the resources may not be incentivized to do as directed; therefore, in the long term, these improvements and projected cost savings may not be realizable.

Due to the complicated nature of the electric power system and the relationship of all market products (e.g., energy, ancillary service, capacity) with each other, it is important that any new modifications to one design do not adversely impact the other. When new designs are made to improve the way that flexibility is incentivized, careful analysis should determine whether this will affect how other markets incentivize other required attributes. Research that goes into new designs should always account for how it may affect other designs. In addition, metrics that are used to show the benefits of any design should be all encompassing. If one design reduces the system production costs, further evaluation should ensure that it either improves or keeps constant the reliability and incentive structure of the system. Therefore, further research into the new designs that may improve incentivizing flexibility should consider all system metrics to the extent that it can, before promoting a new design to be put into practice.

The electricity designs that have been developed in the United States are very sophisticated due to the intricacies with including the physics of the power system within the market mechanisms. Many of the trends of electricity market design evolution, especially with the further improvements with software computational capabilities, have moved toward greater complexity. Another debated topic is whether this complexity is necessary. Should the energy markets be simple with one-part bids and offers? This question will likely continue to be at the center of all market design changes as the thinking continues to evolve.

Many of the mechanisms described in this chapter could have more significant impacts when even greater penetrations of VG are integrated onto the system. Designs such as primary frequency response markets, pay-for-performance ancillary services, and convex-hull pricing are all in their infancy, and their impacts on a changing system should be analyzed further. The research performed in these areas should help all of the market areas find some consistency going forward when determining appropriate market designs with these continually changing systems.

References

1. Ela, E., Milligan, M., Bloom, A., Botterud, A., Townsend, A., Levin, T.: Evolution of Wholesale Electricity Market Design with Increasing Levels of Renewable Generation. Technical Report NREL/TP-5D00-61765, Golden, Colorado (2014)
2. Lannoye, E., Flynn, D., O'Malley, M.: Evaluation of power system flexibility. IEEE Trans. Power Syst. **27**(2), 922–931 (2012)
3. Ma, J., Silva, V., Belhomme, R., Kirschen, D., Ochoa, F.: Evaluating and planning flexibility in sustainable power systems. IEEE Trans. Sustain. Energy **4**(1), 200–209 (2013)

4. Rux, M.: An incremental economic dispatch method for cascaded hydroelectric power plants. IEEE Trans. Power Syst. **8**(3), 1266–1273 (1993)
5. Lu, B., Shahidehpour, M.: Short-term scheduling of combined cycle units. IEEE Trans. Power Syst. **19**(3), 1616–1625 (2004)
6. Ela, E., O'Malley, M.: Studying the variability and uncertainty impacts of variable generation at multiple timescales. IEEE Trans. Power Syst. **27**(3), 1324–1333 (2012)
7. Wan, Y.: A Primer on Wind Power for Utility Applications. Technical Report NREL/TP-500-36230, Golden, Colorado (2005). https://www.nrel.gov/docs/fy06osti/36230.pdf. Accessed 28 Dec 2016
8. Wan, Y.: Analysis of Wind Power Ramping Behavior in ERCOT. Technical Report NREL/TP-5500-49218, Golden, Colorado (2011). https://www.nrel.gov/docs/fy11osti/49218.pdf. Accessed 28 Dec 2016
9. Mills, A., Ahlstrom, M., Brower, M., Ellis, A., George, R., Hoff, T., Kroposki, B., Lenox, C., Miller, N., Stein, J., Wan, Y.: Understanding Variability and Uncertainty of Photovoltaics for Integration with the Electric Power System. Lawrence Berkeley National Laboratory, Report LBNL-2855E, Berkeley, CA (2009). http://utilityscalesolar.lbl.gov/sites/all/files/lbnl-2855e.pdf. Accessed 28 Dec 2016
10. Zhang, J., Hodge, B-M., Florita, A., Lu, S., Hamann, H., Banunarayanan, V.: Metrics for Evaluating the Accuracy of Solar Power Forecasting. Technical Report NREL/CP-5500-60142, Golden, Colorado (2013). https://www.nrel.gov/docs/fy14osti/60142.pdf. Accessed 28 Dec 2016
11. Hodge, B-M., Lew, D., Milligan, M., Holttinen, H., Sillanpää, S., Gómez-Lázaro, E., Scharff, R., Söder, L., Larsén, X., Giebel, G., Flynn, D., Dobschinski, J.: Wind Power Forecasting Error Distributions: An International Comparison. Technical Report NREL/CP-5500-56130, Golden, Colorado (2012). https://www.nrel.gov/docs/fy12osti/56130.pdf. Accessed 28 Dec 2016
12. Ela, E., Kirby, B., Botterud, A., Milostan, C., Krad, I., Koritarov, V.: The Role of Pumped Storage Hydro Resources in Electricity Markets and System Operation. Technical Report NREL/CP-5500-58655, Golden, Colorado (2013). https://www.nrel.gov/docs/fy13osti/58655.pdf. Accessed 28 Dec 2016
13. Milligan, M., Kirby, B., Beuning, S.: Combining Balancing Areas' Variability: Impacts on Wind Integration in the Western Interconnection. Technical Report NREL/CP-550-48249, Golden, Colorado (2010). https://www.nrel.gov/docs/fy10osti/48249.pdf. Accessed 28 Dec 2016
14. Bouffard, F., Galiana, F.: Stochastic security for operations planning with significant wind power generation. IEEE Trans. Power Syst. **23**(2), 306–316 (2008)
15. Meibom, P., Barth, R., Larsen, H., Brand, H., Tuohy, A., Ela, E.: Advanced Unit Commitment Strategies in the United States Eastern Interconnection. Technical Report NREL/SR-5500-49988, Golden, Colorado (2011). https://www.nrel.gov/docs/fy11osti/49988.pdf. Accessed 28 Dec 2016
16. Wang, J., Botterud, A., Bessa, R., Keko, H., Carvalho, L., Issicaba, D., Sumaili, J., Miranda, V.: Wind power forecasting uncertainty and unit commitment. Appl. Energy **88**(11), 4014–4023 (2011)
17. Bertsimas, D., Litvinov, E., Sun, X., Zhao, J., Zheng, T.: Adaptive robust optimization for the security constrained unit commitment problem. IEEE Trans. Power Syst. **28**(1), 52–63 (2013)
18. Price, J., Rothleder, M.: Recognition of extended dispatch horizons in CA energy markets. In: IEEE Power and Energy Society General Meeting, pp. 1–4. IEEE Press, New York (2011)
19. Ela, E., Tuohy, A., Milligan, M., Kirby, B., Daniel, B.: Alternative approaches for incentivizing the PFR ancillary service. Electr. J. **25**(4), 88–102 (2012)
20. Ela, E., Gevorgian, V., Fleming, P., Zhang, Y., Singh, M., Muljadi, E., Scholbrook, A., Aho, J., Buckspan, A., Pao, L., Singhvi, V., Tuohy, A., Pourbeik, P., Brooks, D., Bhatt, N.: Active Power Control from Wind Power: Bridging the Gaps. Technical Report NREL/TP-5D00-60574, Golden, Colorado (2014). https://www.nrel.gov/docs/fy14osti/60574.pdf. Accessed 28 Dec 2016

21. Milligan, M., Kirby, B.: Market Characteristics for Efficient Integration of Variable Generation in the Western Interconnection. Technical Report NREL/TP-550-48192, Golden, Colorado (2010). https://www.nrel.gov/docs/fy10osti/48192.pdf. Accessed 28 Dec 2016
22. FERC: Recent ISO Software Enhancements and Future Software and Modeling Plans. Staff Report prepared by the Federal Energy Regulatory Commission, Washington, D.C (2011)
23. Hirst, E.: Real-Time Balancing Operations and Markets: Key to Competitive Wholesale Electricity Markets. Report prepared for Edison Electric Institute, Washington, D.C. (2001). http://www.consultkirby.com/files/RTMReport.pdf. Accessed 28 Dec 2016
24. FERC: Demand Response Compensation in Organized Wholesale Energy Markets. Order N. 745, Docket N. RM10-17-000, Washington, D.C (2011)
25. Huang, S., Dumas, J., Gonzalez-Perez, C., Wei-Jen, L.: Grid security through load reduction in the ERCOT market. IEEE Trans. Ind. Appl. **45**(2), 555–559 (2009)
26. NYISO: Energy Storage in the New York Electricity Market. New York Independent System Operator White Paper, NY (2010)
27. Ela, E., Edelson, D.: Participation of wind power in LMP-based energy markets. IEEE Trans. Sustain. Energy **3**(4), 777–783 (2012)
28. Ela, E., Gevorgian, V., Tuohy, A., Kirby, B., Milligan, M., O'Malley, M.: Market designs for the primary frequency response ancillary service-part I: motivation and design. IEEE Trans. Power Syst. **29**(1), 421–431 (2014)
29. Liang, J., Grijalva, S., Harley, R.: Increased wind revenue and system security by trading wind power in energy and regulation reserve markets. IEEE Trans. Sustain. Energy **2**(3), 340–347 (2011)
30. Kirby, B., M. Milligan, Ela, E.: Providing Minute-to-Minute Regulation from Wind Plants. Technical Report NREL/CP-5500-48971, Golden, Colorado (2010). https://www.nrel.gov/docs/fy11osti/48971.pdf. Accessed 28 Dec 2016
31. Tuohy, A., Ela, E., Kirby, B., Brooks, D.: Provision of Regulation Reserve from Wind Power: Economic Benefits and Steady-State System Operation Implications. In: 11th International Workshop on Large-Scale Integration of Wind Power into Power Systems, Lisbon, Portugal (2011)
32. FERC: Frequency Regulation Compensation in the Organized Wholesale Power Markets. Docket Nos. RM11-7-000 and AD10-11-000, Washington, D.C. (2001)
33. Rebours, Y., Kirschen, D., Trotignon, M., Rossignol, S.: A survey of frequency and voltage control ancillary services-part II: economic features. IEEE Trans. Power Syst. **22**(1), 358–366 (2007)
34. Makarov, Y., Lu, S., Ma, J., Nguyen, T.: Assessing the Value of Regulation Resources Based on Their Time Response Characteristics. Report of Pacific Northwest National Laboratory, PNNL-17632, Richland, Washington (2008)
35. Ingleson, J., Allen, E.: Tracking the Eastern interconnection frequency governing characteristic. In: IEEE Power and Energy Society General Meeting, pp. 1–6. IEEE Press, New York (2010)
36. Schulz, R.: Modeling of governing response in the Eastern interconnection. In: IEEE Power Engineering Society 1999 Winter Meeting, pp. 1–5. IEEE Press, New York (1999)
37. Virmani, S.: Security impacts of changes in governor response. In: IEEE Power Engineering Society 1999 Winter Meeting, pp. 1–6. IEEE Press, New York (1999)
38. IEEE: IEEE Task Force on Large Interconnected Power Systems Response to Generation Governing: Present Practice and Outstanding Concerns. IEEE Power & Energy Society (2007)
39. Ela, E., Tuohy, A., Milligan, M., Kirby, B., Brooks, D.: Alternative approaches for incentivizing the PFR ancillary service. Electr. J. **25**(4), 88–102 (2012)
40. ERCOT: Future Ancillary Services in ERCOT. Electric Reliability Council of Texas Concept Paper, Austin, TX (2013)
41. O'Neill, R., Sotkiewicz, P., Hobbs, B., Rothkopf, M., Stewart, W.: Efficient market-clearing prices in markets with nonconvexities. Eur. J. Oper. Res. **164**(1), 269–285 (2005)
42. Elmaghraby, W., O'Neill, R., Rothkopf, M., Stewart, W.: Pricing and efficiency in 'Lumpy' energy markets. Electr. J. **17**(5), 54–64 (2004)

43. Gribik, P., Chaterjee, D., Navid, N., Zhang, L.: Dealing with uncertainty in dispatching and pricing in power markets. In: IEEE Power and Energy Society General Meeting, pp. 1–6. IEEE Press, New York (2011)
44. Gribik, P., Hogan, W., Pope, S.: Market-Clearing Electricity Prices and Energy Uplift. Technical Report (2007)
45. Stoft, S.: Power Systyem Economis – Designing Markets for Electricity. IEEE Press and Wiley Interscience (2002)
46. CAISO: Integration of Renewable Resources: Transmission and Operating Issues and Recommendations for Integrating Renewable Resources on the California ISO-Controlled Grid. California ISO Report, Folsom, CA, United States (2007). http://www.caiso.com/1ca5/1ca5a7a026270.pdf. Accessed 14 March 2016
47. CAISO: Integration of Renewable Resources: Operational Requirements and Generation Fleet Capability at 20% RPS. California ISO Report, Folsom, CA, United States (2010)
48. CAISO: Opportunity Cost of Flexible Ramping Constraint: Draft Final Proposal. California ISO Report, Folsom, CA, United States (2011). http://www.caiso.com/Documents/DraftFinalProposal-FlexibleRampingConstraint.pdf. Accessed 28 Dec 2016
49. Abdul-Rahman, K., Alarian, H., Rothleder, M., Ristanovic, P., Vesovic, B., Lu, B.: Enhanced system reliability using flexible ramp constraint in CAISO market. In: IEEE Power and Energy Society General Meeting, pp. 1–5. IEEE Press, New York (2012)
50. CAISO: 2012 – Annual Report on Market Issues and Performance. California ISO Report, Department of Market Monitoring, Folsom, CA, United States (2012)
51. CAISO: Report on Market Issues and Performance. California ISO Report, Department of Market Monitoring, Folsom, CA, United States (2013)
52. CAISO: Flexible Ramping Products. California ISO Report, Folsom, CA, United States (2012)
53. Navid, N., Rosenwald, G.: Market solutions for managing ramp flexibility with high penetration of renewable resource. IEEE Trans. Sustain. Energy **3**(4), 784–790 (2009)

Chapter 6
Long-Term Resource Adequacy, Long-Term Flexibility Requirements, and Revenue Sufficiency

Erik Ela, Michael Milligan, Aaron Bloom, Audun Botterud, Aaron Townsend and Todd Levin

Abstract Variable generation (VG) can reduce market prices over time and also the energy that other suppliers can sell in the market. The suppliers that are needed to provide capacity and flexibility to meet the long-term reliability requirements may, therefore, earn less revenue. This chapter discusses the topics of resource adequacy and revenue sufficiency—that is, determining and acquiring the quantity of capacity that will be needed at some future date and ensuring that those suppliers that offer the capacity receive sufficient revenue to recover their costs. The focus is on the investment time horizon and the installation of sufficient generation capability. First, the chapter discusses resource adequacy, including newer methods of determining adequacy metrics. The chapter then focuses on revenue sufficiency and how suppliers have sufficient opportunity to recover their total costs. The chapter closes with a description of the mechanisms traditionally adopted by electricity markets to mitigate the issues of resource adequacy and revenue sufficiency and discusses the most recent market design changes to address these issues.

This chapter is based on the detailed discussion of existing and evolving market designs to ensure resource adequacy and revenue sufficiency to meet the increased needs from variable generation [1, Sect. 3].

E. Ela (✉)
Electric Power Research Institute, Palo Alto, CA 94304, USA
e-mail: eela@epri.com

M. Milligan · A. Bloom · A. Townsend
National Renewable Energy Laboratory, Golden, CO 80401, USA
e-mail: Michael.Milligan@nrel.gov

A. Bloom
e-mail: Aaron.Bloom@nrel.gov

A. Botterud · T. Levin
Argonne National Laboratory, Argonne, IL 60439, USA
e-mail: abotterud@anl.gov

T. Levin
e-mail: tlevin@anl.gov

F. Lopes and H. Coelho (eds.), *Electricity Markets with Increasing Levels of Renewable Generation: Structure, Operation, Agent-based Simulation, and Emerging Designs*,
Studies in Systems, Decision and Control 144, https://doi.org/10.1007/978-3-319-74263-2_6

6.1 Introduction

In this chapter, we focus on two related topics:

1. Resource adequacy, including newer methods of determining adequacy metrics.
2. Revenue sufficiency and how existing and evolving market designs may enable resources to retrieve sufficient revenue to ensure long-term resource adequacy.

The focus here is on the investment time horizon and the installation of sufficient generation capability. Operational issues, which are closely related, were addressed in the previous chapter.

The topics of resource adequacy and revenue sufficiency—the process of determining the quantity and acquiring that quantity of capacity that will be needed at some future date and ensuring that those resources that offer the capacity receive sufficient revenue to recover their costs—are the focus here. Resource adequacy is generally based on one or more metrics that quantify the long-term reliability of the generation supply (and possibly demand-response resources) and its ability to meet load. When sufficient capacity is acquired—whether through a market, payment, or other incentive, or through a direct regulatory process—resources must have an opportunity to earn sufficient revenue to remain in the market. Without revenue sufficiency to recover both variable and fixed costs, it is likely that resources would retire from the market, potentially compromising long-term reliability. Also, as more variable generation (VG) is brought online, there is a need for greater flexibility attributes from new and existing market participants and technologies. This brings a new dimension to the traditional resource adequacy question of whether there is enough capacity and adds the question of whether there is sufficient flexibility within that capacity.

Several alternative approaches are available to solve the problems that we identify in this chapter. Rather than recommending any single approach, we present an overview of these alternatives. First, we discuss the issues that make ensuring long-term resource adequacy and revenue sufficiency in electricity markets challenging. Then we discuss the current mechanisms for ensuring resource adequacy and the importance of revenue sufficiency. Next, we discuss how the increased penetrations of VG, with its diurnal and seasonal availability patterns, high capital costs and low variable costs, and its increased variability and uncertainty can change the methods and needs of resource adequacy and revenue sufficiency. We present the historical designs that US wholesale electricity markets have used to address these issues. Finally, we present a review of the most recent market design changes to address resource adequacy and revenue sufficiency with a focus on how they are evolving to meet the needs due to increased VG.

6.2 Challenges to Ensuring Long-Term Reliability

In a perfect world, reliability would be bought and sold in a competitive market. This market would feature consumer choice, as exercised by individual demand curves that would be aggregated to the market level, and supply curves that would result in economically efficient market equilibrium. In this perfect world, there is no market power (among buyers or sellers), no free riders, and consumers are free to buy as much (or as little) reliability as they desire at the market price. However, as is well known, most retail consumers purchase electricity via administered prices that are most often a characterization of average total cost plus an administered profit rate. Most consumers are thus insulated from price swings that are, or would be, a function of the relatively volatile cost and marginal wholesale prices of electricity at the bulk system level. In addition, electrical neighbors who wish to purchase different levels of electric reliability must instead purchase the same amount because there is currently no way to differentiate reliability among customers on the same feeder. These issues-that consumers are insulated from actual time-sensitive prices and consumers cannot choose the level of their individual electric reliability-are the two primary demand-side flaws that impact the way that electricity markets are designed.

To be economically efficient—by providing the level of product desired by society at the lowest cost—markets must have functioning supply and demand functions. Market equilibrium is achieved when the plans of buyers coincide with the plans of sellers. In a free market, buyers can choose whether or not to purchase the product at the market price. Price is related to cost, although this relationship may be complex. However, the fundamental principle is that if cost, and therefore price, were to increase, at least some consumers would withdraw from the market.

Well-functioning markets also allow for producers to differentiate among customers—i.e., a customer who is not willing to pay a given price does not receive the product. In the case of electric reliability, it is usually not technically or economically possible to differentiate levels of reliability to customers (especially residential customers) who may be willing to pay more for reliable service—or, conversely, customers who may be willing to receive a lower level of reliability in return for lower prices.

Thus, resource adequacy is not based on a true market outcome. Instead, in most cases it is based on a long-term reliability standard defined by policy. Instead of prices that ration electricity usage, a somewhat-arbitrary reliability standard is introduced along with administered pricing rules that have the effect of muting most forms of price-response from most consumers. There is yet another complication: in some regions, there is only an approximation of a reliability standard, known as a planning reserve margin (PRM). The PRM is usually defined as the percentage by which installed capacity exceeds annual peak demand. In some cases, there is an adjustment to the PRM or to installed capacity that allows for some consideration of the reliability target. This is discussed in more detail in the next section.

In general, markets require the ability of participants to define the product, its quantity, and price. Price is determined by the intersection of the demand and supply curves of buyers and sellers. Reliability targets in the power system exist because the two demand flaws [2] of the market for electricity prevent markets from functioning effectively. Thus, bridging the gap between reliability and electricity markets is challenging. Fundamentally, there are several requirements, including the following:

- A method for choosing and assessing the resource adequacy target.
- Determining whether the resource adequacy target has been, or will be, achieved.
- Determining the contribution of each entity toward meeting the resource adequacy target.
- Utilizing the right time horizon for meeting resource adequacy targets (e.g., 1 year ahead, 3 years ahead, etc.).

When the resource adequacy target is achieved, will the energy and ancillary service markets result in revenue sufficiency? If not, implement measures to ensure the long-term viability of resources that are needed to achieve reliability or other objectives.

In systems with significant levels of VG, additional questions may need to be addressed:

- How does the resource characteristics of VG influence reliability calculations and resource adequacy targets?
- Does the capacity have the right flexibility attributes to effectively handle the increased variability and uncertainty characteristics of VG in grid operations?
- How does VG itself contribute toward the required capacity adequacy requirements?

6.3 Achieving Long-Term Resource Adequacy and Revenue Sufficiency

As described above, because of the limited price elasticity of demand and the inability of suppliers to curtail load based on consumers' reliability preferences, and also because of the length of time involved to build new supply resources, the combined wholesale and retail electricity markets will not function like other commodities markets. Resource adequacy must be considered to ensure that the electricity supply is sufficient to serve load that appears as an inelastic demand. Determining whether resource adequacy targets are achieved is a probabilistic problem. Well-known methods exist and are based on loss-of-load probability (LOLP) and related metrics. PRM, the ratio of installed capacity to peak demand, does not directly address resource adequacy. This is why the existing capacity markets in the United States perform some type of mapping to a reliability-based metric. In nonrestructured areas, there is a mixture of whether and how PRM and reliability-based metrics are used. In this section, we show why the PRM, by itself, is an inadequate tool for measuring reliability and why a probabilistic reliability-based metric is a more rigorous reliability target. This is why some markets derive a PRM from a probabilistic assessment. Loss-of-load expectancy (LOLE), or a related reliability-based metric, is essential to ensuring that the long-term supply is adequate, and it also gives a meaningful way of determining whether resources are needed for long-term reliability and whether they should be incentivized to remain in the market.

Because of the reasons described above, including the large capital costs for generation, effective inelasticity of demand, and regulatory price caps in the wholesale markets, the level of revenue and profit determines whether a resource remains in the market. A resource that is needed to maintain a reliable and secure power system must earn enough revenue to recover both its variable and fixed costs. This concept of revenue sufficiency is the second focus of this section.

6.3.1 Resource Adequacy Calculations

A common reliability target in the United States is a LOLE of 1 day in 10 years. This means that generation supply is sufficient nearly all the time. Alternative targets can be adopted if desired, and the choice of the reliability target is largely one of policy. In modern interconnected systems, it is likely that the LOLE of any one particular area is overstated because if there is insufficient generation within a given balancing authority area, emergency imports may be available from neighboring systems as long as transmission capacity is available. In any case, using reliability metrics provides information regarding how often and/or how much of a generation shortfall might exist.

Reliability models are used by system planners to calculate the LOLE or similar metric for existing or future system configurations. If the calculated LOLE value is higher than the target, alternative resources can be added until the actual LOLE matches the target LOLE. Traditionally, LOLE was calculated using a single data point per day, chosen from the peak hour of the day [3]. To calculate the daily LOLE, each generating unit's capacity and forced outage rate (FOR—the probability the unit would be in an unplanned outage state) are used in a mathematical convolution with forecast demand values [4]. This approach explicitly considers each of these data points to the contribution of a generator to meeting load on a statistically-expected basis. For example, a 100-MW unit with an FOR of 0.10 would have a higher statistically expected output than another 100-MW unit with a 0.20 FOR. Thus, the convolution of multiple units with differing capacities and FORs forms the basis of the reliability calculation and related metrics. Using this simple example, the first generating unit would be considered to have 90 MW of unforced capacity (UCAP) [100 MW $\times$ (1 $-$ FOR)].

Additional metrics may be used instead of LOLE. A relatively simple extension is to apply the basic LOLE convolution algorithm to hourly data instead of to daily data. Interest is increasing in evaluating the performance of this hourly metric, expected loss-of-load hours (LOLH), with the increasing levels of VG that behave differently than more conventional thermal generation.

The relationship between LOLH and LOLE (measured in days) is not straightforward. Unless otherwise specified, LOLE will be taken to mean a LOLE measured

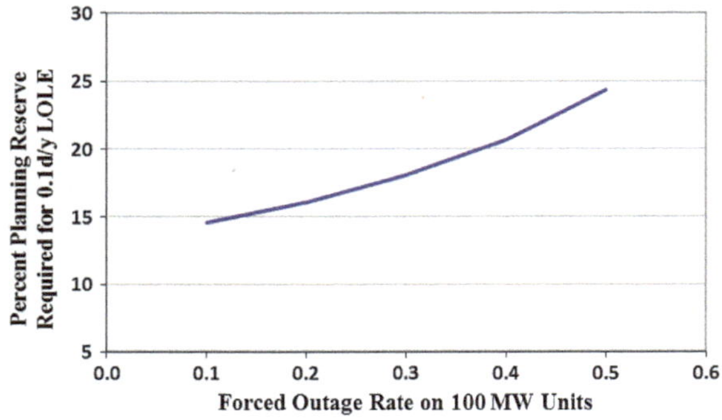

Fig. 6.1 PRM to achieve 0.1 d/y LOLE is a function of FORs (based on [5])

in days. The traditional LOLE calculation accounted for whether an outage might occur in a given day, given the LOLE. Because no hourly data were used in the calculation, there was no way to know, or to calculate, how many hours within the day an outage might occur. Generally, the calculation assumed that if an outage occurred in a given day, there could be anywhere from one hour to many hours of outage. The LOLE days did not have that information. Conversely, LOLH explicitly calculates the number of hours in a year in which there may be insufficient generation supply. For further discussion on the differences between LOLE days and LOLH. From this discussion, we can conclude that an LOLE of 1 d/10 y is not the same as an LOLH of 2.4 h/y.[1]

Another, simpler approach to measuring resource adequacy is the use of the PRM, often expressed as a percentage of capacity above the forecasted peak demand. Because the PRM includes only data regarding capacity and ignores data regarding forced outages, it is easy to see that PRM is not fundamentally a reliability metric-it cannot distinguish between two systems with the same installed capacity and peak loads but with different FORs.

Milligan and Porter [5] provide an example in a parametric study of a system based on the CAISO system. Increasing the FOR on a subset of the thermal units and simultaneously increasing the number of new 100-MW units to maintain a 0.1 d/y reliability target, they found that the PRM required to maintain reliability increased from approximately 15% to nearly 24%, as illustrated in Fig. 6.1.

The implications of this analysis seems clear: when establishing resource adequacy targets, the PRM metric may not prove to be very useful, especially in systems with large penetrations of VG (or any other generation that has a relatively low ratio of capacity value to installed capacity). Although it is possible to calculate the PRM that would result in a 0.1 d/y LOLE target, which is done by some markets, the value of the PRM metric is still questionable because it can no longer be used to compare different systems, nor does it provide consistent information regarding resource adequacy. Common values for a PRM on historical systems range from approximately 13–18% (North American Electric Reliability Corporation, or NERC [6]).

[1] For example, SPP uses 2.4 h/y, which results in a lower reserve margin than 0.1 LOLE (see Astrape Consulting [3]).

The contribution of any resource, or group of resources, to resource adequacy can be calculated using the effective load-carrying capability (ELCC) method, which is built on one of the more fundamental reliability metrics, such as LOLE, LOLH, or expected unserved energy (EUE). This approach can be applied to conventional generation, and it can also be applied to VG. Details can be found in Keane et al. [7] and NERC [8]. The ELCC represents the additional load that can be supplied by the resource being evaluated, holding long-term reliability constant. Historically, this set of calculations has been computationally demanding, so alternative approaches have been developed to rate individual generators. These simplified methods, however, should be benchmarked against the full reliability approach to ensure that they are reasonable (see [5, 7–9]).

For example, a 200-MW gas unit with an FOR of 0.10 would have an ELCC of approximately 180 MW. A 200-MW wind power plant with a capacity factor of 35% might have an ELCC of 30 MW, or 15% of its installed capacity. Note that this example points out a fundamental difference in the ratio of capacity value, as measured by ELCC, to installed capacity when resource types are compared. We discuss the implications of this in more detail later in this section.

The ELCC calculation is graphically illustrated in Fig. 6.2. The example uses a target of 1 d/10 y LOLE. The left curve shows the relationship between the level of peak load that can be served and the LOLE. At the target of 1 d/10 y, a 10-GW load can be served, and as the curve shows, a lower load will have a higher reliability level and a higher load would have a lower reliability level. When a new generator is added to this system, the reliability curve shifts to the right, and the distance of this shift depends on a combination of system and generator attributes. The example diagram shows that the additional load that can be served while maintaining the 1 d/10 y level of reliability is 150 MW; thus, the new generator has a 150-MW capacity value.

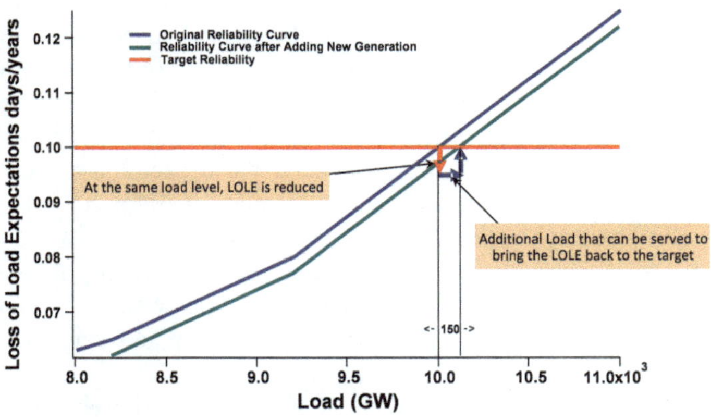

Fig. 6.2 ELCC is the horizontal difference between the reliability curves evaluated at the target reliability level (Source: from NERC)

The usual mathematical formulation for LOLE is based on the daily or hourly estimates of LOLP. The LOLE is the sum of these probabilities, converted to the appropriate timescale. The discussion that follows is based on the calculation of the LOLE in terms of days/year; however, the same procedure can be applied to the LOLH calculation with minimal changes. The annual daily LOLE can be calculated as:

$$LOLE = \sum_{i=1}^{N} P\left[C_i < L_i\right] \tag{6.1}$$

where P denotes the probability function, N is the number of days in the year, C_i represents the available capacity in day i, and L_i is the daily peak load. To calculate the additional reliability that results from adding VG, we can write LOLE′ for the LOLE after the new capacity is added to the system as:

$$LOLE' = \sum_{i=1}^{N} P\left[C_i + g_i < L_i\right] \tag{6.2}$$

where g_i is the power output from the generator under evaluation during hour i. The ELCC of the generator is the additional system load that can be supplied at a specified level of risk (LOLP or LOLE).

$$\sum_{i=1}^{N} P\left(C_i < L_i\right) = \sum_{i=1}^{N} P\left[C_i + g_i < \left(L_i + \Delta C_i\right)\right] \tag{6.3}$$

Calculating the ELCC of any generator amounts to finding the values ΔC_i that satisfy Eq. (6.3). This equation states that the increase in capacity that results from adding a new generator can support ΔC_i more MW of load at the same reliability level as the original load could be supplied (with C_i MW of capacity). To determine the annual ELCC, we simply find the value ΔC_p, where p is the hour of the year in which the system peak occurs after obtaining the values for ΔC_i that satisfies Eq. (6.3). Because LOLE is an increasing function of load, given a constant capacity, we can see from Eq. (6.3) that increasing values of ΔC_i are associated with declining values of LOLE. Unfortunately, it is not possible to analytically solve Eq. (6.3) for ΔC_p. The solution for ΔC_p involves running the model iteratively for various test values of ΔC_p until the equality in Eq. (6.3) is achieved to the desired accuracy. Historically, this calculation was considered to be computationally expensive, and many simplified approaches were developed as shortcuts [10]. However, modern computers can easily manage the computations in a short amount of time.

In modern interconnected power systems, it is likely that during emergency events, such as generation outages, neighboring systems can provide emergency capacity provided there is an unconstrained transmission path and operational procedures in place that allow this response. This may fundamentally alter the LOLE calculation, and this issue was recognized by the Integration of Variable Generation Task force convened by the North American Electric Reliability Corporation. The task force recommends full transparency in reliability assessments with regard to the way interconnected systems are modeled [8]. Examples of the impact that transmission plays in resource adequacy have been shown in recent research [11, 12]. Transmission can enhance long-term reliability, and in some cases it can reduce the need for generating capacity [11].

Capacity contributions of any generator will be subject to interannual variations, although the properties of this variability will differ among technologies. As an example, a thermal plant with an ELCC of 90–95% of its installed capacity, can experience a forced outage event during high-LOLE peak periods, and could conceivably contribute nothing toward meeting load in that year. Similarly, wind and solar generation is a function of the weather and thus may vary from one year to another around the long-term value. More details on LOLE and ELCC for systems with VG can be found in NERC [8].

Resource adequacy is measured in terms of a reliability metric. However, there is no market (or market characteristic) for reliability. Because two otherwise identical plants with different FORs have different ELCC values, using installed capacity as the metric for measuring resource adequacy will result in at least a small, but possibly a relatively large, divergence from the goal of resource adequacy. Conversely, if a target level for installed capacity is utilized, no information about FORs are incorporated in the market. The simplified methods for calculating wind capacity value only include reliability information insofar as the input data represents the time periods of high risk—high LOLP—and that data time series is sufficiently long.

One way to improve this approach is to utilize UCAP. This is done by each of the existing capacity markets in the United States—PJM, MISO, and NYISO—although the mechanisms are different [13]. The various capacity auctions are conducted so as to account for UCAP, even though the market may directly address installed capacity. This means that all capacity acquisitions via the market account for the units' contribution to system reliability and do not simply account for installed capacity. The specific UCAP value will vary based on resource type and by region. This allows two plants of the same size but with different FORs to be differentiated.

Calculating resource adequacy for a future time period is complex. Resource Adequacy targets are predicated on demand forecasts for future time horizons, and the likelihood of over- or underestimating demand as these time horizons increase. The contribution of different types of resources, including demand response, for which characteristics may not be well-known today can be challenging to ascertain. The future is always uncertain. Is it more appropriate to develop a range of targets using multiple scenarios? Can they be combined to form a stochastic expected value, or evaluated subject to a probabilistic hedge? Are there other strategies such as minimizing maximum regret?

6.3.2 Revenue Sufficiency

The missing-money problem [2] is the concern that even robust energy markets may not provide sufficient revenue for at least some generators to earn sufficient revenue to pay for both variable cost and fixed costs. It predates large amounts of VG. It occurs fundamentally because of the market failures described in Sect. 6.2 and the concern that insufficient revenue will result in insufficient installed capacity to serve the load, especially during peak periods, because there is insufficient incentive for generators to build new capacity or even maintain existing capacity. In particular, limited demand-side participation may lead to inadequate scarcity pricing that suppress revenues from the energy market.

Other technical characteristics of electricity markets will impede them from functioning at or near the level of an ideal market. The two demand-side flaws of (1) the inability for buyers to respond to price and (2) the inability for suppliers to supply different levels of reliability imply that there will be some degree of market failure unless a clever approach can be discovered to overcome these obstacles.

Therefore, the market for electricity does not possess the necessary characteristics to perform as a perfectly competitive market; however, markets for specific electricity products, such as bulk energy and ancillary services, can still perform well if there are many buyers and sellers, along with limited congestion and limited market power (i.e., increasing the level of competition). These markets still suffer because the consumer usually cannot respond to price. The system operator is then constrained to provide energy up to an administered level of reliability. When reliability is compromised, or if the operator has concerns that it might be, the threat of penalties or actual reliability events can result in paying high prices for electricity that may exceed the value of the energy from some consumers' point of view. Additionally, it is also well-known that outages are very costly, therefore a strong argument can often be made that the system operator must incur high purchase costs to avoid outages. Thus, the value of lost load (VOLL) has been estimated as high as \$77 000/MWh, although there is a wide variation in estimates for different consumer groups and regions [14].

In the early periods of market design, industry and researchers have made numerous arguments that energy-only markets cannot incentivize appropriate investment. Crampton and Stoft [15] succinctly state several of the issues of energy-only markets and conclude that a forward capacity market is required to ensure reliability:

> "The misconception…is the notion that a cleverly designed 'energy-only' market can induce optimal adequacy, or something close to it, even while the market has insufficient demand elasticity…In an ideal market, with sufficient demand elasticity, the market always clears. This means there can be no adequacy problem because involuntary load shedding occurs only when the market fails to clear and demand exceeds supplyenergy prices do what every economics text says they do, they determine the efficient (not reliable) level of capacity…The concept of an energy-only market solving the reliability problem without selling a reliability product is logically impossible. It suggests that 'the market' can do something 'fairly well' when logic shows that it cannot do it at all."

Long-term reliability needs and short-term economic costs also contribute to revenue sufficiency concerns. When the variable cost of resources unexpectedly goes down (e.g., reduced fuel costs compared to those originally anticipated), the revenues made could be much less in certain years than they are in others. The expected load may be less than anticipated as well. Although this might mean that some of these resources are not needed for reliability, because of the reduced load, they may still be needed in the near future, and building this capacity is a long-term investment. At the same time, long-term markets for electricity are typically few and not very liquid, making it difficult to hedge price and quantity risks for investors.

6.4 Increasing Penetrations of VG Impacts on Long-Term Resource Adequacy and Revenue Sufficiency

The introduction of VG can have an impact on resource adequacy and revenue sufficiency in many ways. First, determining the contribution of VG toward resource adequacy is very different than the method used for conventional generation. Although the FORs of an entire collection of wind turbines or photovoltaic (PV) cells is very rare and not likely to contribute significantly to the resources' unavailability, the availability of the fuel source of VG can be quite variable. Changing weather patterns drive how VG can contribute to meeting long-term reliability needs; it is not is caused by the random forced outages that occur. Second, VG increases the amount of variability and uncertainty on the system, which can require an increased need for flexibility (see the previous chapter). Although certain changes to short-term energy and ancillary service markets may be needed to ensure that the flexibility that is available is provided, this may not guarantee that sufficient flexibility is built or available in the first place. This could lead to the need for new ways in which resource adequacy evaluation is performed. Finally, VG has total costs that are almost entirely fixed capital costs rather than variable operating costs. This can bring down the energy prices further, while potentially increasing (or keeping constant) the total variable and fixed costs in the power system. This could lead to further reliance on markets or incentives other than the energy market to ensure that the resources needed for long-term reliability can recover both variable and fixed capital costs.

6.4.1 Calculating the Capacity Value of VG

For VG, the recommended approach begins with the use of time-synchronized VG data with load. This will implicitly capture the underlying weather that drives load, solar generation, and wind generation. If data from different years are used for load and VG, a situation could be easily envisioned in which the load on a given day is based on hot, sunny weather that induces significant air-conditioning loads, whereas wind data is based on a cloudy, stormy day. Many other similar examples can result in a mismatch between the implicit weather driver of load and the VG resource.

This mismatch would result in an implausible data foundation for the convolution algorithm. The process underlying the calculation is essentially the same as that described above for conventional power plants; the exception is that hourly VG production data (real or simulated, depending on data availability and the specific study requirements) replaces the use of the generator capacity and FOR. Details on the method can be found in Keane et al. [7]. The ELCC of wind power plants typically ranges from approximately 5–40% (see [7]).

Several regional transmission organizations (RTOs) in the United States use simplified methods to calculate the capacity value for wind power. Generally, these methods have been adopted because of their transparency, and they define a peak time period and calculate the capacity factor during that period. For example, PJM calculates the wind capacity factor for the hours ending 3:00 p.m. to 6:00 p.m., June through August for the most recent 3-y period. For wind power plants with at least 3 years of operational data, actual data is used for the calculation. For new wind power plants, a default value of 13% is used initially, which is replaced as operating data become available. Rogers and Porter [9] surveyed methods for calculating wind capacity value during the period from Sept. 2010 to Feb. 2012.

The RTO/ISOs with capacity markets evaluate capacity values for wind resources as described in [9]. Milligan and Porter [5] and Keane et al. [7] recommend periodic analyses and refinements to ensure that the simplified approaches provide good estimates of contributions toward resource adequacy using a more probabilistic approach.

However, note that it is not possible to ensure that these simplified approaches can accurately capture the reliability aspect of resource adequacy. A simple method such as that used by PJM (or other similar approaches) may miss times of significant risk. As an example, Kirby et al. [16] performed ELCC calculations on a 3-y data set from 2001 to 2003, supplied by CAISO. In 2001, the top 20 peak hours all occurred in July or August. In 2002, peak demand in July, September, and June ranked above all hours of August. In 2003, as in 2001, June did not appear in the top 20 peak hours. In the same study, one year experienced an unusually hot period in late September and early October. In the early autumn, some generation was taken out of service for scheduled maintenance as a result of prior planning, and hydro runoff was no longer providing as much energy and capacity as it was during the usual peak season from July to August. Some high LOLP hours thus occurred in the autumn, when load was relatively low compared to levels during the peak season, some generation was out on maintenance, and hydro was not contributing as much as it was during peak periods. Thus, the use of predefined peak windows may miss times of system risk when generating capacity is needed and therefore provide an incomplete picture of the state of reliability of the generation fleet.

Several possible approaches can overcome some of these obstacles, although they may also fall short of providing a true picture of resource adequacy. One approach is to use the top daily or hourly loads. For example, the top 2% of load hours, approximately 175 h, could be evaluated post-hoc, and the VG capacity factor could be calculated for that period.

Table 6.1 ELCC does not always match peak-period capacity factors

Resource	2002		2003		2004		3-Y Average	
	ELCC (% of rated capacity)	Peak capacity factor	ELCC (% of rated capacity)	Peak capacity factor	ELCC (% of rated capacity)	Peak capacity factor	ELCC (% of rated capacity)	Peak capacity factor
Solar	88	97	83	93	79	94	83	95
Wind (Northern California)	24	19	25	20	30	24	26	21
Wind (San Gorgonio)	39	36	24	23	25	28	29	29
Wind (Tehachapi)	26	30	29	24	25	25	27	26

One approach is to identify periods of time during which there is (or may be) high LOLE. These time periods are closely linked to LOLE or related metrics as well as the periods of time during which the capacity value of a resource is determined. Examples of this type of approach have been incorporated into the methods used by NYISO and PJM to determine the capacity value of wind energy. As described in Porter et al. [9], these approaches calculate the capacity factor of the wind resource during the critical time periods and use that as a proxy for the capacity value. In some cases, the capacity factor may match the more rigorous ELCC fairly closely [17], but this is not guaranteed. One example of inconsistent matches was shown in Shiu [18].

In the study, the capacity value (ELCC) of wind and solar generation was evaluated for a 3-y period. Note that the study suffered from some data anomalies; however, the results discussed below are most likely robust against the data concerns.

Table 6.1 illustrates the results of comparing ELCC for three wind power plants and one solar power plant with the capacity factors of these respective plants that were calculated over the peak period, defined as June to September, 12:00 p.m. to 6:00 p.m. The calculations were based on hourly wind, solar, and load data. The utilities and CAISO define this time period as the likely time when the system peak would occur. The table shows that in some cases and for some years, the ELCC matches the time-period factor method reasonably well. San Gorgonio in 2003 is a good match; whereas Northern California in 2004 is not. The implication of these results is that unless a comparison is made between ELCC (which is a reliability metric) and time-period capacity factor (which is not a reliability metric but is used to approximate it), the extent to which the approximation method matches the preferred metric of ELCC is not known. This means that the ability of a capacity market to capture reliability is not likely. This is not surprising, because there is no specific reliability information content in capacity factors in spite of the possibility that high-risk (high-LOLP) hours generally correspond to peak periods.

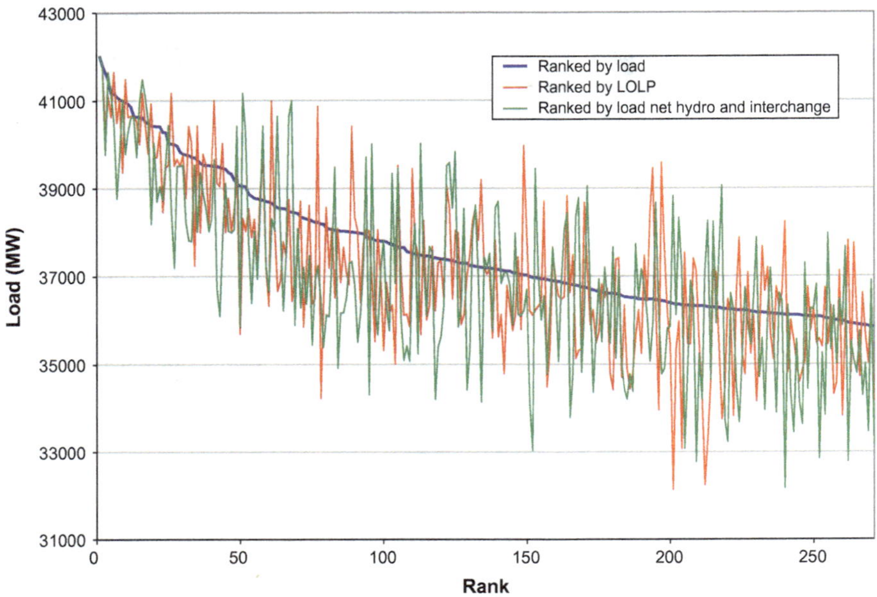

Fig. 6.3 Relationship between rankings of load, LOLP (hourly), and load net of hydro and interchange for 2002 (based on [16])

In fact, the assertion that LOLP and load are highly correlated turns out to be incorrect for at least some regions. Another part of the study by Shiu [18] examined the change in imports and scheduled hydro energy into the CAISO system during various peak and near-peak periods. During many of the highest peak demand periods, additional energy imports were brought in to help support the load. At the same time, to the extent that the hydro could be scheduled to support these peak periods, hydro generation was higher during many of the peak periods. Conversely, at some near-peak, and at many off-peak periods, imports and hydro were lower than at peak times. This means that during near-peak periods when there was less hydro and imports, LOLPs could sometimes be higher than during peak periods.

Figure 6.3 illustrates this condition for the 2002 study year [16]. The black curve is a load duration curve that shows the top 300 h of load. In contrast to the monotonically decreasing load duration curve, the more erratic behavior of the trace that shows the ranking of load by LOLP (hourly) shows that there is only partial correlation between load and LOLP. This partial correlation is clearly not linear, but the downward trend of the LOLP ranking curve follows the trend of the load duration curve. A similarly erratic trace is shown when the curve is ranked by load, net of hydro and imports.

Kirby et al.'s [16] analysis provides evidence that there is only a partial correlation between demand and LOLP. A perfect correlation would imply monotonically decreasing curves for the LOLP-ranking and load duration curve. From this, we can conclude, that a capacity market that relies on peak periods instead of LOLP or similar metrics to calculate the capacity value of VG will have difficulty achieving reliability targets and that approximations of capacity value by the use of capacity factor calculated over peak periods will be unlikely to capture the reliability contribution of wind and solar generation to resource adequacy.

Another view of these results can be taken. Because peak periods generally are the highest risk times for loss-of-load events, system operators and markets will schedule additional generation, including imports, during those times. The impact of increasing imports is that LOLP will be reduced for the period of import. Thus, the system was operated in the way it was intended. Instead, however, some hourly LOLP values were higher during times that they may not have been expected to be significant.

Improvements can be made to the time-period capacity factor approximations. One approach is to perform a ranking of top loads, such as the top 5% or 10% of loads. Multiple years would provide a more robust indicator of possible future critical periods, but predicting the future based on the past is always somewhat problematic. When the sorting exercise is done, it would then be possible to utilize the days/times that are in the top of the ranking as the basis for a modified capacity factor calculation for the VG.

As shown, even a load-based ranking does not necessarily capture reliability. To utilize a simple approach that does not explicitly calculate ELCC, an LOLP ranking of days or hours could be performed, proceeding similarly to that discussed above. Rankings of LOLP could be done for 3 y or more, with the times noted, and VG capacity factors could be calculated over that period. One advantage to either of these ranking approaches is that it provides transparency to market participants with regard to the times that capacity availability can help achieve resource adequacy.

6.4.2 Incorporating Flexibility into Resource Adequacy Needs

As mentioned above, resource performance, in particular the flexibility attributes of a resource, may be significant enough that these attributes should play a role in long-term resource adequacy assessment, and potentially also in forward capacity markets. As experience with VG has increased, there has been a growing recognition that flexibility needs will change in the future, and how to plan for that flexibility has become increasingly relevant. The precise mechanism(s) that would ensure resource adequacy along with the required levels of flexibility are active areas of research.

As an example of one approach to extend the traditional resource adequacy techniques to flexibility analysis, Lannoye et al. [19] adapted LOLP analysis, which is based on the changing levels of load and generation, using the speed of how rapidly resources could respond. The adaptation makes it possible to apply LOLP, ELCC, and related metrics to ramping. Figure 6.4 shows how standard LOLP-related metrics map to ramping. This approach makes it possible to put ramping analysis in the context of reliability because a generator's ability to ramp will depend in part on

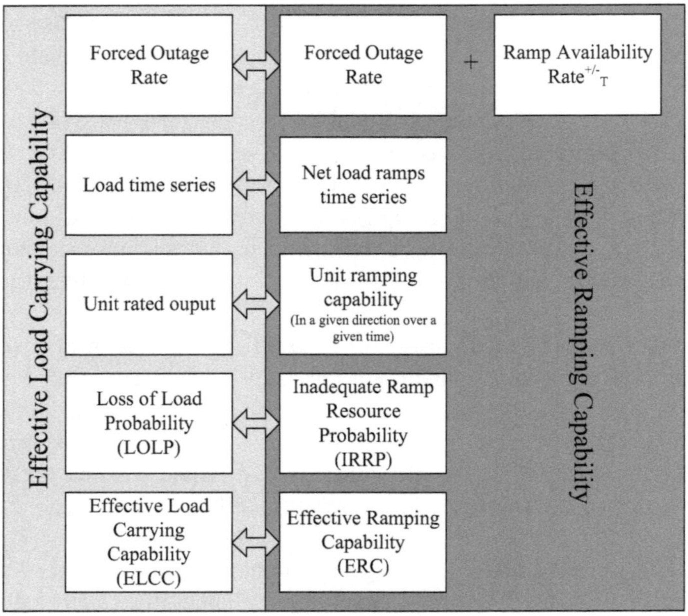

Fig. 6.4 Proposed metrics to understand how effective load-carrying capability can similarly be adapted to understand the capability of resources to meet expected flexibility needs (based on [19])

whether it is on forced outage. Generators that have high FORs will have a lower effective ramping capability compared to a unit with a low FOR, all else equal. It is not clear what target value to use for the inadequate ramp resource probability (IRRP) because this is a new area of research. However, a good starting place is to perform this analysis on existing systems to provide a benchmark that may be useful in setting the target.

In some parts of the electricity market, a mandate to provide a certain service is more useful than a market in which prices incentivize only some of those resources to provide it. For example, some regions require all resources to have synchronous inertia or a capability similar to inertia [20]. Some reasons that might lead to this approach include the cost of such service, such as when it is extremely low, lower than the cost of administering a market to achieve this service, when the market is too complicated, or when there is low diversity in the costs to provide that service (i.e., when it is difficult to innovate and provide the service better or at a lower cost) [21]. If a mandate requires all resources to have a given level of flexibility, then much of the following discussion is not relevant. What may be relevant is how to determine how much flexibility is needed relative to required new capacity and somehow pro-rate that across units. For example, a requirement that all new generation must provide a ramp rate of at least 30 MW/min could be based on an assessment that determines that there will be sufficient ramping if all units have this ramping capability.

However, it is likely that any mandate like this for all resources would be very inefficient because the cost to provide that flexibility would vary extensively between different plants/providers and the required flexibility needs would not be achieved at the least cost.

Ramping is only one part of flexibility. We defined and discussed flexibility in the previous chapter, but for the discussion related to planning, we assume that key flexibility attributes can be grouped into two categories:

- Operating range, the difference between maximum and minimum stable output. A larger operating range suggests a more flexible unit than a small operating range.
- Rate of change from one state to another, including ramping, start-up, shut-down, etc. A high rate of change per minute or per hour denotes a more flexible unit than a low rate of change. Quick-start units that can ramp quickly are more flexible than slow-start units with low ramp rates.

To assess forward flexibility needs, whether for a market or not, some relatively simple approaches can be used as a starting point. One approach to assessing ramp magnitude and timing utilizes hourly (or sub-hourly, if available) data for demand, wind power, and solar power. Similarly to LOLE studies, these should be based on the same weather year so that the often-complex underlying weather impact on each of these variables is consistent. Recognizing that system balance is achieved when the sum of all demand equals the sum of all supply, making maximum use of the installed wind and solar energy means that the net load—demand less wind less solar—must be balanced by the remaining fleet.[2] It remains a simple exercise to calculate this net load. An example graph or up-ramp needs are shown in Fig. 6.5. The y-axis shows each of the 52 weeks of the year, and the x-axis shows the time of day. Ramps in this example are average ramps, but the method can be adapted easily so that the graph shows maximum, minimum, or other ramp metrics, such as mean plus standard deviation, etc.

To read the graph, find a time and then examine the color/legend. For example, the morning hours of approximately 5 a.m. to 11 a.m. in Week 26 show large up-ramp needs. This can be compared to autumn morning up-ramp needs, which are not as high, and do not last as long. These plots can be compared so that alternative scenarios or data views can be analyzed. Some examples can be found in Western Wind and Solar Integration Study Phase 1 [24] and King et al. [23].

Using the same data set as a starting point, various ramp envelopes can be generated and graphed, as illustrated in Fig. 6.6. This example calculates envelopes of up to 12 h, although there is no limit to the number of hours that can be considered in such a graph. Each envelope corresponds to a given percentile boundary or probability. For example, the blue "100% Prob." curve states that all ramps are bounded by this curve. The yellow 99% curve bounds 99% of all ramps.

---

[2]For the discussion, we ignore the possibility of wind/solar curtailment. In reality, some limited curtailment or downward dispatch may help achieve economic and/or reliable system operation.

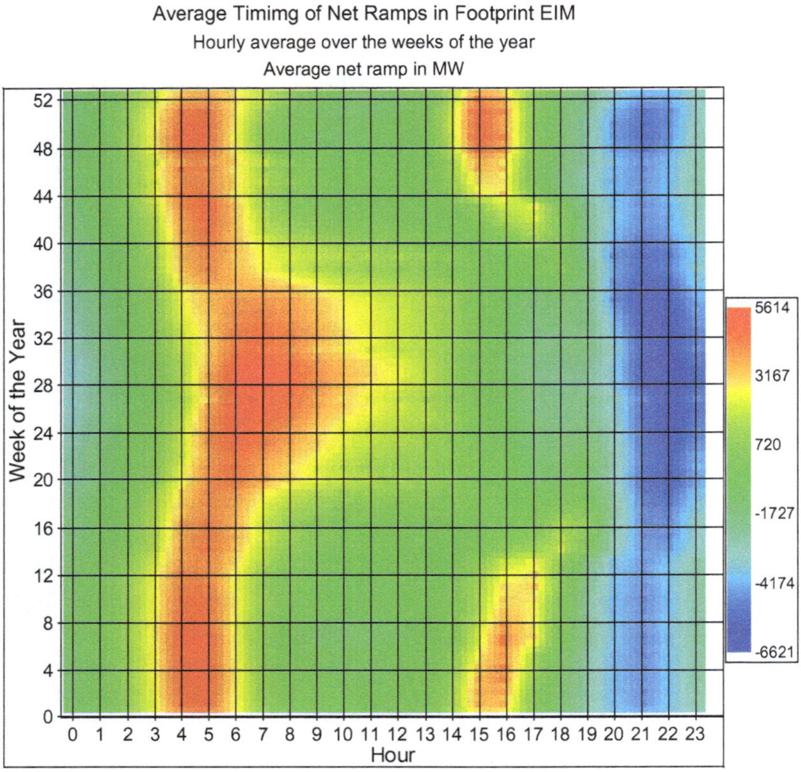

Fig. 6.5 Estimates of ramp timing and magnitude (based on [22])

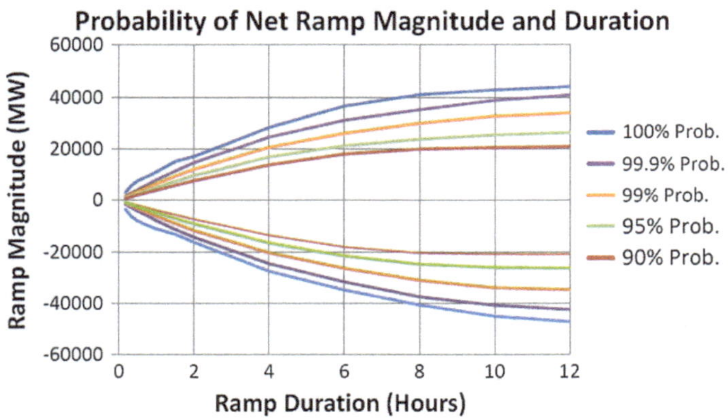

Fig. 6.6 Ramp envelopes can be used to obtain a view of alternative percentage exceedance levels (Source: from [23])

6.4.3 Revenue Sufficiency Challenges as a Result of Increased Amounts of Low-Variable-Cost Resources

Recent analyses of regions with high penetrations of wind generation indicate that the low marginal cost of wind generation may decrease locational marginal prices (LMPs) and thus reduce revenue of all suppliers in the energy market [25]. This may add to the revenue insufficiency problem for some resources, preventing them from recovering both variable and fixed costs. This impact of wind power on market prices is not confined only to the United States but has been apparent in other countries as well [26]. In addition to the empirical evidence regarding lower LMPs resulting from VG, the Western Wind and Solar Integration Study Phase II (WWSIS-2) [27] found that prices were suppressed under VG penetrations of approximately 33%. In addition to reducing electricity prices, VG displaces other resources via the merit-order effect, such that capacity factors for other generator types are also reduced. The question is what the combination of lower energy prices and lower capacity factors of existing plants implies for revenue sufficiency. It may be possible that even though the majority of prices are being depressed, the occasions of high price spikes may increase, helping to capture needed revenue. This may depend, however, on price caps, market mitigation procedures, and the price levels of administratively-set scarcity pricing. Finally, the existence and design of forward capacity markets can have a large impact on the level of revenue sufficiency.

To demonstrate the reduction in operating profits as a result of price suppression and reduced capacity factors caused by widespread deployment of wind and solar generation, we examined archived results from WWSIS-2. WWSIS-2 used a unit-commitment and dispatch model to analyze how the system would operate under different penetration levels of wind and solar generation. For this analysis, revenues and costs for each generator type (nuclear, coal, gas combined cycle, gas steam boiler, and gas combustion turbine) from electricity sales were calculated from the model results. WWSIS-2 was not intended to study price formation, and the results included penalty prices that significantly impacted the conclusions from this analysis of revenues and operating profits. After controlling for the penalty prices by applying a price cap, the analysis found that capacity factors of certain generator categories decline and electricity prices are suppressed, leading to lower revenues and operating profits. This analysis emphasizes the importance of a model providing accurate price formation in addition to operations, especially in the context of high penalty prices, such as reserve shortage penalties. The full analysis is presented in [1].

In a separate study, summarized in [1], capacity adequacy and revenue sufficiency were analyzed using a generation expansion model that finds the optimal portfolio of thermal power plants for a given wind penetration level considering the variability in the wind resource and the increased need for operating reserve due to wind power forecast uncertainty. After the optimal expansion plan was determined for a given wind level, expansion variables were fixed, and the model was solved again to derive the prices for energy and spinning reserve that would result under the optimal expansion plan. A case study found that increasing wind penetration reduces energy prices while the prices for operating reserve increase. Moreover, scarcity pricing for operating reserve through reserve shortfall penalties significantly impacts the prices and profitability of thermal generators. This was the case regardless of the wind penetration level. Without scarcity pricing, no thermal units are profitable; however, scarcity pricing can ensure profitability for peaking units, also at high wind penetration levels. Capacity payments can also ensure profitability, but the payments required for baseload units to break even increase with the amount of wind power. The results indicate that baseload units are most likely to experience revenue sufficiency problems when wind penetration increases.

The study builds on a number of simplifying assumptions. For instance, expansion decisions were based on a system-wide least-cost objective function assumed to represent a fully competitive market. Also, spinning reserve up was the only reserve product considered in the analysis. Moreover, similar to the study of revenue sufficiency for the WWSIS-2 results, it was found that the level of reserve shortfalls and corresponding frequency of scarcity prices were very sensitive to minor changes in parameters inputs, as discussed in more detail in [1]. However, the study still provides some insights into what a future electricity market with high levels of renewable generation may hold in terms of prices and revenue sufficiency. A more detailed summary of assumptions and results are presented in [1].

6.4.4 Revenue Sufficiency Challenges as a Result of Increased Flexibility Needs

With increasing penetrations of VG anticipated during the next several years, it has become clear that a system design perspective is needed to determine the best resource mix for the non-VG fleet. Markets that incentivize both the investment in and use of flexible generation should be designed from a perspective of how they can best achieve an economically optimal solution subject to various reliability constraints. What is needed is a clear market signal to investors that communicates how much flexibility is needed at some future date. Of course, there are many uncertainties around this question and about how the need for flexibility will change over the lifetime of a new generator. This chapter provides a discussion of potential market structures that can incentivize investment in new flexible resources. This can be broken into two areas of exploration:

- Whether the market design provides incentives for new resources entering the market to have the flexibility attributes that are needed.
- Whether the market design provides incentives for existing resources to increase their ability to provide flexibility (e.g., through retrofits), subject to technical barriers and economic trade-offs.

This discussion focuses on long-term investment in flexibility.

There is no widespread agreement as to whether volatile energy prices will induce suppliers to invest in the needed level of flexibility, or whether an explicit market for flexibility is required. Another open question is whether existing capacity markets (whether modified to better capture resource adequacy aspects) should be conducted in tranches of differing flexibility needs or whether there should be separate, linked markets for capacity (resource adequacy) and flexibility. Adding to these complications is that the electric power system is not in a steady state, but is rather in a transition between a low-VG past and a potentially high-VG future. Thus, it is unlikely that current price signals can provide a good indication of the flexibility requirements 5 or 10 years from now. Therefore, investors in flexibility will need to determine the needed level of flexibility over the lifetime of potential new flexible technologies, such as demand response. This is a difficult determination to make because there is a large number of variables that can influence the need for flexibility that cannot easily be determined over the asset life of the flexible resource.

The question as to whether a given suite of flexibility attributes should be required from all new generators is an important one. Proponents of setting a requirement that applies to all new generation argue that this way there is no discrimination between different types of units, and that markets may result in unintended arbitrage that, coupled with market power, may needlessly increase costs. Proponents of the market view argue that specifying a requirement for all new market entrants will result in some types of technology that may be able to provide the required flexibility at a very high cost compared to others. This can result in getting the flexibility at a much higher cost compared to a least cost solution or getting more flexibility than is really needed, driving up costs further without apparent benefit.

6.5 Traditional Market Design Elements to Ensure Resource Adequacy and Revenue Sufficiency

Several potential approaches can be used to help address potential revenue sufficiency issues. Revenue sufficiency has been discussed since the initial stages of electricity market design [2, 28]. Market designs had traditional ways of meeting this issue from the inception of wholesale electricity markets. The extent to which existing market designs provide revenue to recover the fixed costs of enough resources to remain available in the market for long-term reliability is still an ongoing debate. Also, the way in which they incorporate the changing resource adequacy needs of systems with increasing VG penetrations is also somewhat unclear. Two distinct directions in terms of market designs for long-term resource adequacy have emerged. First, scarcity pricing, both through administrative prices as well as offered prices, may provide prices that go above the variable costs of the most expensive operating resources for short periods of time, when capacity is scarce, such that those prices can help recover the fixed costs of the peaking units. Second, forward capacity markets have been a part of several US wholesale electricity markets for a long time. These markets look ahead to ensure that enough available capacity will be available to meet load in peak periods and aim to provide incentives for new capacity to be built in locations where it is most needed. Below, we provide a brief summary of these two designs that many of the US wholesale markets have traditionally adopted to mitigate the issues of revenue sufficiency.

6.5.1 VOLL or Ancillary Service Scarcity Pricing

In markets that approach perfectly competitive markets, price volatility provides signals to both buyers and sellers. In the absence of market power and with the other attributes of perfectly competitive markets, prices would be free to fluctuate. The interplay between buyers and sellers in the market would result in an economic profit of zero in the long-run. In cases such as this, there is no concern regarding market power or a level of profit that is above and beyond what is needed to elicit the economically efficient level of supply.

The VOLL evaluates the potential cost of supply shortages where load must be involuntarily curtailed. It may be used to determine price caps in the energy market. The importance of the VOLL concept in electricity markets stems from the limited demand response, which may prevent end-users' preferences to be reflected in prices during times of scarcity. In typical power system operations, shortages will first result in insufficient reserve while load is maintained. Therefore, ancillary service scarcity prices rather than VOLL are more frequently used to determine prices during scarcity. Stoft [2] shows that the system operator can induce the same optimal capacity expansion as under VOLL pricing by setting an administratively determined price cap for operating reserve. Sometimes the administratively-set ancillary service scarcity prices use the VOLL in its calculation so that VOLL is still reflected in the resulting prices. Research studies have looked at many new ways of incorporating VOLL in the ancillary service pricing explicitly [29, 30], accounting for the probability of not being able to meet load (e.g., [28, 31]). The VOLL and associated administratively-set ancillary service scarcity prices are typically very high because of the large economic losses that are usually associated with outages, such as food spoilage, failure of expensive industrial processes, etc. For example, ERCOT VOLL estimates vary from \$110/MWh to \$6 979/MWh for residential and commercial customers, respectively. However, estimated VOLL ranges even more widely elsewhere, up to \$42 256 for MISO's small commercial/industrial consumers [14]. In New Zealand, VOLL has been estimated as high as \$77 687/MWh. In practice, a price cap based on VOLL exists only to protect the purchaser from market power, which could potentially cause prices to move even higher than VOLL pricing.

These administratively-set ancillary service scarcity prices will set the price for both operating reserve and energy when the system is short of reserve because it means there is an overall capacity shortage (or ramping capability shortage). The scarcity prices are set such that they are triggered only during rare instances when capacity (or ramping capability) is unavailable and therefore no supply resource is marginal to set the price. The number of times these prices are triggered combined with their high price are intended to provide the revenue needed to make the peaking units (those that have the highest variable cost, but still nonzero capital cost) recover their capital costs over the lifetime of the resource. Thus, these administratively-set scarcity prices, when triggered an appropriate number of times to reflect the reliability target of how often the reserve requirements should be scarce, would help resources recover sufficient revenue to recover their capital cost. In practice, operators still attempt to reduce the number of times scarcity pricing occurs, and it is extremely difficult to predict the number of occurrences. Hence, for investors in new generation capacity, it is difficult to secure financing based on revenues from volatile scarcity prices and it is being questioned whether these prices alone provide sufficient investment incentives.

In practice, the administratively-set scarcity prices throughout the United States vary significantly. The prices for various services have different meanings, and the ways in which these are triggered are also very different. Some regions have fixed stepwise curves, in which the greater the scarcity, the higher the price. Others have more dynamic prices depending on the system conditions. For example, Midwest Independent System Operator (MISO) bases its regulation reserve scarcity price on the monthly peaker proxy price, whereas it bases its spinning reserve scarcity price according to a formula that evaluates the VOLL and the probability of an outage for resources that are online. Other scarcity prices are very low, thereby not strictly in place to assist in capital cost recovery. For example, the ISO-NE spin scarcity price, at $50/MW-h, may be in place to reflect that a system operator sometimes prefers to be scarce by a small amount rather than make additional costly commitment decisions. Table 6.2 shows a summary of some of the scarcity prices in the market areas in the United States.

In theory, VOLL pricing in the energy markets can be shown to support the optimal mix of generation capacity, because the generators recover their capital and operating costs from the resulting market-clearing prices as a result of the infrequent periods of very high prices at VOLL (e.g., [2]). The optimal level of reliability is also obtained if the price cap is set equal to the true VOLL. However, VOLL pricing gives rise to extreme price volatility, high investment risks, and the potential exercise of market power. Therefore, in practice, whether VOLL pricing will appropriately compensate peaking units to provide sufficient revenue to cover fixed costs when operating only during a limited number of hours is in question.

In some ways, the introduction of demand response for energy provision can provide a similar effect to scarcity pricing. During times when otherwise the system would be in scarcity conditions, demand response can be utilized to avoid the scarcity condition. When this demand response is used, the price can rise above the cost of the peaking units but below that of VOLL or even ancillary service scarcity price. For example, in both NYISO and PJM, the real-time LMP will be set to $500/MWh whenever emergency demand response is called upon [32]. Depending on frequency of occurrence of the emergency demand response and the level at which the price is set, the affect can similarly provide additional revenue for capital cost recovery.

Depending on how it is done, VOLL energy pricing or ancillary service scarcity pricing may directly link reliability to economics. For example, a price cap may be established so that it corresponds to a target LOLE, LOLH, or EUE. One example

Table 6.2 Administrative ancillary service scarcity prices for selected US markets

Market	Regulation	Spin reserve	Total contingency	Other
NYISO[a]	• $400 if scarcity greater than or equal to 80 MW • $180 if scarcity between 25 and 80 MW • $80 if scarcity is less than 25 MW	• $500	• $450	30-min reserve • $200 if scarcity greater than or equal to 400 MW • $100 if scarcity between 200 and 400 MW • $50 if scarcity is less than 200 MW
ISO-NE[b]	• N/A	• $50	• $850	30-min operating reserve[c] • $500
MISO[d]	Monthly peaker proxy price[e]	• $98 if scarcity is greater than 10% of requirement • $65 if scarcity less than 10%	System-wide operating reserve[f] • Min: $1 100 • Max: VOLL-RegDC	• N/A
CAISO[g,h]	Regulation up • $200 Regulation down • $700 if scarcity greater than 84 MW • $600 if between 32 and 84 MW • $500 if scarcity less than or equal to 32 MW	• $100	• $700 if scarcity greater than 210 MW • $600 if scarcity between 70 and 210 MW • $500 if scarcity less than or equal to 70 MW	• N/A

[a]NYISO Market Services Tariff
[b]ISO-NE Market Services Tariff, Sect. III – Market Rule 1
[c]These results represent the system-wide thirty-minute operating reserve. Local rules exist for this product as well, not shown here
[d]MISO Business Practices Manual-002, Energy and Operating Reserve Markets
[e]The monthly peaker proxy price is equal to the average cost per MW of committing and running a peaking unit for an hour. The price is updated on a monthly basis
[f]The scarcity price for operating reserve is determined based on the product of VOLL and the conditional probability that a resource contingency will occur. The minimum price of $1 100 is set based on the sum of the energy and reserve offer price caps ($1 000 + 100). The maximum price is set based on VOLL (~$3 500) minus the regulating reserve demand curve price
[g]California ISO, Final Draft Proposal: Reserve Scarcity Pricing Design
[h]The scarcity prices depend on the bid cap. The numbers presented here assume a $1 000 bid cap. A lower bid cap would result in lower scarcity prices

is the NEM in Australia, which establishes the price cap for energy at an EUE of 0.002%. Assuming that the VOLL is precisely calculated, this would appear to allow prices to rise up to the cap so that the reliability target is achieved. The price cap is applied to energy prices so it does not explicitly account for ancillary service pricing or reserve pricing, which could potentially be used as substitutes to VOLL pricing.

6.5.2 Forward Capacity Markets

In the United States, PJM, ISO-NE, and NYISO all have mandatory forward capacity markets, which establish the level of needed capacity and allow for fixed cost recovery in a transparent market environment. The mechanisms to procure the capacity are somewhat different, but several common steps are undertaken.

1. *Establish a capacity target.* A forward capacity market is designed to procure capacity for a future period that is above and beyond what exists today. Existing resources and bilateral power purchase agreements can be used to count toward a load-serving entity's capacity obligation.
2. Determine the future capacity need for the period(s) in question.
3. Determine how existing resources will count toward the target. When this has been accomplished, the difference between the future capacity need in (2) and the existing resources in (3) is the new capacity that must be acquired via the forward capacity market.

A recent white paper [33], published by the Federal Energy Regulatory Commission (FERC), categorizes some of the key features of the capacity markets currently operating in the United States. The main design elements include:

1. Demand curves for capacity.
2. Forward and commitment periods.
3. Definition of the capacity product.
4. Performance requirements.
5. Market power mitigation [33].

The overall objective of a forward capacity market is to provide a mechanism to send signals to investors regarding the need for new capacity at a future date. As noted above, the existing markets in the United States have different forward and commitment periods, and the shorter periods can provide a challenge because they may not be long enough to adequately stimulate the needed investment and construction time required for needed resources in the future.

The ultimate objective of resource adequacy, and the associated capacity markets, is to achieve some level of long-term reliability that is consistent with society's preferences. The existing capacity markets use differing measures of the product and different methods of "trueing up" or adjusting for the difference in what was acquired by the market compared to what was delivered in the key operating periods during which capacity resources are expected to perform. Thus, the market attempts to link resource adequacy/long-term reliability of supply to a market.

In general, the existing forward capacity markets have a somewhat general definition of capacity that is resource neutral, although this may not be strictly true in all cases. Demand response, for example, does not always qualify or is limited in how it is able to participate. However, several types of resources can participate in these markets, including VG. The methods for calculating the contribution of these resources may be different than those used for conventional generation. Also capacity needs are differentiated by constrained transmission zones, which may result in different capacity prices even within the same overall market.

Two aspects of generator performance may be relevant for capacity markets. The first aspect relates to the capability that the unit can provide in terms of ramping, minimum up/down times, and generally flexible operations. It is likely that the increase in renewable generation will increase the need for flexibility in the system. Current markets do not explicitly recognize this issue, which is the subject of intense interest, research, and debate although some concepts are being developed.

The second aspect of generator performance is how the unit responds as a capacity resource during the critical system times. Generally, this is the question of how the level of installed capacity relates to delivered capacity, and markets can handle this issue in different ways. The US markets address this by using either installed capacity (ICAP) or UCAP—i.e., accounting for a resource's likelihood of forced outage at a certain time—in their capacity auctions [13]. This linking of the capacity market to a reliability metric is key, and it results from the absence of total price transparency from generation to end user. In general, system operators in the United States use a probabilistic assessment to assess how much capacity is needed to meet the required reliability standard (1 event in 10 y), but the specific approaches and software packages used differ [13].

The details regarding the size and composition of the jurisdiction that establishes and operates the capacity market, along with the rules for how and what capacity counts toward the target, may have significant impacts. For example, consider an RTO market that includes multiple load-serving entities (LSEs). Each LSE is allowed to acquire its own capacity via long-term power purchase agreements or other similar mechanisms. Thus, a long-term purchase of 100 MW would result in 100 MW, derated as per its ELCC, to count toward the capacity that is needed to serve the load obligation of the LSE. Consider the case that multiple LSEs secure long-term capacity in this manner, each acquiring capacity toward its own target. Generally, such LSEs will have noncoincident peak loads, and absent significant transmission constraints between them, they will be able to share contingency reserve under some circumstances. As shown in [11], there is capacity value that can be acquired via aggregation: if a pooled area considers its combined adequacy needs, it will often be lower than the required capacity needed if each area undertakes its own individual assessment. Therefore, the assessments of resource adequacy, even when considered on a reliability basis, are affected by the level of aggregation in establishing the appropriate target, crediting LSEs for capacity acquired via power purchase or other mechanisms, and establishing a resource gap that must be provided via a capacity

Table 6.3 Design features for US capacity markets

Market	Longest forward period	Longest commitment period	Demand curve	Auction product
ISO-NE	3 y	5 y	Vertical with descending clock auction	ICAP
NYISO	30 d	6 mo	Downward sloping	UCAP
PJM	3 y	3 y	Downward sloping	UCAP

market. The ultimate level of reliable supply that is "in the ground" is what it is. However, whether there is a need for additional capacity to be secured via a market mechanism can be driven in part by the assessment level. This means that the market may incorrectly assess the quantity and price of new capacity unless a regional (such as within an RTO) target is developed first and then allocated to the appropriate load-serving entities.

Table 6.3 outlines general differences of the mandatory capacity markets including time horizon, resource qualification, slope of the demand curve, and type of capacity product (ICAP or UCAP). The forward period is defined as the length of time the auction takes place ahead of when the capacity resource is needed. The commitment period is the length of time that the capacity resource must provide capacity. The administratively-established demand curve, discussed further below, is used to determine the amount of capacity that is procured by the market. The auction product is either installed or unforced capacity.

The table shows some of the key characteristics of these markets and illustrates that there is significant variation. For example, the NYISO capacity market considers periods of 30 d and required selected resources to continue providing its capacity service for 6 months. Conversely, both ISO-NE and PJM use a 3-y forward period and require selected resources to perform for 5 y and 3 y, respectively. A forward period of 30 d is clearly insufficient to incentivize the development of new resources that have not been built, thus the NYSIO mechanism secures operating capacity over a short forward horizon. A period of 3 y is sufficient to develop some forms of generation, such as natural gas combined-cycle or peaking units, along with wind or solar generation.

Thus, the longer forward periods aim to balance the time necessary to construct new resources with the risk of over-procurement. If the forward period is too short, a resource may incur significant costs before being able to participate in an auction to cover those costs. If the forward period is too long, there is a risk that an inefficient level of capacity will be procured. NYISO has elected to use a comparatively short forward period; whereas ISO-NE and PJM have longer periods and realignment auctions closer to the commitment period [33]. The final two columns in the table illustrate the different approaches for setting the capacity target and valuing capacity additions as well as the type of capacity product being auctioned.

6.6 Evolving Market Design Elements to Ensure Resource Adequacy and Revenue Sufficiency

The impact of renewable energy on resource adequacy and future capacity needs is receiving increasing attention in US electricity markets. In regions with existing capacity markets, there are discussions regarding the potential need to revise market designs to ensure that sufficient flexible capacity is procured in the capacity auctions [33]. Other developments include improved scarcity pricing through dynamic operating reserve demand curves (ORDCs) in ERCOT and the introduction of specific requirements for flexible capacity in the centralized resource adequacy program in CAISO. We discuss these two developments, which represent two different approaches to address different challenges, in more detail below.

6.6.1 Dynamic Demand Curves for Operating Reserve

ERCOT is the only electricity market in the United States that relies on an energy-only market design to ensure capacity adequacy. Under the energy-only design, prices in the electricity market need to reach high levels during periods of scarcity to ensure that generators recover their fixed and variable costs, as discussed above. Therefore, scarcity pricing therefore becomes particularly important to obtain revenue sufficiency and reduce the missing-money problem, because there are no additional incentives for generation investment.

ERCOT has recently taken two steps to improve scarcity pricing. First, offer caps in the energy market are gradually being increased (e.g., to $5 000/MWh in June 2013, $7 000/MWh in June 2014, and $9 000/MWh in June 2015). Increasing offer caps mean that energy prices can increase to higher levels during periods of extreme scarcity. Second, a demand curve for operating reserve is being introduced in the real-time market to better reflect the marginal value of reliability from reserve on electricity market prices [34]. The ORDC influences the prices for both reserve and energy because the two products are closely linked through opportunity costs.

The proposed ORDC is derived based on a probabilistic assessment of the LOLP for different reserve levels and an estimate of VOLL. Ideally, the ORDC should be used in a co-optimization of energy and reserve markets. However, because there is currently no co-optimization in the ERCOT real-time market, a post calculation that mimics co-optimization and derives prices for reserve and price adders for energy is used instead. The procedure involves the following main steps [35]:

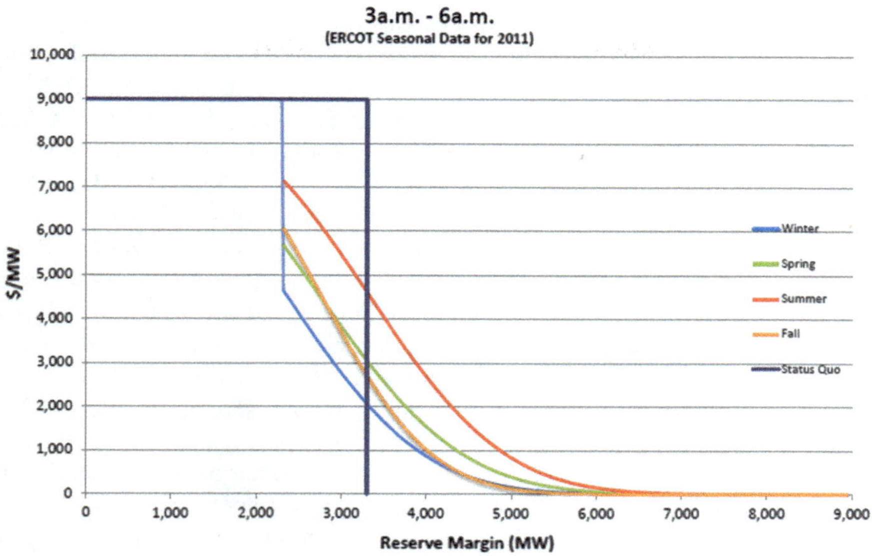

Fig. 6.7 Illustration of ORDCs for different seasons in ERCOT for the time period from 3 a.m. to 6 a.m. based on data from 2011 (Source: from: [36])

- Estimate probability distributions for LOLP based on historical differences between hour-ahead scheduled reserve and available reserve in real time. These estimates are done for 4 seasons and 6 time-of-day blocks, a total of 24 time segments. The resulting LOLP probability distributions, assumed to take the shape of a normal distribution, reflect the likelihood of forced outages, load forecasting errors, and wind power forecasting errors for the 24 different time segments.
- An adjusted LOLP distribution, LOLP′, is derived by assuming that a certain minimum level of contingency reserve, X, is required to avoid load shedding. LOLP′ is assumed equal to 1 whenever the available reserve is below X. Moreover, the LOLP′ probability distribution is shifted to the right by X compared to the original LOLP distribution. Separate LOLP′ curves are also derived for spinning and nonspinning reserve to reflect the different response times.
- The price on the demand curve for operating reserve for a given reserve level, R, equals LOLP′(R) multiplied by (VOLL minus LMP).
- Price adders for spinning and nonspinning reserve are derived based on the actual available reserve in real time.
- The LMP is increased by the calculated price adder for spinning reserve. ERCOT does not currently co-optimize energy and reserve in the real-time market, so this adder must be manually added to the energy price.
- An ancillary service market imbalance settlement ensures that resources are indifferent between energy and reserve in real time.

Figure 6.7 compares the proposed ORDCs for four different seasons to the current practice of using a fixed reserve requirement and scarcity price, which is equivalent to a vertical demand curve (in black). The downward-sloping demand curves will result in higher prices whenever the reserve margin is above the current requirement of 3 300 MW, but it may actually give lower prices when the reserve level drops below 3 300 MW. Simulation results for ERCOT [35] indicate that if the ORDC had been in place in 2011—i.e., a year with several extreme weather events—the average energy price would have increased in the range of $7/MWh to $26/MWh, depending on the VOLL parameter and the minimum contingency reserve, X. In contrast, in 2012 the average energy price increase would have been much more modest, in the range of $1/MWh to $4.5/MWh. Overall, any increases in the prices for energy and reserve would increase the incentives to invest in new system resources. Improved scarcity pricing could also provide improved operational incentives for both supply and demand resources. The ORDC concept does not prevent the introduction of additional incentives for capacity and flexibility. In fact, several other ISO/RTOs have simple ORDCs in place already, along with other capacity adequacy incentives. Current demand curves for ancillary services in other markets are versions typically not based on the same dynamic, rigid, probabilistic assessment as the one proposed for ERCOT. For instance, other ORDCs do not account for the actual probability of load shedding for that particular time. However, improved scarcity pricing through the ORDC and increased offer caps are currently the main new resource adequacy initiatives in ERCOT, which continues to rely on the energy-only market design to provide adequate investment incentives as wind energy penetration continues to rise.

An ORDC is not a new idea in the academic literature (see, e.g., [28]). However, the current ERCOT initiative represents the first rigorous implementation based on a first-principles probabilistic assessment of the reliability impacts and marginal value of reserve. Still, the ERCOT implementation could be enhanced in several ways. For instance, full co-optimization of energy and reserve in the real-time market would ensure more efficient operation and pricing. Moreover, the ORDC could also be considered in the day-ahead market to avoid inconsistencies and to achieve better price convergence between day-ahead and real-time markets. The ORDCs could be derived based on a dynamic assessment of relevant information in real time rather than using predefined LOLP distributions for different time segments. Zhou and Botterud [31] propose to calculate ORDCs dynamically, accounting for the uncertainty in wind power through a probabilistic wind power forecast along with load forecasting uncertainty and the probability of forced outages from thermal generators. Hence, the expected forecast uncertainty, which is likely to change depending on the weather situation, would be factored into the ORDC. Zhou and Botterud [31] conducted simulations of scheduling, dispatch, and market-clearing prices in co-optimized day-ahead and real-time markets with different reserve strategies. The results also indicate that the ORDC would lead to higher prices in many hours, whereas extreme price spikes are less likely to occur because of the downward-sloping demand curve (see Fig. 6.8).

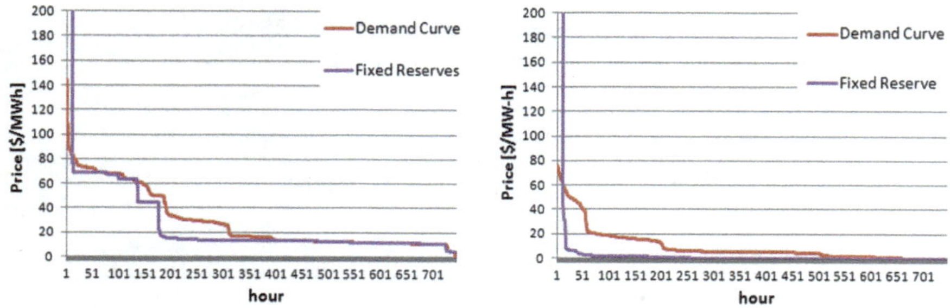

Fig. 6.8 Simulated price duration curves in a day-ahead market for energy (left) and spinning reserve (right) with an ORDC and a fixed reserve requirement for one month (based on [37])

6.6.2 *Forward Flexible Capacity Requirements*

CAISO is proposing a different approach to ensure resource adequacy in the long term. Since 2005, a resource adequacy program has been in place in which load-serving entities are required to meet a local PRM. Certain deliverability criteria are in place for resources to qualify and count toward the PRM, which must be documented at monthly and yearly levels. Moreover, qualified resources must make themselves available to the system operator. There is no centralized capacity market, but market participants can use bilateral trading to ensure that sufficient capacity is available to meet reserve margins [38]. The bilateral resource adequacy market provides generators with a potential source of income to help ensure revenue sufficiency.

Until recently, the focus of the resource adequacy program has been to ensure that sufficient capacity was available to meet peak load. However, several developments in recent years have prompted an ongoing revision and extension of the resource adequacy program. In particular, California's renewable portfolio standard, which requires the state to meet 33% of its load with renewable resources by 2020, has raised concerns about whether or not there will be sufficient system flexibility to efficiently operate the system with the rapid increase in renewable energy. Therefore, changes to the resource adequacy program are being introduced so that LSEs will be required to not only procure sufficient capacity to meet forecasted peak load but also to meet additional flexibility requirements with their capacity. In short, the current proposal, scheduled to be introduced in 2015 if approved, requires that LSEs in aggregate have sufficient flexible capacity available to meet forecasted system needs. The new flexible capacity initiative includes six measures [39].

1. Flexibility requirement determination for the system for the upcoming year. This is based on net load forecasts (Fig. 6.9) considering the most current RPS contracts. In the figure, Category 1 is base flexibility, Category 2 is peak flexibility, and Category 3 is super-peak flexibility. "A" refers to the maximum 3-h net load ramp in a month, and "C" refers to the largest secondary 3-h net load ramp for the month [39].

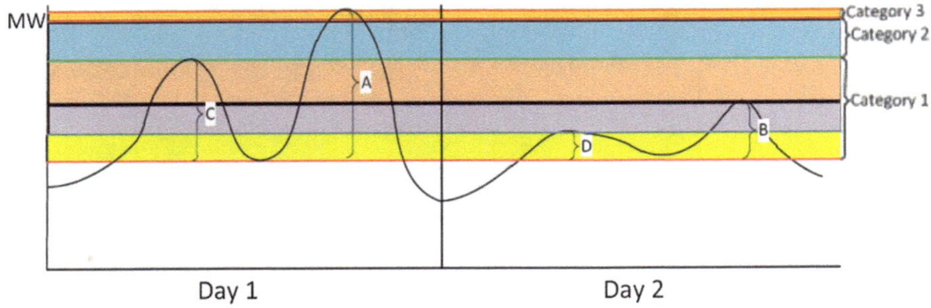

Fig. 6.9 Capacity requirements for different flexibility categories as a function of forecasted net load ramps (based on [39])

2. An allocation methodology that translates the system flexibility requirements to the individual LSEs based on their historical contributions to net load ramps.
3. Flexible capacity showings in which LSEs are required to demonstrate adequate flexible capacity procurement. Flexible capacity can come from several sources (generation, storage, demand).
4. Assessment of the capacity showings in which the ISO uses a flexibility counting methodology to ensure that LSEs' flexible capacity requirements have been met.
5. Must-offer obligations—i.e., flexible capacity must provide economic bids to the ISO's day-ahead and real-time markets from 5 a.m. through 10 p.m.
6. Backstop procurement, which allows the ISO to procure flexible capacity on a one-year forward basis if there is a deficiency in the aggregate supply of flexible capacity to the system.

Overall, the approach taken by CAISO relies on administrative and centralized planning procedures to ensure both the amount of future capacity and level of flexibility in the future resource mix. This is in contrast to the approach in ERCOT, which relies on price incentives in the short-term markets for energy and reserve to ensure capacity adequacy and revenue sufficiency.

6.7 Conclusion

At this point, it should be clear that designing a market for the reliability of supply—resource adequacy—is a very difficult proposition. The large-scale expansion of VG adds further complexities to resource adequacy assessments and market design to ensure revenue sufficiency. Whether current market designs provide the incentives that are needed to ensure adequacy in the long run or if new approaches are needed is still an open debate.

If electricity market designs rely on the energy-only approach, the key challenge is to ensure that the short-term prices for energy and reserve provide sufficient incentives for investments in a resource mix with sufficient capacity and flexibility. Appropriate scarcity pricing has been and still is the main solution that is required to ensure that resources recover both capital and operating costs in an energy-only market. This will require clever regulatory market interventions that allow prices to rise during scarcity conditions without creating opportunities for market manipulation. Renewable resources may lead to higher frequency of low and negative energy prices and increase the variability in market prices, increasing the reliance on scarcity pricing under the energy-only designs. At the same time, prices for operating reserve may increase because of higher reserve requirements associated with the increased variability and uncertainty of renewables. However, it is an open question whether future prices for energy and reserve will adequately compensate ramping capabilities to meet system flexibility needs as well as the required need for available capacity. Analyzing revenue sufficiency is difficult because it depends on having an accurate model of price formation in a future system, which is in itself a challenging problem that deserves more research. Finally, stochastic programming approaches have been proposed to more efficiently handle the challenges of renewables in electricity market operations, but setting prices for energy and reserve under stochastic scheduling is not straightforward.

If the solution is to use an additional incentive mechanism for resource adequacy, such as a capacity market, several other questions need to be addressed:

- What reliability metric should be used at the system level: LOLE, LOLH, EUE, or other?
- What target reliability level should be used? Is the traditional target of 1 d/10 y the right target?
- Is a separate metric required to ensure sufficient flexibility, in addition to capacity?
- How many years should be used in rolling reliability and capacity assessments?
- Do time-period capacity factor approximation methods sufficiently capture the link between performance attributes and resource adequacy?
- Does the combination of metric and reliability target exhibit consistency across generator types and their contribution to resource adequacy?
- What type of performance assessment should be required to ensure and incentivize resources to be available when needed?
- How should capacity market auctions be designed to minimize the potential exercise of market power?

In this chapter, we have argued that it may be useful to consider ELCC itself as a candidate metric for a capacity/resource adequacy market, because it would better capture a resource's contribution to reliability than current metrics such as unforced capacity. However, there are several challenges to implementing such a probabilistic ELCC metric. As illustrated, ELCC is highly nonlinear and potentially sensitive to several other influences. Clearly, more work would be needed to develop an ELCC-based auction for resource adequacy. As an alternative, more rigorous mapping between ELCC and UCAP or a similar metric may result in achieving the goal of retaining some reliability information in the market, yet perhaps overcome some of the concerns regarding the nonlinearity and other issues, such as sensitivity to ordering, that may exist.

A variety of market designs are considered and being introduced in different ISO/RTO markets to address the issues of long-term flexibility needs, resource adequacy, and revenue sufficiency. We believe that the industry will go through several iterations before consensus emerges on the specific best practices on these complex topics of resource adequacy and revenue sufficiency in electricity markets with large-scale penetrations of renewable resources.

Several other open research questions remain, many of which exist independently of the future structure of long-term resource adequacy and flexibility markets.

- What is the behavior of VG ELCC over multiple years, and what is the appropriate number of years of data to use in a resource adequacy study to achieve statistically-expected results and/or behavior and quantification of statistical tails?
- What metric, or family of metrics, can best describe future flexibility needs? What is an appropriate choice for a target level of flexibility, especially given that there is some uncertainty surrounding the levels of VG that will be experienced during the next several decades?
- What are the desired properties of these flexibility metrics, and how do they perform when considering multiple technologies? Are the metrics robust enough to provide consistent evaluations of new, possibly unknown, technologies that may emerge?
- How can multiple flexibility metrics be established within an incentive mechanism? Will it incentivize both new and existing resources to have flexibility capabilities?
- Is there a right combination of scarcity pricing and forward capacity markets that can take the best attributes of both concepts?
- Is it important to have standard resource adequacy and flexibility definitions and markets across all regions? If there are differences in product definitions and/or assessment algorithms, will that create gaming from entities that could potentially sell into multiple markets?

As these markets undergo changes and potentially new markets evolve to address long-term issues described here, it will also be critical to identify and correct any unintended consequences that undermine one or more markets and the way they interact. And, as always, the potential for market power must be assessed and managed.

References

1. Ela, E., Milligan, M., Bloom, A., Botterud, A., Townsend, A., Levin, T.: Evolution of wholesale electricity market design with increasing levels of renewable generation. Technical Report NREL/TP-5D00-61765, Golden, Colorado (2014)
2. Stoft, S.: Power Systyem Economis – Designing Markets for Electricity. IEEE Press and Wiley Interscience (2002)
3. Carden, K., Wintermantel, N.: The Economic Ramifications of Resource Adequacy White paper. A Report of Astrape Consulting Funded by the U.S, Department of Energy (2013)

4. Billinton, R., Allan, R.: Reliability Evaluations of Power Systems. Springer, New York (1996)
5. Milligan, M., Porter, K.: Determining the Capacity Value of Wind: An Updated Survey of Methods and Implementation. Technical Report NREL/CP-500-43433, Golden, Colorado (2008) http://www.nrel.gov/docs/fy08osti/43433.pdf (Cited on 10 January 2017)
6. NERC: Long-Term Reliability Assessment. North American Electric Reliability Corporation, Atlanta, GA (2013)
7. Keane, A., Milligan, M., D'Annunzio, C., Dent, C., Dragoon, K., Hasche, B., Holttinen, H., Samaan, N., Söder, L., O'Malley, M.: Capacity value of wind power. IEEE Trans. Power Syst. **26**(2), 564–572 (2011)
8. Milligan, M.: Methods to Model and Calculate Capacity Contributions of Variable Generation for Resource Adequacy Planning. Presentation at Joint NERC-UWIG Workshop, Kansas City (April 2011) https://www.nrel.gov/docs/fy11osti/51485.pdf (Cited on 10 January 2017)
9. Rogers, J., Porter, K.: Summary of time period-based and other approximation methods for determining the capacity value of wind and solar in the United States. Technical Report NREL/SR-5500-54338, Golden, Colorado (March 2012) https://www.nrel.gov/docs/fy12osti/54338.pdf (Cited on 10 January 2017)
10. Garver, L.: Effective load carrying capability of generating units. IEEE Trans. Power Appar. Syst. **8**, 910–919 (1966)
11. Ibanez, E., Milligan, M.: Impact of transmission on resource adequacy in systems with wind and solar power. Technical Report NREL/CP-5500-53482, Golden, Colorado (Feb. 2012) https://www.nrel.gov/docs/fy12osti/53482.pdf (Cited on 10 January 2017)
12. Zavadil, R.: Eastern Wind Integration and Transmission Study. Report prepared for NREL by EnerNex Corporation, NREL/SR-5500-47078, Golden, Colorado (Feb. 2011) https://www.nrel.gov/docs/fy11osti/47078.pdf (Cited on 10 January 2017)
13. Pfeifenberger, J., Spees, K., Carden, K., Wintermantel, N.: Resource adequacy requirements: reliability and economic implications. Report Prepared by the Brattle Group and Astrape Consulting for FERC (Sept. 2013) https://www.ferc.gov/legal/staff-reports/2014/02-07-14-consultant-report.pdf (Cited on 10 January 2017)
14. LE: Estimating the Value of Lost Load. Report Prepared by London Economics to OFGEM, London, UK (July 2011)
15. Cramton, P., Stoft, S.: Forward reliability markets: less risk, less market power, more efficiency. Util. Policy **16**(3), 194–201 (2008)
16. Kirby, B., Milligan, M., Makarov, Y., Hawkins, D., Lovekin, J., Jackson, K., Shiu, H.: California Renewables Portfolio Standard: Renewable Generation Integration Cost Analysis-Phase III: Recommendations for Implementation. Consultant report, California Energy Commission, Sacramento, CA (2004)
17. Piwko, R., Bai, X., Jordan, G., Miller, N., Zimberlin, J.: The Effects of Integrating Wind Power on Transmission System Planning, Reliability, and Operations. Report Prepared by GE Energy and Energy Consulting for The New York State Energy Research and Development Authority (March 2005)
18. Shiu, H., Milligan, M., Kirby, B., Jackson, K.: California Renewables Portfolio Standard Renewable Generation Integration Cost Analysis: Multi-Year Analysis Results and Recommendations (Final Report). Report for The California Energy Commission, Sacramento, CA (May 2006)
19. Lannoye, E., Milligan, M., Adams, J., Tuohy, A., Chandler, H., Flynn, D., O'Malley, M.: Integration of variable generation: capacity value and evaluation of flexibility. In: IEEE Power and Energy Society General Meeting Proceedings, pp. 1–6. IEEE (2010)
20. Brisebois, J., Aubut, N.: Wind farm inertia emulation to fulfill hydro-québec's specific need. In: IEEE Power and Energy Society General Meeting, pp. 1–4. IEEE Press (2011)
21. Ela, E., Tuohy, A., Milligan, M., Kirby, B., Brooks, D.: Alternative approaches for incentivizing the frequency responsive reserve ancillary service. Electr. J. **25**(4), 88–102 (2012)
22. Milligan, M., Clark, K., King, J., Kirby, B., Guo, T., Liu G.: Examination of benefits of an energy imbalance market in the western interconnection. Technical Report NREL/TP-5500/57115. Golden, Colorado (2013) http://www.nrel.gov/docs/fy13osti/57115.pdf (Cited on 10 January 2017)

23. King, J., Kirby, B., Milligan, M., Beuning, S.: Flexibility reserve reductions from an energy imbalance market with high levels of wind energy in the western interconnection. Technical Report NREL/TP-5500-52330, Golden, Colorado (2011) http://www.nrel.gov/docs/fy12osti/52330.pdf (Cited on 10 January 2017)
24. GE Energy: Western Wind and Solar Integration Study Phase 1. Report prepared for NREL by GE Energy, NREL/SR-550-47434, Golden, Colorado (May 2010) https://www.nrel.gov/docs/fy10osti/47434.pdf (Cited on 10 January 2017)
25. Maggio, D.: Impacts of wind-powered generation resource integration on prices in the ERCOT nodal market. In: IEEE Power and Energy Society General Meeting, pp. 1–5. IEEE Press (2012)
26. Milligan, M., Ela, E., Lew, D., Corbus, D., Wan, Y., Hodge, B.: Assessment of simulated wind data requirements for wind integration studies. IEEE Trans. Sustain. Energy **3**(4), 620–626 (2012)
27. Lew, D., Brinkman, G., Ibanez, E., Florita, A., Heaney, M., Hodge, B.-M., Hummon, M., Stark, G., King, J., Lefton, S.A., Kumar, N., Agan, D., Jordan, G., Venkataraman, S.: The western wind and solar integration study phase 2. Technical Report NREL/TP-5500-55588, Golden, Colorado (2013)
28. Hogan, W.: On an "Energy Only" Electricity Market Design for Resource Adequacy. Report. Harvard University, Cambridge (2005)
29. Ortega-Vazquez, M., Kirschen, D.: Optimizing the spinning reserve requirements using a Cost/Benefit analysis. IEEE Trans. Power Syst. **22**(1), 24–33 (2007)
30. Wang, J., Wang, X., Wu, Y.: Operating reserve model in the power market. IEEE Trans. Power Syst. **20**(1), 223–229 (2005)
31. Zhou, Z., Botterud, A.: Dynamic scheduling of operating reserves in co-optimized electricity markets with wind power. IEEE Trans. Power Syst. **29**(1), 160–171 (2014)
32. Walawalkar, R., Fernands, S., Thakur, N., Chevva, K.: Evolution and current status of demand response (DR) in electricity markets: insights from PJM and NYISO. Energy **35**(4), 1553–1560 (2010)
33. FERC: Centralized Capacity Market Design Elements. Staff Report of the Federal Energy Regulatory Commission, AD13-7-000, Washington, D.C. (Aug. 2013)
34. Hogan, W.: Electricity Scarcity Pricing Through Operating Reserves: An ERCOT Window of Opportunity. Working paper, Harvard University, Cambridge, MA (2012)
35. Hogan, W.: Back Cast of Interim Solution B+ to Improve Real-Time Scarcity Pricing. White paper, Electric Reliability Council of Texas, Austin, TX (March 2013)
36. Anderson, K.W.: Memorandum to Public Utility Commission of Texas. Memo for Project No. 40000 – Commission Proceeding to Ensure Resource Adequacy in Texas, USA (2013)
37. Zhou, Z., Botterud, A.: Price Responsive Demand for Operating Reserves in Co-Optimized Electricity Markets with Wind Power. Presentation at Federal Energy Regulatory Commission (FERC) Technical Conference, Washington DC (2013) http://www.ferc.gov/CalendarFiles/20140411130626-T1-B%20-%20Botterud.pdf (Cited on 10 January 2017)
38. CAISO: Business Practice Manual for Reliability Requirements. California ISO Report, Folsom, CA, United States (2013)
39. CAISO: Flexible Resource Adequacy Criteria and Must-Offer Obligation (Market and Infrastructure Policy Draft Final Proposal). California ISO Report, Folsom, CA, United States (2014)

Chapter 7
Requirements for Strategic Reserves in a Liberalized Market with Wind Power

Lennart Söder

Abstract The requirements concerning the reliability in power supply are high. This chapter addresses the issue of system adequacy, i.e., the need of enough installed capacity in each area to meet the load with an acceptable reliability. The challenge in liberalized markets is that the utilization time of rarely used peak units is so low that they will require extreme prices in order to be profitable. This has led to different methods including the creation of different types of capacity markets. The aim of this chapter is to analyze the connection between peak prices, system adequacy, needed size of a strategic reserve, i.e., the volume of the capacity market, and the impact of wind power on strategic reserves. The chapter uses data from Sweden to perform a detailed analysis of the influence of renewable generation on the capacity adequacy requirements for three different situations.

7.1 Introduction

A power market involves producers and traders who supply power at prices that are set by the market. The basic role of the system operator (SO) is to keep the technical continuous consumption-production balance. This means that it is not fair from the market point of view if some companies keep a high generation adequacy when they sell power while others neglect the risk of high demand and rely on "system resources". The basic principle should be: *if you sell power you should have power to sell.*

The challenge for the market designer and the SO is on the border between "system security" and "system adequacy" [1–4]. Keeping the continuous physical production-consumption balance means that there must be resources available and such resources should be (in the basic approach) financed by market participants, i.e., not by the SO. But in many places there are some pressure and expectations that the SO should be involved in this business. One reason is associated with blackouts or adequacy

L. Söder (✉)
KTH, the Royal Institute of Technology, Stockholm, Sweden
e-mail: lennart.soder@ee.kth.se

F. Lopes and H. Coelho (eds.), *Electricity Markets with Increasing Levels of Renewable Generation: Structure, Operation, Agent-based Simulation, and Emerging Designs*,
Studies in Systems, Decision and Control 144, https://doi.org/10.1007/978-3-319-74263-2_7

problems: in case of a blackout or adequacy problem the physical balance is kept by the SO (and not by the market), so journalists and politicians normally ask the SO: why this "does not work"? At the same time, there is a criticism when prices are "too high", since they can be an incentive to the exercise of market power. So, there are several challenges in order to get an economically efficient and market based way to get enough capacity in the market.

There are several drawbacks of a system with a large role of the SO concerning financing and operation of peak power plants. These include:

1. The liberalization is based on a clear separation between production and transmission. In a system with transmission system operators (TSOs), as in Europe, a lot of questions including cross subsidization can be raised, with one of the actors involved in both areas.
2. It is not a trivial task to define how large the role of the SO should be: 1000 MW? 3000 MW?
3. For rarely used units the solution of demand flexibility is probably the cheapest solution (cheaper than open cycle gas turbines, OCGTs) [5]. But in a market framework, the demand is supplied by retailers/traders/suppliers and these actors have probably the best contacts and the best possibilities to introduce a system with flexible consumers. And not the SO/TSO.
4. If the SO is involved in peak power investments, there will be a negative incentive for "ordinary market players" (e.g., producers) to be involved in peak power investments. Why should they invest, e.g., in open cycle gas turbines to be used during 10 h/year with high prices, if they know that the system operator takes the responsibility for this and finances the investment in a socialized way?

But there is still another challenge when the SO is a central player. The question is, what happens to the prices for imbalances when there is a "lack of power"? If there is "a lack of power" (i.e., demand > available capacity) then, in reality, there is at least one trader/retailer who sells power without having power to sell. In the legal framework they do not have to do this (they are only "balance responsible" = economically responsible to keep the balance), but it is important that the imbalance price be so high in order to have an incentive to not come into this situation, i.e., an incentive to invest in peak capacity.

To obtain a relevant system adequacy a *capacity market* can be created. If there is not a specific capacity market, then this is denoted: (a) *energy-only market* [6]. There are different set-ups of capacity markets including: (b) *long-term contracts or options for energy*, (c) *payment mechanisms for capacity*, (d) *quantity requirements for capacity* and (e) *demand curves for capacity* [6].

According to the economic theory a capacity market is not required, but some practical difficulties arise in real markets from the generators' viewpoint that prevent the straightforward results from this theory [4]. One consequence is, e.g., the impact of extreme prices, which causes the need of price-caps if the load price sensitivity is low [7].

The method of long-term contracts or options for energy implies a regulatory requirement that those who sell power to consumers should hold long-term contracts of options for energy [6]. An alternative is forward reliability markets [8], where physical capacity bundled with financial options supplies energy above a strike price.

Payment mechanisms for capacity were introduced, e.g., in Sweden where every year, in a tender process, the SO pays for originally up to 2000 MW of capacity to be used during the coming winter in peak load situations. According to a parliament decision, this level is going to be decreased and about the year 2020 there should be no SO responsibility. The capacity includes both production and reduction of consumption. For the winter of 2014/2015, the levels are 626 MW of production and 874 MW of demand [9]. This type of market is defined as *selective capacity market* [3] or *strategic reserve* [1].

Quantity requirements for capacity include the ICAP (installed capacity) market which implies a target level of system generating reserves and the allocation of responsibilities for meeting that target [6, 10]. France is on the way of starting up a capacity market for 2016–2017 [11], and the idea is that obligations will be assigned to suppliers based on the actual consumption of their customers during peak periods.

Demand curves for capacity mean that the system operator creates a downward sloping demand curve that pays more for capacity if reserves are short and provides some payment even when there is significantly more capacity than the amount needed [6, 12–14].

An important issue for all these different setups is the connection between the price-cap, the system reliability and the amount of needed reserves, i.e., the volume of the capacity market. The amount of needed capacity is treated in [15], but here this issue will get a more detailed description which will also be expanded and commented for systems with large amounts of wind power. The critical role of the SO concerning purchase or investments in reserve power in a liberalized market is treated in Sects. 7.2–7.5. The situation of a system with wind power is considered in Sect. 7.6. In order to draw conclusions for wind power, a comparison is made with a case with the same yearly energy production in nuclear power (Sect. 7.7). Finally, the main conclusions are drawn in Sect. 7.8.

7.2 Peak Power Requirements

The aim of this section is to analyze the connection between risk of capacity deficit, power prices and the function of the market. The section will show that there are three central variables: the amount of subsidized reserve power (= strategic reserve), R, the risk of capacity deficit, $LOLP$, and the maximal accepted price, λ_{max}. For a certain system, one of these variables can be estimated from the other two.

An important question is as follows: what is an "acceptable" supply reliability (= generation adequacy)? It will be shown that this question is directly related to the amount of strategic reserves and the maximal accepted price.

First, assume a simplified future power system involving:

1. One area, i.e., no bottlenecks.
2. An assumed distribution of future power consumption, available as a duration curve.
3. An assumption that the consumption is price independent.
4. Power stations, assumed to be 100% reliable, i.e., the installed capacity is always available.
5. A power production consisting only of a certain type of open cycle gas turbines.
6. An assumption of perfect information and perfect competition (if a power station can get its total costs—both investments and operation costs—covered, then it is built by some producer).

We consider first an analysis based on these assumptions and then we comment the assumptions themselves. The analysis is illustrated with an example with the following data, denoted **Data–1**:

- The future power consumption is the Swedish consumption for the studied period (the years 1996–2001 and 2007–2013). This period was chosen because good wind data are available for it.
- The cost of an OCGT is assumed to include a capitalized investment cost (discount rate of 6%) of 36 k€/MW-year, an availability of 95% and an operation cost of $c_G = 0.12$ k€/MWh. This means that the cost per MW is $\alpha_G = 36/0.95 = 37.89$ k€/MW-year. The cost data are from a Swedish report [16] and an assumed exchange rate of 10 SEK/€. This means that the limited availability is, for simplicity, not assumed to be stochastic, but 95% of the installed capacity is always available.

Figure 7.1 shows the yearly load duration curve, $F(x)$, and the needed market price, λ_x, if this price alone should cover the total costs for investments and operation of the OCGT. Assume that, e.g., the OCGT is used 60 h/year and has an availability of 100%. The needed price is then:

$$\begin{aligned}\lambda_x(60) &= c_G + \frac{\alpha_G}{60} \\ &= 120 + 36{,}000/60/0.95 = 120 + 632 = 752\,\text{€/MWh}\end{aligned} \tag{7.1}$$

The load duration curve, $F(x)$, is defined as follows:

$$F(x) = \text{number of hours per year that the load is} \geq x \tag{7.2}$$

It can be noted that, e.g., the level of 0.5 h/year in reality means, perhaps, 2 h every 4th year, since the power consumption varies between different years. This is illustrated in Fig. 7.2, where the situations with high load, during the studied period, are shown.

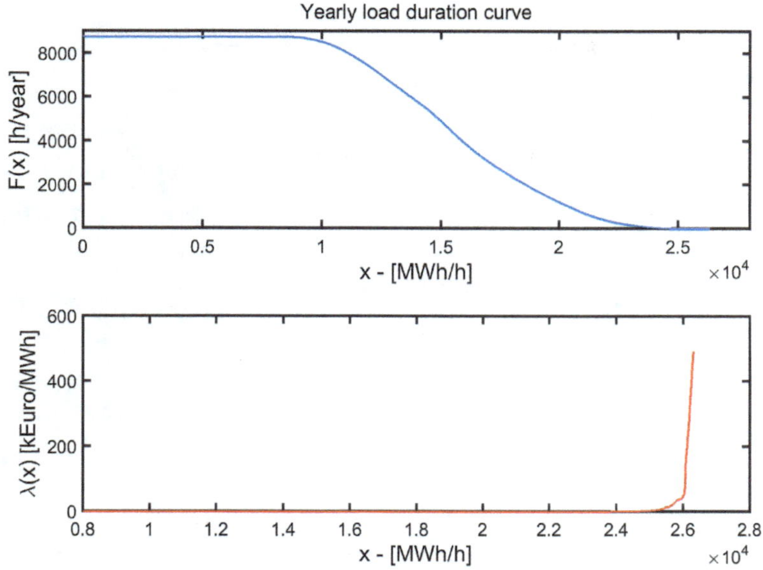

Fig. 7.1 Duration curve for the consumption (top) and needed price level for a profitable investment in OCGT (bottom)

The needed price corresponds to an energy price that covers the total cost of the power plant (including investment costs) at an utilization time that the power consumption has at this level.

The needed price level, $\lambda(x)$, can be calculated as follows:

$$\lambda(x) = c_G + \frac{\alpha_G}{F(x)} \tag{7.3}$$

Table 7.1 shows some of the values depicted in Fig. 7.1. In Fig. 7.3, some new variables are introduced and the same data as in Fig. 7.1 are shown, but for the interval when the load is larger than 23,500 MW, i.e., high load situations.

Now we introduce some new variables and use Fig. 7.3 for an illustration (i.e., Data–1): P = total installed capacity, λ_{max} = maximal accepted price level, M = load level corresponding to λ_{max}(= capacity installed by the market), and $R = P - M$ = subsidized reserve capacity.

In Fig. 7.3, this corresponds to the following:

- P = installed capacity = $x2$ = 25,689 MW. This is not the same level as the maximum load, and the load increases this level during 2 h/year. This means that the *LOLP* is 2 h/year.

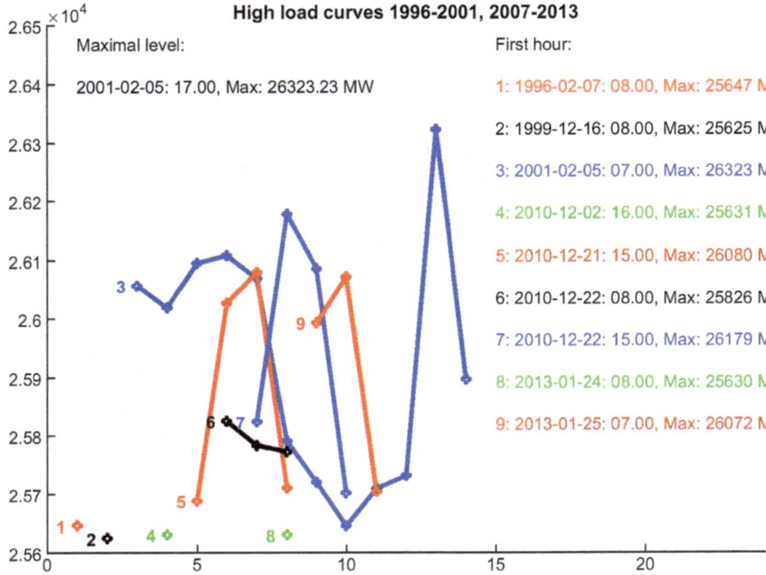

Fig. 7.2 Occasions with high load during 13 years in Sweden

- λ_{max} = maximal accepted price level = 0.75 k€/MWh. This means that the market will only perform investments for OCGTs which have an utilization time higher than 60 h/year. If there are no other power plants, then there will be a *LOLP* of 60 h/year.
- M = load level corresponding to the maximal accepted price level λ_{max}. This means that the market will only make investments which cover the load up to the level $M = 24{,}028$ MW.
- $R = P - M = x2 - x1$ = need of subsidized reserve capacity since the market will only invest in M (MW) and P (MW) if a *LOLP* of 2h/year is requested.

Table 7.1 Utilization time and needed price depending on load level in Fig. 7.1

x: load level (MW)	$F(x)$: duration (h/year)	$\lambda(x)$: needed price (k€/MWh = €/kWh)
>23,000	191.3	>0.32
>23,500	118.1	>0.44
>24,000	62.6	>0.73
>24,500	29.3	>1.41
>25,000	11.6	>3.38
>25,500	2.7	>14.20
>26,000	0.85	>44.9
>26,250	0.08	>492.75

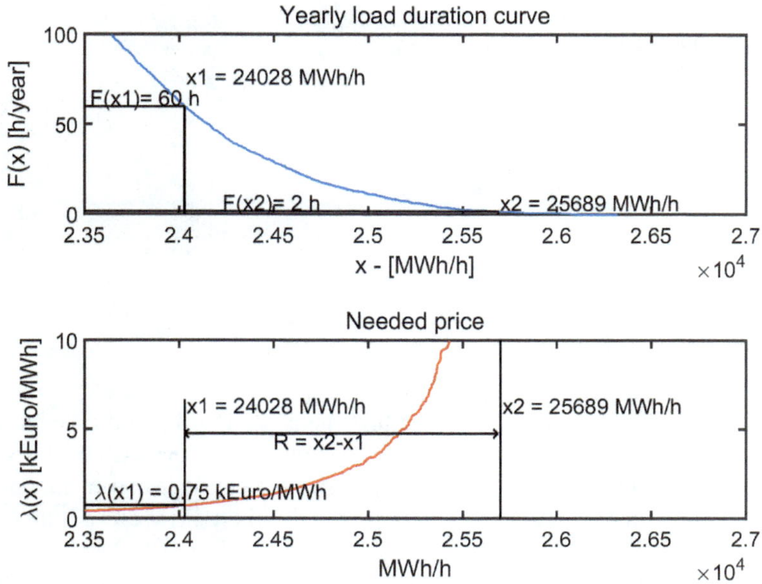

Fig. 7.3 Duration curve for the consumption (top) and needed price level for profitable investment in OCGT (bottom)

Relevant questions are now the following:

1. How much installed capacity (P) is needed?
   This is the same question as: what reliability level, $LOLP = F(P)$, is required?
2. How high prices (λ_{max}) can be accepted?
   This is the same question as: how much power (R) has to be subsidized (but see below).

Considering the first question, if one has $P = 25{,}689$ MW installed, then there will be a deficit of power during 2 h/year ($P \Rightarrow LOLP$). If one accepts a $LOLP$ during 2 h/year then it is necessary to install 25,689 MW ($LOLP \Rightarrow P$). The lower the risk, the more power has to be installed. This example is illustrated in Fig. 7.3.

Now assume that one, in an analysis, have come to the conclusion that 25,689 MW of capacity is enough, i.e., accepts a capacity deficit with a mean value of 2 h/year. In reality, the amount of hours per year with capacity deficit varies.

If one assumes that all power stations should only be paid with the current power price, then the price has to be $\lambda(25{,}689) = 20.65$ k€/MWh, at this consumption level, in order to make it profitable to invest in the last MW of an OCGT which is only used 2 h/year. This is outside the figure.

Table 7.2 Needed amount of reserve power for different accepted power prices and an accepted *LOLP* of 2 h/year

Accepted price λ_{max} (k€/MWh)	Needed amount of reserve power (MW)
0.32	2689
0.44	2189
0.73	1689
1.41	1189
3.38	689
14.20	189

7.3 Maximum Price

Now assume that the society considers that there are too many problems if one accepts a price higher than λ_{max}. If this is the case, then only $M = 24{,}028$ MW will be installed by market participants without any extra payment since power stations with lower utilization time will not be profitable. Figure 7.3 (bottom) shows a combination of $\lambda_{max} = 0.75$ k€/MWh which corresponds to $M = 24,028$ MW. More power will not be installed since the total cost (investment + operation cost) is higher than the revenue. If the risk of capacity, $LOLP = F(M)$, is considered too high, $F(24{,}028) =$ 60 h/year, then investments in more power stations have to be subsidized in some way. If the assumption that $LOLP = 2$ h/year is acceptable, as in Fig. 7.3, then $R = P - M = 1661$ MW has to be subsidized.

Considering now the second question (above), if a price higher than 0.75 k€/MWh ($\lambda_{max} = 0.75$) is not accepted, then this implies that one has to subsidize $R = P - M = 25{,}689 - 24{,}028 = 1661$ MW. This means: (λ_{max} and *LOLP*) $\Rightarrow R$. If one, on the other hand, has decided how much reserve power can subsidize, then it is possible to estimate which price can be accepted: (R and *LOLP*) $\Rightarrow \lambda_{max}$. The higher the price, the lower the amount of reserve power that has to be subsidized. Table 7.2 shows the required amount of reserve power as a function of the accepted price. There are some important conclusions from this discussion:

1. If one has a maximum price, λ_{max}, but no subsidized power stations, $R = 0$, it is then possible to estimate the resulting *LOLP*. In the example: $\lambda_{max} = 0.75$ k€/MWh $\Rightarrow LOLP = 60$ h/year ($\lambda_{max}, R = 0 \Rightarrow LOLP$).
2. If one sets a maximum price, λ_{max}, and a certain amount of reserve power, R, then it is possible to estimate the resulting risk of capacity deficit. With $\lambda_{max} = 0.75$ k€/MWh and $R = 1661$ MW the load up to 24,028 MW will be covered (without subsidies) by the market since the costs are covered, while the load between 24,028 and 25,689 MW will be covered by subsidized plants. Higher loads cannot be covered: $LOLP = 2$ h/year ($\lambda_{max}, R \Rightarrow LOLP$).
3. If one assumes a maximum price (λ_{max}), and accepts the concept of strategic reserve (i.e., subsidized power plants), and also considers an acceptable risk of capacity deficit, then it is possible to estimate how much reserve capacity (= volume of capacity market) that has to be subsidized. With $\lambda_{max} = 0.75$ k€/MWh and $LOLP = 2$ h/year, it is necessary to subsidize $R = 1661$ MW of reserve power,

(λ_{max}, *LOLP* $\Rightarrow$ *R*). It can be noted that if the reserve power is bid into the market at a lower price than λ_{max}, then a larger amount of reserve power is needed to keep down the *LOLP* at 2 h/year (see below).

4. If one assumes a certain amount of reserve power (*R*) and a given risk of capacity deficit, then its is possible to estimate the bidding price of the reserve power into the market in order to avoid replacing market financed power stations. In the example, this means that with $R = 1661$ MW of reserve power and $LOLP = 2$ h/year, it is possible to bid this power into the market at a price $\lambda_{max} = 0.75$ k€/MWh (*LOLP*, $R \Rightarrow \lambda_{max}$).

As shown in this analysis, there are three central variables: size of the strategic reserve, *R*, risk of capacity deficit, *LOLP*, and maximum accepted price, λ_{max}. For a certain system, one of the variables can be estimated based on information about the other two.

Data–1b. This considers a slight change in the data in order to make a comparison with wind and nuclear power. Assume that the maximum accepted price is 1.4 k€/MWh, the accepted *LOLP* is 0.2 h/year, i.e., 1 h every 5th year, and OCGTs are the most expensive market financed power plants. Also, assume that the price to the market for the strategic reserve is not lower than the needed market price to finance the last market financed MW of power. The load is the Swedish load for period 1 (as above). The amount of strategic reserve is shown in Fig. 7.4.

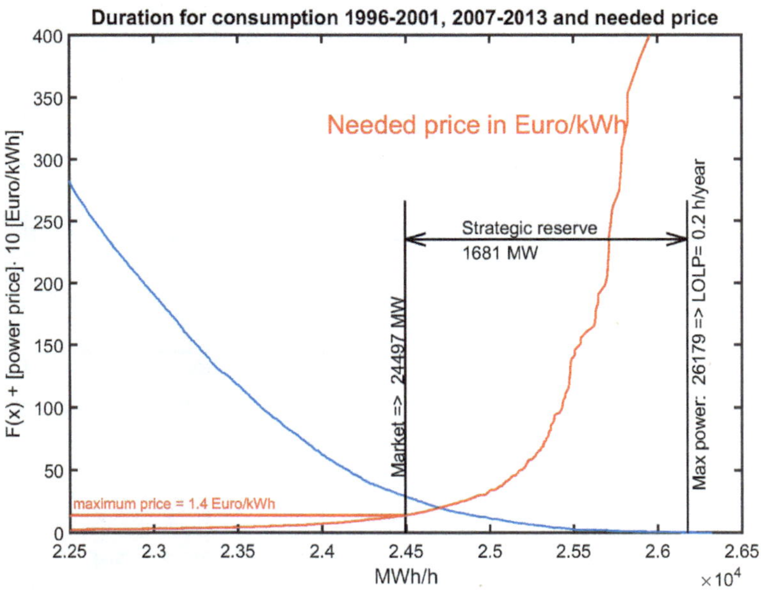

Fig. 7.4 Amount of needed strategic reserve in a system with $LOLP = 0.2$ h/year and a max price of 1.4 k€/MWh

7.4 On the Bid Price for Reserve Power

Assume now that there is no formal maximum price, but a decision on which price that the subsidized reserve power will use when it is bid into the market. An example based on Data–1 is shown in Fig. 7.5. In this figure, there are two types of reserve power: R_S = subsidized reserve power and R_M = other reserve power financed only by the market, i.e., by market price. The following is valid:

1. The subsidized reserve power ($R_S = 1000$ MW) is bid at 0.75 k€/MWh. This means that it is only when this is not enough that more reserve power is needed. This occurs at the following consumption level: 24,028 + 1000 = 25,028 MW.
2. For load levels above 25,028 MW, the reserve power plants have to be financed by the market, since the subsidized reserve power plants are already used. This corresponds to the following: the first extra MW that comes in at load level 25,028 MW has to get a price of $\lambda(M_1 = M + R_S) = 3.47$ k€/MWh. For more power even higher prices are needed, since the utilization time decreases.
3. The amount of subsidized reserve power (R_M) that is needed depends on the accepted price level or required reliability level (corresponding to a certain *LOLP* level). In the example in Fig. 7.5, market finance reserve power $R_M = P - R_S - M = x2 - x_{rs} = 611$ MW. This corresponds to a *LOLP* of 2 h/year and an accepted price of 20.65 k€/MWh.

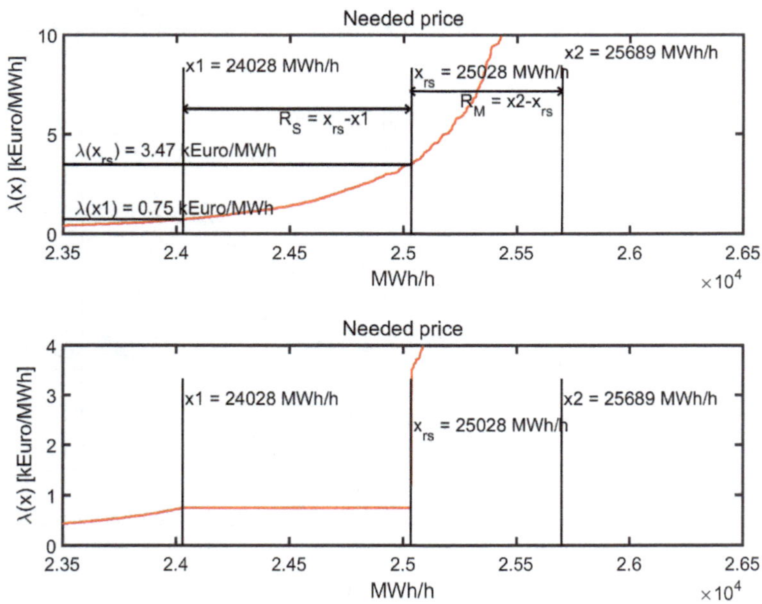

Fig. 7.5 $\lambda(M) = 0.81$ k€/MWh. Needed price for a market financed gas turbine (top) and actual marginal price to the market at different load levels (bottom)

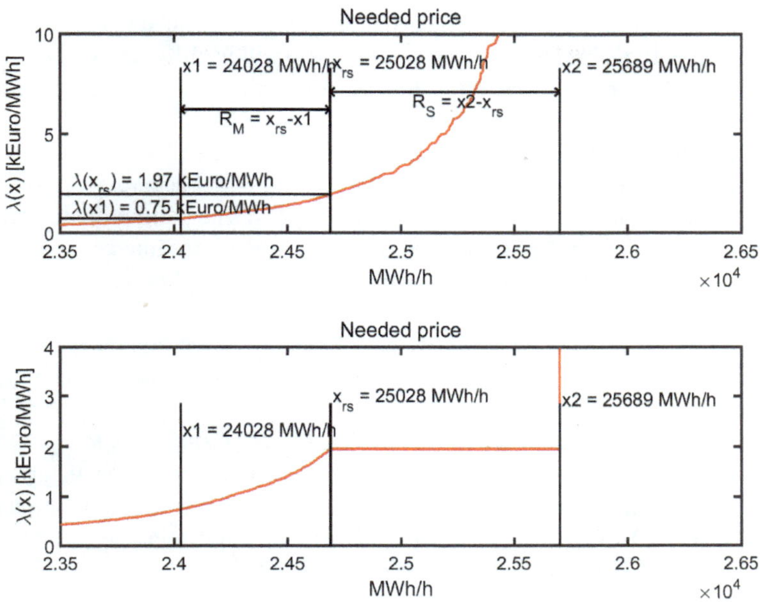

Fig. 7.6 Needed price for a market financed gas turbine (top) and actual marginal price to the market at different load levels (bottom)

In this example, there is a need to accept the price of 20.65 k€/MWh in order to get a *LOLP* of 2 h/year. If one instead has set the price of subsidized reserve power to 1.97 k€/MWh, $\lambda(P - R_S) = \lambda(25{,}028) = 1.97$ k€/MWh, then the maximal price becomes 1.97 k€/MWh (and not 20.65 k€/MWh) for an accepted *LOLP* of 2 h/year. This case is illustrated in Fig. 7.6. The conclusion is that pricing of subsidized reserve power is essential, since subsidized power on the market will compete with fully market financed power plants. If subsidized plants lower the price on the market, this will reduce the interest of the market to invest in market financed power plants. This can lead to very high prices and/or a high risk of capacity deficit and/or requirements of more subsidized reserve power plants.

7.5 Comments on the Assumptions of the Illustrative Examples

Important comments concerning the basic assumptions used in the examples in Sects. 7.2–7.4 are as follows:

1. *One area, i.e., no bottlenecks*: the analysis can be made for a whole system involving bottlenecks. But this means that there will be different price levels and different risks of capacity deficit in different areas, since load and transfer capacity varies.
2. *A known distribution of future power consumption available as a duration curve*: the results are naturally dependent on the structure and accuracy of this curve. If it is not appropriate then the results will be poor. The main problem is: if one overestimates the duration curve then one overestimates the interest of the market to construct power plants with low utilization times, which has an impact on the risk of capacity deficit.
3. *An assumption that the consumption is price independent*: price dependent load can be modeled as a production source, where the load decreases at a certain price instead.
4. *Power stations assumed to be 100% reliable, i.e., the installed capacity is always available*: it is possible to include outages in the duration curve by the use of the theory of probabilistic simulation, including the equivalent load duration curve (*ELDC*) [17]. However, this method is based on an area approach, which needs to be extended if a multi-area system is to be studied.
5. *The power production consists only of a certain type of gas turbines*: in a real power system there are a lot of types of power plants with different operation and investment costs. This has no principal impact on the price curve $\lambda(x)$. At high utilization times (x is small), high investment costs per MW are not the dominant problem, since there are many MWh that can share the investment cost. But at low utilization times (x is large), low investment costs are essential, since the investment costs can only be distributed to comparatively few produced MWh.
6. *If a power station can get its costs covered, then it is built*: this is probably not true since the future is uncertain and there needs to be an expected profit (including risk premium) before an investment decision should be made. Probably, this can be included by raising the costs of new power plants, i.e., using a higher discount rate.

7.6 Peak Power Requirements in the Presence of Wind Power

In this section, we analyze the impact of wind power on the requirements of reserve power. We start by defining two new data-sets, which change the basic set up considered in Data–1b:

- **Data–2a**. The same production system and load duration curve as in Data–1b, but adding 4000 MW of wind power installation. The first part of the data set consists of wind power data, namely synthetic data from an assumed installation of 4000 MW of wind power in Sweden for the period 1996–2001 [18]. For this period, both hourly load and hourly wind power production are available.

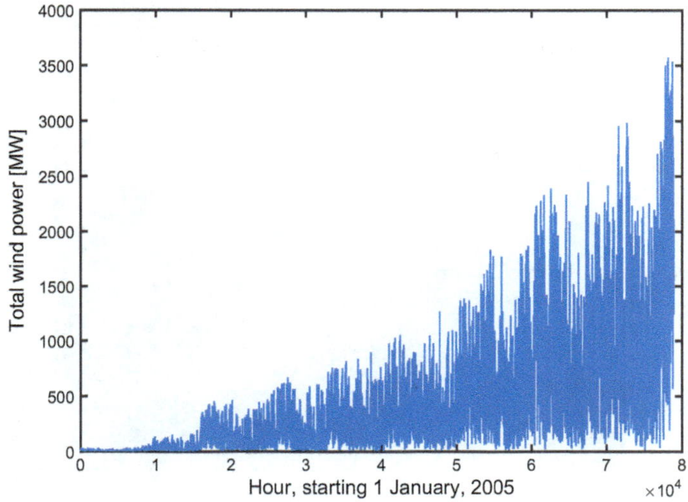

Fig. 7.7 Production of Swedish wind power in the period 2005–2007

- **Data–2b**. The second part of the data set consists of real Swedish wind power production data for the period 2007–2013. A time curve of Swedish wind power is shown in Fig. 7.7. The idea is to use the data in order to estimate a possible variation of 4000 MW of wind power to get a longer time series of both load and 4000 MW of wind power. The amount of Swedish wind power is shown in Table 7.3. The hourly data are then scaled assuming a linear expansion of the installed amount of power during the year. We assume data for 07:00–08:00 on March 15, 2011, i.e., hour $31 \times 24 + 28 \times 24 + 14 \times 24 + 7 = 1759$. If we assume a total amount of 4000 MW of wind power then the formula becomes the following (using data from Table 7.3):

$$\text{scaling} = \frac{4000}{2899} \times \frac{8760 - 1759}{8760} + \frac{4000}{3745} \times \frac{1759}{8760} = 1.3172 \qquad (7.4)$$

which means a weighted value of the installed capacity at the beginning of the year (4000/2899 = 1.3798), and at the end of the year (4000/3745 = 1.0681). For years 2005–2006, the data in Fig. 7.7 are not so reliable since there were a lot of wind power in the distribution grid that was not reported. So, these two years are not used. Data–2b then consists of the true wind power production data for the period 2007–2013 scaled to 4000 MW.

Table 7.3 Installed wind power capacity in Sweden at Dec. 31, for the period 2004–2013

Year	2004	2005	2006	2007	2008
MW	442	525	580	788	1021
Year	2009	2010	2011	2012	2013
MW	1560	2163	2899	3745	4470

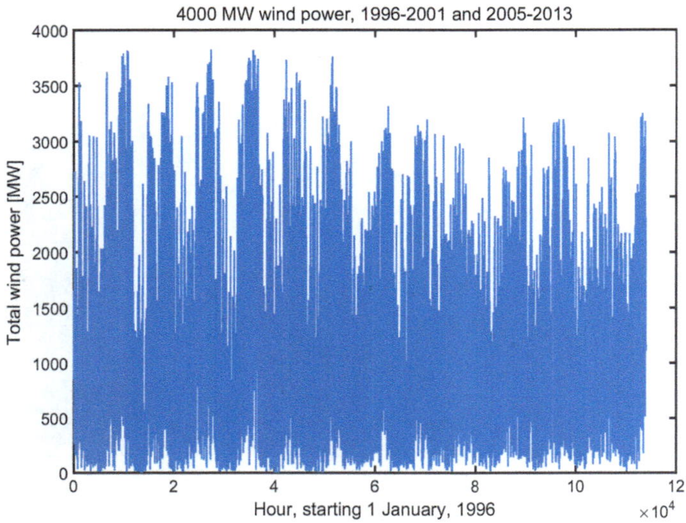

Fig. 7.8 Data–2: production of 4000 MW of Swedish wind power for the period 1996–2001 and 2005–2007

- **Data–2**. If one then applies this method to the whole period 2007–2013 and add the data from Data–2a above, then the whole Data–2 series can be obtained, which is shown in Fig. 7.8. One can note that Data–2a provides a slightly higher production (which depends on good wind sites and 2 MW units with rather high towers), while the real data for Data–2b also includes slightly smaller units (and not so good wind sites). The duration curve for this wind power is shown in Fig. 7.9 (top). The mean yearly wind power production is 9.0 TWh, corresponding to a mean value of 1022 MW, a capacity factor of 0.26 and an utilization time of 2240 h. It is assumed that this wind power is base loaded which means that the other units have to meet the excess demand. For the whole period then both wind power and load are available as hourly values. This means that it is possible to calculate the net load (net load = load − wind power) which has to be covered by other units. This is shown in Fig. 7.9 (bottom).

Data–2b. Now assume the same as for Data–1b, i.e., maximum accepted price is 1.4 k€/MWh, accepted *LOLP* is 0.2 h/year, i.e., 1 hour every 5th year, and OCGTs correspond the most expensive market financed power. It is also assumed that the price to the market for the strategic reserve is not lower than the needed market price to finance the last market financed MW of power. The load is the Swedish load for the selected period and the wind power is as indicated for Data–2 (as described above and shown in Fig. 7.9).

We now have a new situation where the OCGTs should cover the load that is not covered by wind power, i.e., the net load. This situation is shown for the peak load situation together with the resulting amount of strategic reserve in Fig. 7.10.

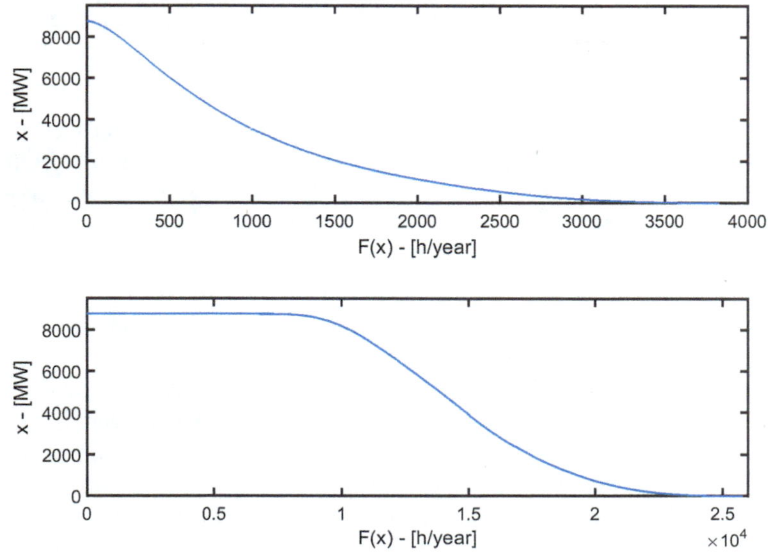

Fig. 7.9 Duration curve for 4000 MW of wind power (top) and net load duration curve for non-wind power plants (bottom)

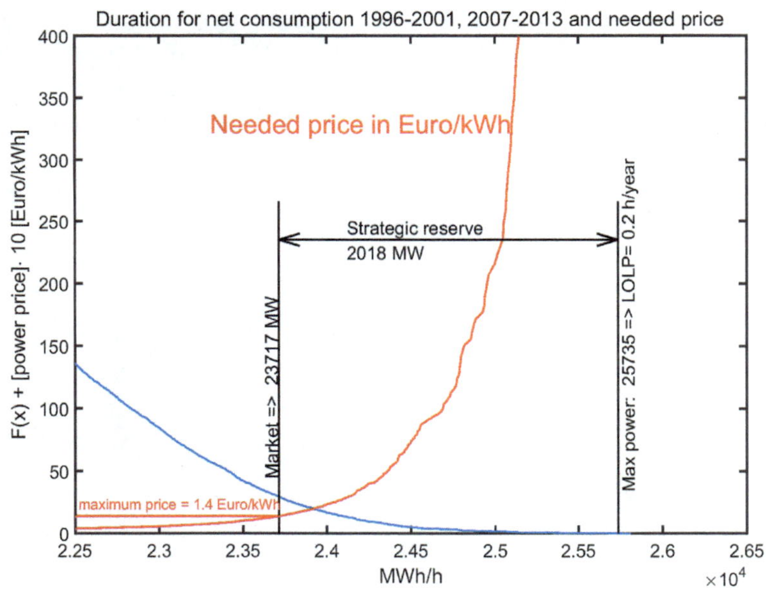

Fig. 7.10 Amount of needed strategic reserve in a system with $LOLP = 0.2$ h/year, max price = 1.4 k€/MWh and 4000 MW wind power

7.7 Peak Power Requirements in the Presence of Nuclear Power

In this section, we analyze the impact of nuclear power on the requirements of reserve power. The basic idea is to compare the impact on the strategic reserve from both wind power and nuclear power. Specifically, we compare the impact of two set-ups with the same expected yearly energy. The 4000 MW of wind power above produced 9.0 TWh, as a mean value during the 13 years. We will then use a set-up with 2 nuclear stations with an expected production level of 9.0 TWh/year. Hourly data have not been available for this period for the nuclear stations, so another approach is applied.

An annual production of 9.0 TWh corresponds, roughly, to the annual production of the Swedish nuclear reactors Oskarshamn–1 (473 MW) and Ringhals–1 (854 MW). This gives a total installed capacity of $473 + 854 = 1327$ MW, i.e., about 1/3 of what is needed in the form of wind power. With an assumed utilization time of 6780 h/year, these units will deliver 9.0 TWh/year. Assuming 90% availability and seven week summer maintenance/refuel, they will receive an expected annual production of O1: $45/52 \times 0.9 \times 473 \times 8760 = 3.2$ TWh, and R1: $45/52 \times 0.9 \times 854 \times 8760 = 5.8$ TWh, respectively. This gives a total annual production of $3.2 + 5.8 = 9$ TWh. Annual production of these stations for the same years as wind power, i.e., both older and some newer data, are shown in Table 7.4.

It may be noted that some power upgrades have been made. The above numbers for installed power (473 and 854) are from 2012, but in 2001 the installed capacity of these power plants were 445 MW (O1) and 835 MW (R1). As shown in Table 7.4, the above assumptions are optimistic, since we have not achieved an average of 90% availability and/or 7 weeks of maintenance. However, in the analysis below, the data are used to obtain a size of the contribution.

It is assumed that maintenance can be neglected, as it cannot have any influence over the hours of interest, since it is performed during low load periods. What is

Table 7.4 Yearly production in nuclear stations Oskarshamn–1 and Ringhals–1 for the period 1996–2001 and 2007–2013

Yearly production	2007	2008	2009	2010	2011	2012	2013
Oskarshamn–1	2.6	3.5	2.8	3.2	3.0	0.0	0.5
Ringhals–1	6.0	4.5	3.9	5.0	4.2	4.0	6.1
Total	8.6	8.0	6.7	8.2	7.2	4.0	6.6
Yearly production	1996	1997	1998	1999	2000	2001	Mean
Oskarshamn–1	2.4	2.9	1.3	3.3	3.1	3.1	2.4
Ringhals–1	6.5	2.2	5.6	4.9	3.2	5.0	4.8
Total	8.9	5.1	6.9	8.2	6.3	8.9	7.2

Table 7.5 Probability for different nuclear production levels during high demand

	O1: 473 MW	R1: 873 MW	Total power (MW)	probability = share of time
State 1	Available	Available	1346	$0.9 \times 0.9 = 0.81+ \Rightarrow 81\%$
State 2	Available	Not available	473	$0.9 \times 0.1 = 0.09+ \Rightarrow 9\%$
State 3	Not available	Available	873	$0.1 \times 0.9 = 0.09+ \Rightarrow 9\%$
State 4	Not available	Not available	0	$0.1 \times 0.1 = 0.01+ \Rightarrow 1\%$

important during high load hours is that plants may not be available. During high load, plants may have four different possible states (see Table 7.5).

If we now start with the load duration curve, $F(x)$, as shown in Fig. 7.1, then we can calculate the net load duration curve, i.e., the load-nuclear power:

$$F_{net-load}(x) = 0.01 \times F(x) + 0.09 \times (x + 473) + 0.09 \times F(x + 873) + 0.81 \times F(x + 1346) \quad (7.5)$$

The load duration curve and two net load duration curves are shown in Fig. 7.11, for peak load situations, when nuclear is 100 and 90% available. We will now define one new data-set, for the case with nuclear power, which changes Data–1.

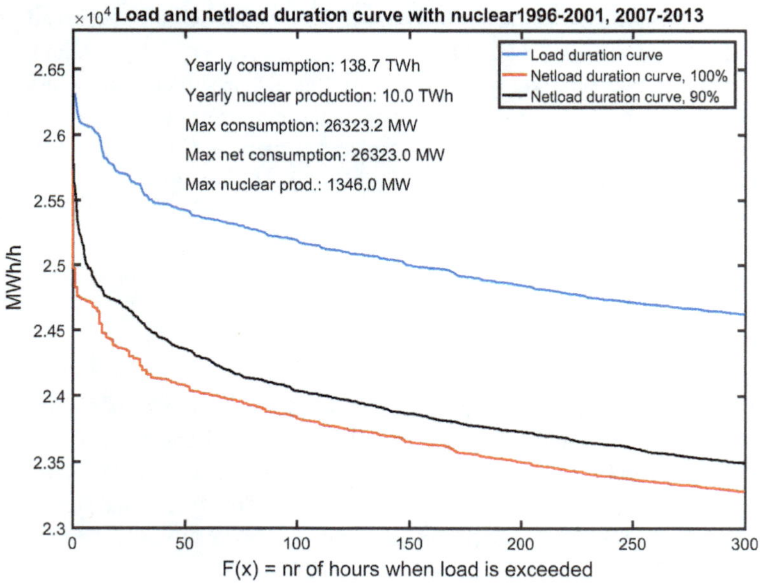

Fig. 7.11 Load and net load duration curves with nuclear power availability of 100 and 90%

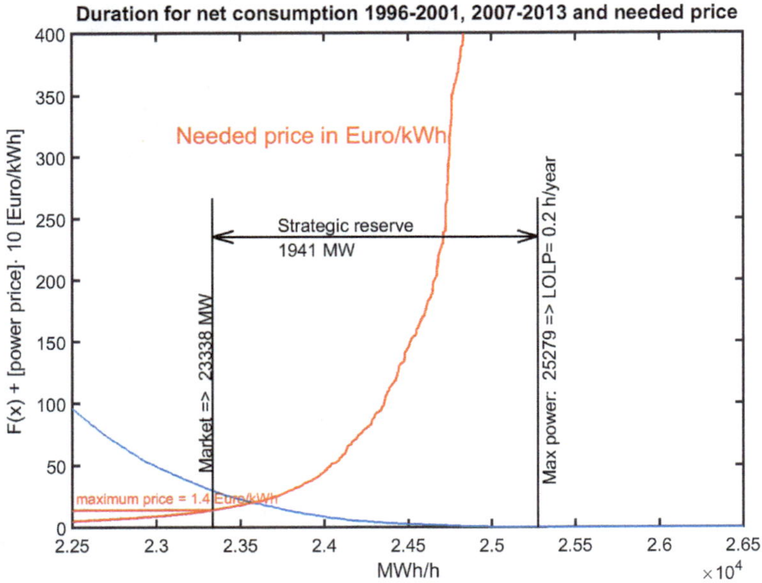

Fig. 7.12 Amount of needed strategic reserve in a system with *LOLP* = 0.2 h/year, max price =1.4 k€/MWh and 2 nuclear stations

Data–3b. Assume the same as for Data–1b, i.e., the maximum accepted price is 1.4 k€/MWh, the accepted *LOLP* is 0.2 h/year, i.e., 1 h every 5th year, and OCGTs are the most expensive market financed power. It is also assumed that the price to the market for the strategic reserve is not lower than the needed market price to finance the last market financed MW of power. The load is the Swedish load for the selected period and the nuclear power is as described above. The amount of strategic reserve is shown in Fig. 7.12.

7.8 Conclusion

This chapter analyzed three different cases about the amount of strategic reserve and how that amount is influenced by the installation of wind and nuclear power. The results are shown in Figs. 7.4, 7.10 and 7.12. We now comment on the capacity credit for the different cases, defined as a small modification of the traditional "effective load carrying capability" [19]. The version used here is the "equivalent firm capacity", calculated as follows:

Table 7.6 Amount of strategic reserve and capacity credit for 9 TWh of wind or nuclear power (maximum price = 1.4 €/kWh and accepted *LOLP* = 0.2)

Example	Figure	Power/Energy	Market investment	Strategic reserve	Total capacity	Capacity credit (MW)
Only load	7.4	—	24,497	1681	26,178	—
Wind power	7.10	4000 MW, 9 TWh	23,717	2018	25,735	443
Nuclear power	7.12	1346 MW, 9 TWh	23,338	1941	25,279	899

1. Assume that a power plant has the capacity X (MW).
2. Calculate the *LOLP* of the system without this power plant $\Rightarrow LOLP_1$.
3. Install the X (MW) power plant and calculate the *LOLP* $\Rightarrow LOLP_2$.
4. Decrease the amount of installed capacity in the system with Y (MW) until $LOLP = LOLP_1$.
5. This means that the X (MW) power plant has a capacity credit of Y (MW).

The data from the figures can be summarized and the results are presented in Table 7.6. From the table, the following can be summarized:

- 4000 MW of wind power decreased the needed amount of installed capacity from 26,178 to 25,735 MW while keeping the same $LOLP = 0.2$ h/year. This means a capacity credit of $26{,}178 - 25{,}735 = 443$ MW, which is 11% of the installed capacity.
- Wind power decreases the amount of market financed power with $24{,}497 - 23{,}717 = 780$ MW, but increases the needed strategic reserve with 337 MW. This depends on the net-load curve, which is slightly "sharper" than the load curve.
- 1346 MW of nuclear power decreased the needed amount of installed capacity from 26,178 to 25,735 MW while keeping the same $LOLP = 0.2$ h/year. This means a capacity credit of $26{,}178 - 25{,}279 = 899$ MW, which is 66% of the installed capacity.
- Nuclear power decreases the amount of market financed power with $24{,}497 - 23{,}338 = 1159$ MW, but increases the needed strategic reserve with 260 MW. This depends on the net-load curve (still slightly "sharper" than the load curve). But nuclear power varies slightly less (81% of time is installed capacity available), so the increase of strategic reserve is smaller than the increase caused by wind power (260 MW instead of 337 MW).
- When one compare the capacity credit values, it looks that wind power capacity credit is 6 times lower than nuclear (11% compared to 66%). If one compares the MW level, i.e., the capacity credit for the same yearly energy production, then wind power capacity credit is around 2 times lower (443 instead of 889 MW).

We now recalculate all data in Table 7.6 using $LOLP = 0.1$, i.e., one hour in 10 years, and an accepted maximum price of 3 €/kWh. The results are shown in Table 7.7. From the table, the following can be summarized:

Table 7.7 Amount of strategic reserve and capacity credit for 9 TWh of wind or nuclear power (maximum price = 3 €/kWh and accepted *LOLP* = 0.1)

Example	Power/Energy	Market investment	Strategic reserve	Total capacity	Capacity credit (MW)
Only load	—	24,936	1387	26,323	—
Wind power	4000 MW, 9 TWh	24,127	1683	25,810	513
Nuclear power	1346 MW, 9 TWh	23,807	1735	25,542	781

- More capacity is needed in all cases. This depends on a higher *LOLP* requirement.
- The market will make higher investments in all cases. This depends on the higher accepted price.
- The needed amount of strategic reserve decreased in all cases. This depends on the higher accepted price, which had a higher impact on the requested volume than the stricter *LOLP* requirement.
- 4000 MW of wind power increased the capacity credit from 443 MW (11%) to 513 MW (13%). This result indicates the challenge of having long time series in order to get a realistic estimation of the wind power availability during high load.
- 1346 MW of nuclear power decreased the capacity credit from 899 MW (66%) to 781 MW (58%). This depends on the stricter LOLP requirement, which in general decreases the capacity credit for thermal stations (see, e.g., [20]).

This chapter evaluated the basic principles for the volume of the strategic reserve and how this is strongly connected to the required reliability level and accepted maximal price. Also, the chapter showed that the price that the strategic reserve of the market uses when it bids into the market has an important impact on the interest of the non-subsidized part of the market to make investments. Furthermore, it analyzed the impact of wind power on the volume of needed market investments and strategic reserve.

References

1. Bhagwat, P, De Vries, L.: The effect of German strategic reserves on the Central European electricity market. In: 10th International Conference on the European Energy Market (EEM), pp. 1–7. IEEE (2013)
2. Söder, L.: Who should be responsible for generation capacity addition? In: International Conference on Electric Utility Deregulation and Restructuring and Power Technologies (DRTP 2000), pp. 296–301. IEEE (2000)
3. Söder, L.: Analysis of pricing and volumes in selective capacity markets. IEEE Trans. Power Syst. **25**(3), 1415–1422 (2010)
4. Vázquez, C., Rivier, M., Pérez-Arriaga, I.: A market approach to long-term security of supply. IEEE Trans. Power Syst. **17**(2), 349–357 (2002)

5. IEA: Implementing agreement on demand-side management technologies and programmes. 2007 Annual Report, International Energy Agency Demand-Side Management Programme (2008). http://www.ieadsm.org/. Accessed 10 Feb 2017
6. Baldick, R., Helman, U., Hobbs, B., O'Neill, R.: Design of efficient generation markets. Proc. IEEE **93**(11), 1998–2012 (2005)
7. Zhou, H., Tu, Z., Talukdar, S., Marshall, K.: Wholesale electricity market failure and the new market design. In: IEEE Power Engineering Society General Meeting 2005, pp. 503–508. IEEE (2005)
8. Crampton, P., Stoft, S.: Forward reliability markets: less risk, less market power, more efficiency. Util. Policy **16**, 194–201 (2008)
9. Svenska kraftnät: purchase of capacity for winters up to 2015 (in Swedish). http://www.svk.se. Accessed 10 Feb 2017
10. Chao, H.: Capacity markets in NYISO and ISO-NE. In: IEEE Power Engineering Society General Meeting 2007, pp. 1–4. IEEE (2007)
11. RTE: French capacity market report accompanying the draft rules. Technical report, Réseau de Transport d'électricité (2014). http://www.rte-france.com/sites/default/files/2014_04_09_french_capacity_market.pdf. Accessed 10 Feb 2017
12. Hobbs, B., Hu, M., Inon, J., Stoft, S., Bhavaraju, M.: A dynamic analysis of a demand curve-based capacity market proposal: the PJM reliability pricing model. IEEE Trans. Power Syst. **22**(1), 3–14 (2007)
13. Hobbs, B., Inon, J., Hu, M., Stoft, S.: Capacity markets: review and a dynamic assessment of demand-curve approaches. In: IEEE Power Engineering Society General Meeting 2005, pp. 514–522. IEEE (2005)
14. Sener, A., Kimball, S.: Reviewing progress in PJM's capacity market structure via the new reliability pricing model. Electr. J. **20**(10), 40–53 (2007)
15. Liu, G., Xu, H., Xiao, L., Zeng, M.: The optimal generation capacity and investment incentives of peak units. In: International Conference on Electric Utility Deregulation and Restructuring and Power Technologies (DRPT 2008), pp. 1293–1297. IEEE (2008)
16. Nohlgren, I., Svärd, S., Jansson, M., Rodin, J.: Electricity from new and future plants 2014. Elforsk report 14:45, Stockholm, Sweden (2014)
17. Baleriaux, H., Jamoulle, E., Guertechin, F.: Simulation de l'exploitation d'un Parc de Machines Thermiques de Production d'électricité Couplé á des Stations de Pompage, Revue E (edition SRBE) **v**(7), pp. 225–245 (1967)
18. Magnusson, M., Krieg, R., Nord, M., Bergström, H.: Effektvariationer av vindkraft (in Swedish), Elforsk report 04:34. Stockholm, Sweden (2004)
19. Garver, L.: Effective load carrying capability of generating units. IEEE Trans. Power Appar. Syst. **85**(8), 910–919 (1966)
20. Amelin, M.: Comparison of capacity credit calculation methods for conventional power plants and wind power. IEEE Trans. Power Syst. **24**(2), 685–691 (2009)

Part III
Agent-based Simulation of Electricity Markets with Increasing Levels of Variable Generation: Traditional and New Design Elements

Chapter 8
MATREM: An Agent-Based Simulation Tool for Electricity Markets

Fernando Lopes

Abstract This chapter presents the key features of an agent-based simulation tool, called MATREM (for Multi-Agent TRading in Electricity Markets). The tool allows the user to conduct a wide range of simulations regarding the behavior and outcomes of electricity markets (EMs), including markets with large penetrations of renewable energy. In each simulation, different autonomous software agents are used to capture the heterogeneity of EMs, notably generating companies (GenCos), retailers (RetailCos), aggregators, consumers, market operators (MOs) and system operators (SOs). The agents are essentially computer systems capable of flexible, autonomous action and able to interact, when appropriate, with other agents to meet their design objectives. They are able to generate plans and execute actions according to a well-known practical reasoning model—the belief-desire-intention (BDI) model. MATREM supports two centralized markets (a day-ahead market and an intra-day market), a bilateral market for trading standardized future contracts (a futures market), and a marketplace for negotiating the terms and conditions of two types of tailored (or customized) long-term bilateral contracts: forward contracts and contracts for difference. The tool is currently being developed using both JADE—the JAVA Agent DEvelopment framework—and Jadex—the BDI reasoning engine that runs over JADE, enabling the development of BDI agents. A graphical interface allows the user to specify, monitor and steer all simulations. The human-computer interaction paradigm is based on a creative integration of direct manipulation interface techniques with intelligent assistant agents. The target platform for the system is a 32/64-bit computer running Microsoft Windows.

F. Lopes (✉)
LNEG–National Laboratory of Energy and Geology,
Estrada do Paço do Lumiar 22, 1649-038 Lisbon, Portugal
e-mail: fernando.lopes@lneg.pt

F. Lopes and H. Coelho (eds.), *Electricity Markets with Increasing Levels of Renewable Generation: Structure, Operation, Agent-based Simulation, and Emerging Designs*,
Studies in Systems, Decision and Control 144, https://doi.org/10.1007/978-3-319-74263-2_8

8.1 Introduction

The computer-controlled operating processes at most real-world applications (e.g., automated factories and telecommunication centres) are complex by nature. As this complexity grows, it will be increasingly difficult to control such applications with centralized management and scheduling policies that are both robust in the face of unexpected events and flexible at dealing with operational and environmental changes that might occur over time. The inherent distribution of competence, control, and information, as well as the complexity of the theoretical problems underlying such applications call for new ways of domain modeling and problem-solving. The ability of different autonomous components to interact, exchange information, and coordinate their actions—that is, interoperability—is often considered a central requirement. Accordingly, a promising approach to model and analyze the behavior of real-word applications consists in distributing—along such dimensions as space and function—the control of operations to a number of software agents. Intelligent agents and multi-agent systems (MAS) are indeed a relatively new way to conceptualize and implement complex systems (see, e.g., [1–3]).

Traditionally, the electric power industry was dominated by a few, often publicly owned, entities with monopoly control over both large power generation plants and power grids. Increasingly, however, this centralized paradigm has been replaced by deregulated and unbundled markets with a large number of stakeholders involved in power generation, transmission, distribution, and network development. Policy makers have introduced wholesale and retail markets and encouraged different forms of transaction that compelled market entities to compete both for generation and for serving customers (see, e.g., [4, 5]). Conceptually, the agent-based approach is an ideal fit to the naturally distributed domain of EMs. As a result, agent-based simulation of EMs has attracted considerable attention over the last years and a number of simulation tools have emerged (e.g., SEPIA [6], EMCAS [7], NEMSIM [8], AMES [9], PowerACE [10], MASCEM [11, 12], and GAPEX [13]; see also [14–17] for comprehensive reviews and critical surveys).

The share of variable generation (VG), such as wind and photovoltaic solar power, in the energy mix, has increased significantly in recent years. The European Union (EU) has been one of the major drivers of the development of renewable energies. The Renewable Energy Roadmap [18] kick-started the real growth in the renewable energy sector. The EU Renewable Energy Directive [19] has laid the ground for a policy framework on renewable energy sources until 2020. The core of the legislative framework involves binding targets for the production of renewable energy (RE) of individual member states, and also requires priority grid access for such production. However, the rules regulating the integration of RE in the marketplace are neither mandatory nor quantified. Even though there is potential for a large penetration of RE in the European grid infrastructure, the electricity markets of individual members are still designed for power systems based on predictable production from centralized, conventional sources, like coal and nuclear power.

At present, it is unclear whether or not existing market designs cater well to the needs of the increasingly complex and interdependent power systems with high penetrations of variable generation [20]. There is, therefore, a growing need to monitor the potential impacts of RE to determine if existing designs are still effective. Furthermore, and although the agent-based approach presents itself as an advanced modeling approach to simulate the behavior of power markets over time, most existing agent-based tools for EMs were arguably developed to simulate traditional market models, proposed when the vast majority of generation units were controllable and fuel-based, meaning that production could be shifted in time with limited economic impact.

Against this background, this chapter presents the key features of an agent-based simulation tool for EMs, called MATREM (for Multi-Agent TRading in Electricity Markets). MATREM allows the user to conduct a wide range of simulations regarding the behavior and outcomes of energy markets, including markets with large penetrations of variable generation. The tool is currently being developed using the JAVA Agent DEvelopment framework (JADE) [21], an agent-oriented middleware, offering a framework for developing multi-agent systems, and fully integrated with the JAVA programming language. The remainder of the chapter is organized into the following three major parts:

1. *Overview of the tool.* This part begins by describing the six types of market entities currently supported by MATREM: generating companies, retailers, aggregators, consumers, market operators and system operators. It then presents the two exchanges simulated by the tool: a power exchange, comprising a day-ahead market and an intra-day market, and a derivatives exchange, comprising a futures market for trading standardized bilateral contracts. Finally, the part describes the marketplace supported by MATREM: a bilateral marketplace for negotiating the details of tailored (or customized) long-term contracts, particularly forward contracts and contracts for difference.
2. *The user interface.* This part is devoted to the graphical user interface of the system. First, the major interface functions are discussed. Next, the two main styles of human-computer interaction are briefly introduced: direct manipulation and indirect management via interface agents. Following this introductory material, some important design choices are presented and analyzed. Finally, the human-computer interaction paradigm adopted within this work is described in detail.
3. *Autonomous software agents.* This part begins by describing the various types of agents defined in MATREM, notably assistant agents and market agents, and then presents a Unified Modelling Language (UML) class diagram showing the relationships among the different agent types. Next, it delves into the technical details of the conceptual agent model. First, the model initially adopted—a "traditional" deliberative model—is presented and then the model currently being implemented—a simplified belief-desire-intention (BDI) model—is described in detail. The part closes with details of the implementation of the BDI model.

8.2 Overview of the Tool

MATREM relies on multiple autonomous agents to simulate the commercial functions of wholesaling (essentially sales to resellers) and retailing (basically sales to final customers). The current version of the tool supports two centralized markets (a day-ahead market and an intra-day market) and a bilateral market for trading standardized future contracts (a futures market).[1] Accordingly, the tool supports a (power) exchange in which supply bids and demand offers are aggregated to find a clearing price at which supply and demand are equal. Also, the tool supports a (derivatives) exchange where private parties can trade standardized contracts. This exchange uses an electronic trading system that automatically matches the bids and offers from various market participants. Worthy of mention is the possibility of market participants to negotiate the details of tailored (or customized) long-term bilateral contracts, specifically contracts designed to cover the delivery of large amounts of energy over long periods of time. The parties are able to negotiate the terms and conditions of both forward contracts and contracts for difference (see Fig. 8.1).

Furthermore, the tool supports four key types of agents participating in the aforementioned markets: generating companies (GenCos), retailers (RetailCos), aggregators and consumers.[2] It also supports two key types of agents responsible for all markets: market operators (MOs) and system operators (SOs). The target platform for MATREM is a 32/64-bit computer running Microsoft Windows. A graphical interface allows the user to specify, monitor and steer all simulations. The remainder of this section is structured as follows. Section 8.2.1 describes the entities that take part in the different markets, Sect. 8.2.2 presents the centralized day-ahead and intraday markets, Sect. 8.2.3 introduces the market where standardized bilateral contracts are traded (referred to as futures market), and finally Sect. 8.2.4 introduces the main phases involved in the negotiation of tailored long-term bilateral contracts.

8.2.1 Key Market Entities

Generating company agents (GenCos) own the power plants that are in charge of producing electrical energy. They may own a single plant or a portfolio of plants of different technologies. Non-dispatchable GenCos are agents with non-dispatchable sources, such as solar-thermal or wind power plants (see Fig. 8.1).

---

[1]Other markets are expected to be available soon. In particular, a market for trading standardized option contracts is currently under development. Also, future work aims at extending the simulation tool by incorporating a market to match the imbalances caused by the variability and uncertainty present in power systems.

[2]To date, our focus has been on four key types of market participants. One important area for future work is to consider other types of traditional power industry agents, including transmission company agents (TransCos) and distribution company agents (DistCos).

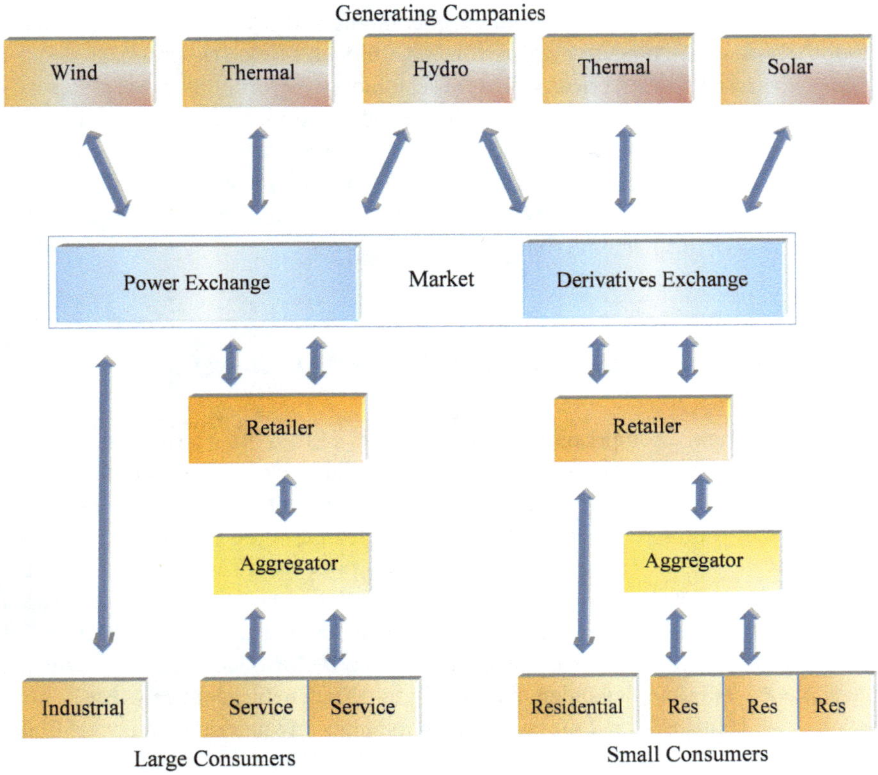

Fig. 8.1 Simulation tool: power and derivatives exchanges and key market entities

GenCos strive to maximize their profit from selling energy either in the organized markets or directly to retailers and consumers through bilateral contracts. Retailer agents (RetailCos) do not own production units and need to purchase all the electric power needed to provide energy to their clients. They buy energy in the organized markets and re-sell it to end-use customers who are not allowed, or do not want, to participate in these markets. Also, they may deal indirectly with end-use customers through aggregators—that is, agents that support groups of customers in trading electrical energy. The key objective of RetailCos is to maximize the profit obtained by selling energy to their clients. Since clients can change retailer when they are offered a better price, RetailCos try to buy energy as cheap as possible to provide clients with the lowest possible prices. Large consumers can take an active role in the market by buying electrical energy through the centralized markets, or alternatively they may sign customized (or tailored) bilateral contracts with producers or may be supplied by retailers. They aim at minimizing their procurement cost (or maximizing the utility obtained from electricity usage). Small consumers either buy energy (directly) from retailers or deal (indirectly) with retailers through aggregators (see Fig. 8.1).

Aggregators are essentially entities that combine end-use customers into buying groups pursuing the objective of buying large blocks of electric power at cheaper prices (when compared to prices for single customers). Accordingly, the tool supports coalitions of end-user customers—that is, various customers ally into coalitions to enhance the likelihood of achieving their individual outcomes (i.e., covering their needs at better energy prices). In other words, two or more customers intentionally form a coalition—represented by an aggregator—who interacts with a retailer in search for a superior outcome (but see [22, 23]).

8.2.2 *Day-Ahead and Intra-day Centralized Markets*

The power exchange comprises a day-ahead market and a shorter-term market known as intra-day market. The day-ahead market (DAM) buys energy from sellers and sells energy to buyers in advance of time when the energy is produced and consumed (see, e.g., [24, 25]). For a given day d, the DAM is cleared in the previous day $(d-1)$, typically at 12 noon.[3] The intra-day market sets prices and schedules a few hours ahead to facilitate balancing on advance of real time. For day d, the intra-day market is cleared several times once the DAM has been cleared (e.g., at 7 p.m. and 10 p.m. of day $d-1$, and 2 a.m., 5 a.m., 9 a.m. and 1 p.m. of day d).[4] Producers that deal with the intermittency and time-dependent nature of non-dispatchable sources—that is, non-dispatchable GenCos—tend to rely more on the intra-day market than conventional producers, since the time that elapses between the closure of this market and the delivery of electrical energy allows a higher accuracy on actual power production forecasts.

Most energy transactions take place in the day-ahead market, i.e., the intra-day market is mainly used to make adjustments in the positions of market participants as delivery time approaches. Accordingly, we present next a description of the operation of the DAM only.[5] Generating company agents submit bids to sell energy at some price for every hour of the market horizon. Also, retailer agents and large consumers submit offers to purchase electricity at some price for every hour of the market horizon. A market operator agent then collects the bids and offers and sort them according to the price. In a next step, the marker operator clears the market and determines the prices and traded quantities. All bids to sell energy with prices lower than or equal to the market-clearing price are accepted. Likewise, all offers to buy energy with prices greater than or equal to the market-clearing price are accepted.

---

[3]The gate closure time of the DAM is a simulation parameter. Thus, the tool allows the user to specify market simulations involving different gate closures (e.g., 10 a.m. or 6 p.m. of the day before the day of operation).

[4]The number of intra-day market sessions is also a simulation parameter, meaning that the tool allows the user to phase the intra-day market into different sessions (e.g., only two sessions a day, or twenty-four sessions a day, one for each hour).

[5]The operation of the intra-day market is essentially identical to that of the day-ahead market, and is therefore omitted.

Both system marginal pricing (SMP) and locational marginal pricing (LMP) are supported. Under SMP, a more detailed description of the market operation is as follows:

- GenCos submit bids to sell energy in the form of price and quantity pairs—that is, bids to supply a certain amount of energy at a certain price for each of the 24-hours of the day of operation (i.e., 24 price/quantity pairs).
- RetaiCos and large consumers submit offers to buy a certain quantity of energy at a specific price for each of the 24-hours of the day under consideration.[6]
- The market operator collects the two-part bids, ranks them from the lowest price to the highest price, and builds a supply curve (i.e., a curve showing the price as a function of the cumulative quantity). Also, if the previous step was not omitted, this agent collects the two-part purchase offers, ranks them from the highest energy price to the lowest energy price, and builds a demand curve.
- The market operator agent clears the market and determines the market-clearing prices and the traded volumes. For each hour, the market-clearing price is defined by the intersection of the supply and demand curves. The traded volume is the sum of all energy quantities specified in the purchase offers that are satisfied at the market-clearing price.
- The market operator instructs GenCos to produce the energy corresponding to their accepted bids. Similarly, this agent informs RetailCos and large consumers of the amount of energy that they are allowed to draw from the system.
- The market operator manages a simple invoicing process. All trades between GenCos and RetailCos or large consumers are invoiced as if the agreed quantities have been delivered exactly.[7]

Thus, SMP involves simple bids/offers consisting of price and quantity pairs and does not take into consideration transmission constraints.

Locational marginal pricing is more complex than system marginal pricing. Basically, LMP involves the pricing of electrical energy according to the location of its injection or withdrawal from the transmission grid. In practical terms, the grid is modeled as a balanced three-phase network with $M \geq 1$ branches and $N \geq 2$ nodes. Each pair of nodes is assumed to be connected by a linked branch path consisting of one or more branches. Kirchoff's Current Law is also assumed to hold for the grid for each hour of operation (see, e.g., [26] for details about this law).[8]

[6]The simulation tool allows the user to omit this step if necessary or desired. Specifically, the user may assume that demand is (highly) inelastic and set according to a load forecast. The demand curve is then a vertical line defined by simply considering the value of the load forecast.

[7]Future work aims at extending the tool by incorporating a detailed invoicing and settlement process. The user will then be able to simulate forward payments from buyers to sellers following the delivery of energy. GenCos will be paid the market-clearing price for every megawatt-hour that they will produce, whereas RetailCos and large consumers will pay this price for every megawatt-hour that they will consume.

[8]Under LMP, the user needs to specify the location of the various GenCos and RetailCos at the various nodes of the transmission grid.

GenCos submit bids to sell energy in the form of marginal cost functions defined over feasible production intervals—that is, bids consisting of a linear function ($a' + 2b' \times P$) defined over a feasible interval ($LP' \leq P \leq UP'$) for each of the 24-hours of the day of operation, where a' and b' are the reported cost coefficients, P is the hourly production level, and LP' and UP' are the reported lower production limit and upper production limit, respectively.[9] These bids can be strategic in the sense that a' and b' can deviate from the true cost coefficients (a and b). Also, LP' and UP' can deviate from the true feasible production interval (defined by LP and UP).

The offers to buy energy of the RetailCo agents may involve two different parts: a fixed demand part and a price-sensitive demand part. The fixed demand part is essentially a 24-h load profile. The price-sensitive demand part consists in demand functions for each of the 24-hours of the day under consideration. Each function takes a linear form ($c' - 2d' \times D$) and is defined over a purchase capacity interval ($LD' \leq D \leq UD'$), where c' and d' are the demand coefficients, D is the hourly power load, and LP' and UP' are the lower and upper load limits, respectively.[10] The market operator collects all bids to buy energy and all offers to sell energy and runs an Optimal Power Flow (OPF) procedure that defines the optimal power commitments and the locational marginal prices for the day of operation. "Optimal" here means that the total net surplus is maximized.

8.2.3 *Futures Market: Standardized Bilateral Contracts*

The derivatives exchange comprises a futures market for trading standardized bilateral contracts. The futures market provides both financial and physical products that span from days to several years and allow market participants to hedge against the financial risk inherent to day-ahead and intra-day prices. The products are base and peak energy contracts, where base indicates 24-hours of each day of the corresponding derivative time span, and peak indicates peak hours, normally from 8 a.m. to 8 p.m., of each weekday of the time span. Typical examples are as follows: base and peak daily future contracts (one-day delivery period, starting on one of the next 3 days), base and peak weekly future contracts (for one of the next 3 weeks), base and peak monthly future contracts (for one of the next 6 months), base and peak quarterly future contracts (for one of the next 6 to 7 quarters), and base and peak yearly future contracts (for one of the next 3 or 4 years).

---

[9]The supply bids of the GenCo agents are modeled as linear functions, relating money and power. Although other functions are discussed in the literature on energy markets (e.g., quadratic functions), we note that the implications of the supply bid format for the operation of EMs is an important topic that requires further research (but see Chap. 2).

[10]For simplicity, the tool allows the user to model RetailCos as non-strategic agents servicing price-insensitive loads only—that is, the demand serviced by RetailCos may exhibit a negligible price sensitivity and thus the price-sensitive demand part of the offers to buy energy may be omitted. Alternatively, the user may omit the fixed demand part of the offers.

Financial and physical standardized contracts involve the trade of electrical energy, in a standardized quantity and quality, on a predefined date and place, at a price agreed in the present.[11] Both types of contracts are market products that may include the following basic specifications: (i) nominal, (ii) form of quotation, (iii) tick and tick value, (iv) trading period (first and last trading days), (v) method for determining the settlement price (during the trading period), (vii) price change limits, and (v) delivery period (first and last delivery days). The nominal (or nominal value) corresponds to the "fixed" energy quantity (e.g., $7 \times 24\,\text{h} \times 1\,\text{MW} = 168\,\text{MWh}$, for a base week contract). The price quotation is € per MWh (although $ per MWh can also be considered). The tick (or tick size) is the minimum price change (typically, 0.01 €/MWh) and the tick value is the minimum contract's value change, i.e., the nominal times the tick size (e.g., $168 \times 0.01 = 1.68$ €, for a base week contract).

The trading period is the period comprised between the first and the last day on which contracts are admitted for trading on a continuous basis.[12] During this period, market participants may introduce, modify or cancel orders on contracts. Buy and sell orders likely to interfere with each other, immediately and individually, generate transactions and give rise to specific prices (known as transaction prices or, less frequently, contract prices). Transactions are essentially trades involving particular contracts and generate positions after registration in the system. On a daily basis, the system considers a settlement price (SP) for each traded contract (the settlement price is also referred to as the trading reference price). The SP of financial contracts (e.g., base weekly financial future contracts) is equal to the SP of the corresponding physical contracts (e.g., base weekly physical future contracts). Trades on individual contracts are subject to price variation limits—that is, the price of any transaction must not exceed a value (positive and negative) when compared with the previous SP. At present, contracts are not subject to a mark-to-market procedure during the trading period, i.e., a daily settlement of the profit and losses due to price movements (but see, e.g., [27]).

The delivery period is the period following the trading period, for the settlement of positions, and in case of physical contracts, for the physical delivery of electrical energy (see below). Contracts specify the trading and delivery periods by defining the following days: first trading day (FTD), last trading day (LTD), first delivery day, and last delivery day. These four days are typically defined by specifying some of them and then determining the others according to a set of normative rules. For example, base daily future contracts specify the delivery day (the delivery period starts at 00:00 and ends at 24:00 of the delivery day). The trading period is then determined as follows: LTD is the trading day preceding the delivery day and FTD is the last trading day of the previous week (the trading period is the period comprised between FTD and LTD, both included).

[11] Financial future contracts involve the notional supply of electrical energy—that is, the delivery is purely financial based on a reference price. On the other hand, physical future contracts involve the real supply of electricity at constant power (e.g., 1 MW) during all the hours of the delivery period.

[12] Contracts are traded in a continuous mode. Future work aims at extending the tool to support auctions during the trading period to achieve more flexibility.

Market participants enter orders involving either bids to sell energy or offers to buy energy in the trading platform. Sellers and buyers do not have any information about the identity of each other, although some details of the submitted orders may be observed by all participants (e.g., prices and quantities).[13] Typically, to introduce orders, participants indicate a particular type of contract (e.g., a base weekly futures contract), the respective quantity, and the price. More specifically, at least two of the following elements should be specified for each order: (i) the nature (buy or sell), (ii) the contract, (iii) the type, (iv) the price conditions, and (v) the validity period. Both buy and sell orders relate to the standard contracts admitted to trading on the market and are of the following type: limit orders. This means that orders can be executed at their specified limit price or at a best price—that is, a price greater than the specified price, if they involve bids (to sell), or lower than the specified, if they involve offers (to buy). Orders remain active according to one of the following validity periods: good for day (orders are valid until the end of the activity period of the trading day in which they were introduced), or good till date (orders are valid until the date and time indicated).

The introduction of orders is time stamped after validation by the trading platform.[14] Also, orders may be modified with respect to the price and the quantity.[15] A change in the price or an increase in the quantity leads to a new introduction time. On the other hand, a reduction in the quantity does not affect the time at which the orders are introduced. The trading platform automatically and continuously matches the buy and sell orders for each contract likely to interfere with each other. The operation of the platform complies with the following two criteria for the execution of orders: the time-price criterion and the time-time criterion (which is subordinated to the previous one). Firstly, the time-price criterion, according to which, either in the buy or in the sell side, the orders are executed at the most favorable price. Secondly, the time-time criterion, according to which all the orders at the best price are chronologically executed in accordance with the time assigned by the trading platform. All orders submitted and the subsequent changes are registered by the trading platform until they are executed, expired or canceled.

Matching buy and sell orders generate transactions—that is, trades executed on contracts—and open positions after registration. A position is essentially a set of rights and obligations inherent to a transaction. Positions opened on specific contracts may be closed by carrying out the opposite operations—that is, if market participants initially sold, then they need to buy, or conversely, if participants initially bought, then they need to sell—on the same contracts.

---

[13]The trading platform supports anonymous operations on contracts only. The underlying anonymity model is widely used and often considered very useful (see, e.g., [27]). Market participants can formulate expectations relative to price variations based on specific strategies without discriminating between different agents, creating conditions for determining fair energy prices. Also, since all bids and offers are public, participants may exploit eventual disparities resulting from the "gap" between the supply and demand of electricity.

[14]Orders specifying a price out of the price variation limits are not accepted by the platform.

[15]Price is the main negotiable element of the standardized future contracts. However, if desirable or even necessary, the parties may increase or reduce the energy quantity.

The positions that remain opened at the end of the activity period of the last trading day are considered firm and definitive for settlement during the delivery period. Financial positions are subject to a purely financial settlement. To this end, the market operator agent determines a delivery settlement value (DSV) daily, by considering a formula analogous to the formula used in the MIBEL Derivatives Market [28].[16] Briefly, for each delivery day, the system defines a spot reference price (SRP) based on the PTEL base index, which is equivalent to the arithmetic mean of MIBEL's clearing prices (but see [29]). DSV then results from the difference between the SRP and the SP (the settlement price) of each contract on the last trading day, for the number of hours of each day under consideration. Physical positions are subject to a similar financial settlement as well as a physical settlement.

8.2.4 Tailored Long-Term Bilateral Contracts

The simulation tool allows market participants to negotiate all the details of two different types of tailored (or customized) long-term bilateral contracts: forward contracts (see, e.g., [30, 31]) and contracts for difference (see, e.g., [32, 33]).[17] Such contracts are frequently designed to cover the delivery of large amounts of energy (hundreds of MW) over longs period of time (several months to several years). Their terms and conditions are very flexible and can be negotiated privately to meet the objectives and needs of the negotiating parties.

Both types of contracts include several basic items, notably: (i) starting date and time, (ii) duration or length, (iii) constant price over the length of the contract, (iv) fixed amount not to exceed (or in some special cases, variable amount with minimum and maximum limits), and (v) length of commitment for energy. In a more general form, the price and quantity may be time-varying over the contract duration—that is, the contract specifies the provision of different amounts of energy for different blocks of time, at somewhat different prices. This generalization accounts for time-varying tariffs that reflect the value and cost of electricity in different periods of a 24 hour day—that is, a two-rate tariff (peak/off-peak), three-rate tariff (peak/mid-peak/off-peak), four-rate tariff (peak/mid-peak/off-peak/super off-peak), or even more refined tariffs (e.g., a hour-wise tariff, involving different prices for each hour).

---

[16]The financial settlement based on a DSV value applies exclusively to existing positions on daily, weekly and monthly financial future contracts.

[17]Arguably, most real-world long-term contracts are forward contracts between retailers and end-use customers. Standardized long-term contracts for differences (CFDs) have recently started to be used as a mechanism to support renewable generation (see, e.g., [34]). This work aims at going one step beyond by considering tailored long-term CFDs. Accordingly, the current version of the tool allows market participants (e.g., GenCos, RetailCos and large customers) to negotiate any CFD terms that are deemed appropriate. Tailored CFDs allow market participants to take part in the centralized day-ahead market, while insulating them from the market-clearing prices (i.e., they provide a hedge against price volatility in the DAM).

Forward bilateral contracts are agreements between two parties to exchange a specific amount of electric power at a certain future time for a specific price. Such contracts are typically negotiated weeks or months prior to their delivery and involve physical obligations—that is, they specify the physical participants that generate and consume the power agreed to as well as the buses of injection and consumption.

Contracts for difference are agreements in which each party ensures the other against discrepancies between the contract price (or strike price) and the market-clearing price. Such contracts operate in parallel with the day-ahead market. The trading parties negotiate their terms and conditions (notably the strike price and the energy quantity), and then in a separate transaction they take part in the day-ahead market (like any other participant). Once trading on the DAM is complete, the parties compensate each other for the difference between the strike and the market-clearing prices. Specifically, if the strike price is higher than the market price, the buyer pays the seller the difference between these two prices times the agreed quantity. Conversely, if the strike price is lower than the market price, the seller pays the buyer the difference between these two prices times the agreed quantity.

For both types of contracts (i.e., forward contracts and CFDs), the parties can privately negotiate any terms and conditions in accordance with their objectives and preferences. To this end, buyer and seller agents are equipped with a negotiation model that handles two-party and multi-issue negotiation (see, e.g., [35–37]). The negotiation process involves an iterative exchange of offers and counter-offers. An offer is, essentially, a set of issue-value pairs—such as "energy price = 45 €/MWh", "contract duration = 12 months", and so on—and a counter-offer is an offer made in response to a previous offer. Negotiation proceeds as follows:

- One of the trading parties ("Party A") submits an offer to the other party ("Party B or Opponent") in the first time period.
- Party B receives the offer and can either accept it, reject it and opt out of the negotiation, or reject it and continue bargaining. In the first two cases, negotiation ends. Specifically, if the offer is accepted, negotiation ends successfully and the agreement is implemented. In the last case, negotiation proceeds to the next time period, in which the Opponent makes a counter-offer.
- Party A receives the counter-offer and the tasks described in the previous step are repeated, i.e., Party A may either accept the counter-offer, reject it and opt out of the negotiation, or reject it and continue bargaining.
- The agents continue to bargain in this manner and negotiation may end with either agreement or no agreement. In particular, negotiation may end in a compromise agreement (an agreement on some middle ground on an obvious dimension that links the two parties' initial offers) or an integrative agreement (an agreement that reconciles the two parties' interests). Failure to agree can occur in two ways: (i) either party decides to opt out unilaterally (or the two agree to break off), or (ii) a deadline is reached and the parties do not agree to any offer.

The negotiation of long-term contracts represents a novel and powerful tool for bilateral contracting in competitive electricity markets and, we believe, the main contribution of this work.

8.3 The User Interface

This section is organized as follows. Section 8.3.1 introduces the graphical user interface of the simulation tool and discusses the major interface functions. Section 8.3.2 is divided into three further parts. The first part deals with the design of intelligent user interfaces in general. It introduces the two main styles of human-computer interaction: direct manipulation and indirect management via interface agents. The next two parts are devoted to the design of the MATREM user interface. First, some important design choices are analyzed in the context of existing interaction styles. Next, the paradigm for the interaction between the user and the simulation tool is described in detail.

8.3.1 Main Interface Functions

The user interface is characterized by a collection of widgets—such as tool bars, menus and pop-up windows. The common (or default) screen layout of the system is shown in Fig. 8.2.[18] The top of the screen is reserved for the menu bar (displayed under the title bar of the system window). On the left side, it contains several drop down menus, namely the agents menu, the markets menu, the participants menu and the simulation menu (as well as a menu item allowing the user to exit the simulation tool). The right side contains menu extras that display information relevant to the user (for example the system clock).

According to the menus of the menu bar, the interface incorporates the following key functions:

- agent management: allows the user to create new agents (e.g., new GenCos), load existing agents (e.g., previously created RetailCos) and kill active agents;
- scenario construction: allows the user to select particular markets (e.g., a day-ahead or a forward market) and to specify their participants (e.g., GenCos and RetailCos)[19];
- simulation management: allows the user to start new simulations, steer and monitor active simulations, and save completed simulations;
- report analysis: allows the user to view and analyze simulation results. A variety of simulation-generated data can be displayed in tables and live graphs and/or exported to log files for offline analysis.

[18] The term "common screen layout" refers to the primary windows displayed on the screen when the system starts running. Each window is associated with a specific area (or part) of the screen.

[19] The current version of the system allows the user to indicate the agents participating in a particular market by using the participants menu—that is, the user selects the agents sequentially, one at time, and confirms their bids/offers. Future work will focus on developing a graphical editor for constructing and/or modifying electric power industry scenarios. The editor will allow the user to specify market agents graphically and to define relationships between them (i.e., interconnecting market agents by links representing the power grid, ownership and money flow).

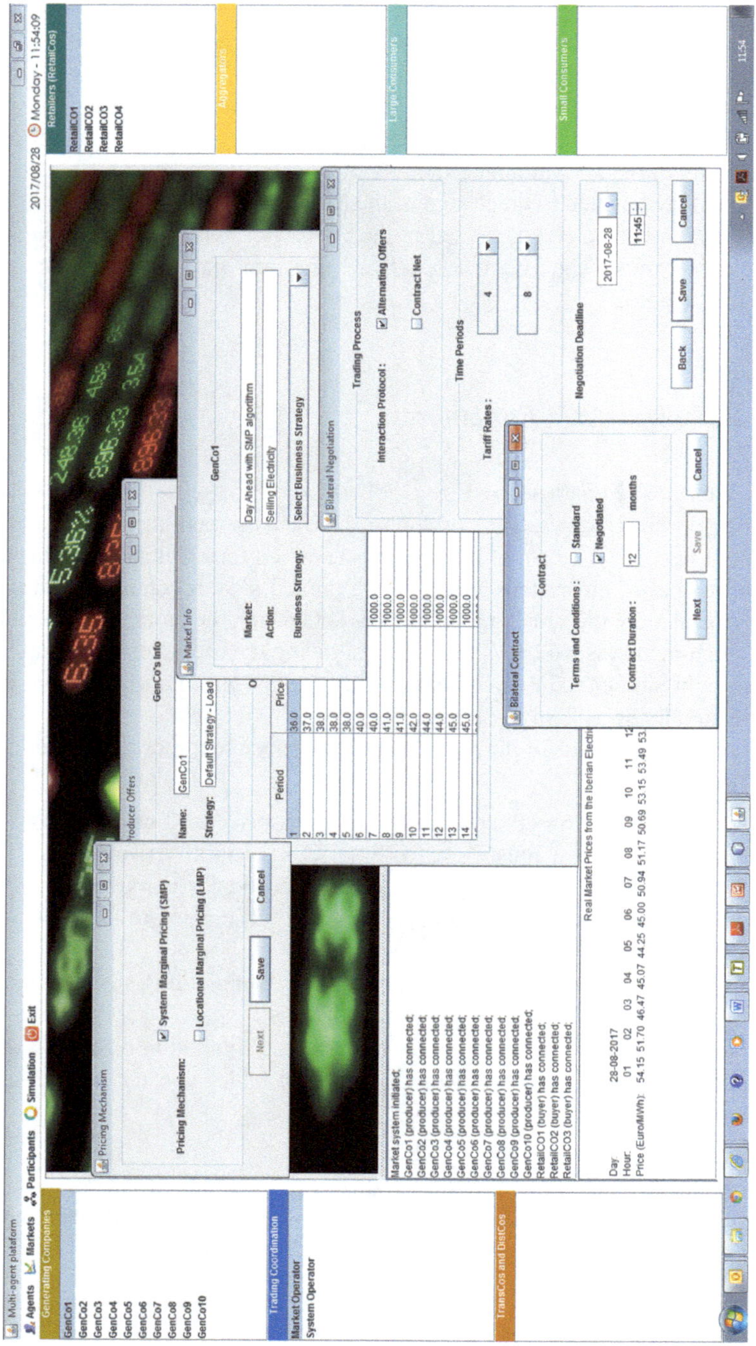

Fig. 8.2 Snapshot of the agent-based system for simulating electricity markets

The area of the screen under the menu bar is divided into ten different parts corresponding to the primary windows of the system. Both the three left-hand windows and the four right-hand windows display active agents—that is, agents that can be selected by the user to participate in a specific market (as well as the market operator and the independent system operator). Specifically, the top left window (GenCo window) shows a list of active generating company agents, the middle left window (trading coordination window) displays the market operator and the system operator, and the bottom left window (TransCo and DistCo window) is reserved for transmission and distribution company agents.[20]

Likewise, the top right window (RetailCo window) shows a list of active retailer agents. The other three right-hand windows are reserved for aggregators and end-use customers. In particular, the middle right window (under the RetailCo window) displays active aggregators—that is, agents that support groups of consumers in trading energy. The other middle right window displays large consumers, i.e., agents that can take an active role in the market. The bottom right window displays small consumers.

The top middle area of the screen contains an image that offers a real-life perspective of the flow of information (and money) typical of competitive markets. Below that, the middle window summarizes the tasks performed by active agents and displays information about the content of key messages exchanged between agents (e.g., the energy prices offered by buyers and sellers during bilateral contracting of electricity).[21] The bottom middle window displays both problem-specific and general-purpose information acquired by software agents acting as "information assistants" (e.g., the daily market-clearing prices of the Iberian Electricity Market).[22] For illustrative purposes, Fig. 8.2 also shows several pop-up windows related to both centralized trading and bilateral contracting of electricity.

8.3.2 *Interface Design: Intelligent Assistance*

8.3.2.1 Human-Computer Interaction: Two Prominent Styles

Generally speaking, the design, implementation and evaluation of a user interface is inextricably linked with a natural and productive human-computer interaction—that is, the quality of the communication between the user and the computer. Two main styles have emerged [38]: direct manipulation and indirect management.

---

[20] An noted earlier, one important area for future work is to develop TransCo and DistCo agents responsible for operating the transmission and distribution systems, respectively.

[21] The information displayed in the middle window is essentially complementary to that provided by the Sniffer agent or the Java Sniffer application (see [21] for details of these two tools).

[22] The simulation tool supports not only market agents (e.g., GenCos and RetailCos), but also a special type of agent referred to as assistant agent (but see below).

Direct manipulation [39] is often considered the most successful style (typical examples include editors and video games). The central ideas are as follows [40]: continuous representation of the objects of interest, rapid and reversible operations whose impact on the objects of interest is immediately visible, labeled button presses or physical actions (e.g., movement and selection by mouse), and a layered or spiral approach to learning that permits usage with minimal knowledge. Systems based on these principles have several beneficial attributes, including [41]: (i) the user can immediately see if his/her actions are furthering the goals (and if not, he/she can change the direction of the activity), (ii) the user gains confidence and mastery because he/she initiates actions, feels in control, and can predict system responses, and (iii) the user experiences less anxiety because the system is comprehensible.

Simply put, the success of direct manipulation relies mainly on the direct and constant feedback about the tasks being performed at any moment. This feedback allows the user to realize changes or corrections in the operations being executed. However, despite its success, direct manipulation is not free of problems. Significant weakness include [41]: (i) the use of spatial or graphic representations of the problems does not necessarily improve performance, (ii) the user must learn the meaning of the components of the graphic representations, and (iii) graphic representations may be misleading and take excessive screen display space. Also, the requirement that the user should perform every task and control all events arising from interaction may lead to cognitive overcharge, particularly in complex applications, resulting in a reduction of the usability of the system [38].

Indirect management [42] is closely related to the concept of software agent. The metaphor used is that of a *personal assistant* who collaborates with the user in the same work environment [43]. Instead of exercising complete control and taking responsibility for every move the computer makes, the user is engaged in a collaborative process in which both him/her and software agents initiate communication, monitor events and procedures, and perform tasks to achieve specific goals.

Software agents playing the role of personal assistants are often able to exhibit goal-direct behavior and take the initiative when they detect situations that are believed to be relevant to the user.[23] They may be endowed with extensive domain-specific knowledge about both the application domain and the user. At run time, they may employ that knowledge to recognize the user's plans and find opportunities for contributing to them. Furthermore, incorporating artificial intelligence (AI) techniques, such as reasoning, planning and learning, into such agents has been considered very helpful (see, e.g., [46, 47]). Some key design principles for successful integration of AI are [48]: analyze what the user is doing (take advantage of the information implicit in the actions of the user to infer his/her goals and interests), suggest rather than act, operate when the user is busy (e.g., thinking about what input to provide next), and do not disturb the user's interaction (the user should always have the possibility to ignore an agent).

[23] *Software assistants* are computer programs that provide assistance to users dealing with computer-based applications [44]. *Autonomous interface agents* are agents capable of operating the interface—or at least part of the interface—in an autonomous way and also act in parallel with the user [45].

A central issue associated with software agents is that the user might feel a sense of loss of control caused by agent autonomy [38]. Other key issues related with the use of software agents include [49]: poor guessing about the goals and interests of the user, inadequate consideration of the costs and benefits of agents' actions, poor timing of agents' actions, and inadequate attention to opportunities that allows the user to guide the invocation of agents' services. Furthermore, little attention has been paid to the following issues [50]: how to best interact with different users and how to provide them assistance of the right sort at the right time.

8.3.2.2 Strategic Choices

The advantages and disadvantages of direct manipulation interfaces and agent interfaces have been the subject of a somewhat heating debate [51]. A fundamental question is whether agents should be presented to the user in the interface of real-world applications. Furthermore, should the user give up control of his/her interaction with the interface? These and other problems associated with intelligent user interfaces can by no means be regarded as solved (see, e.g., [52]). However, as pointed out by some researchers, agents can be considered a complementary technique to well-designed interfaces—visualization and direct manipulation—not a substitute for the tools allowing the user to personally interact with the application [43]. Accordingly, this work considers a creative integration of direct manipulation interface techniques with assistant agents, providing adequate control and responsibility to the user—that is, letting the user be responsible for all the important decisions.

Now, different forms of assistance suit different real-world applications, depending on the balance of expertise and knowledge between the user and the agent-based system [47]. When human problem-solving skills are weak or compromised in some way, intelligent personal assistants can be helpful by intervening to provide guidance when the user reaches an impasse or makes specific mistakes. In situations where human skills are an essential part of problem solving, a more appropriate form is to have assistants relieving the user of common tasks to enable him/her to focus on strategic decision making. Finally, when there is a distribution of problem-solving skills between the user and the system, collaborative assistants can work in conjunction with the user to achieve a common goal in a way that exploits their complementary capabilities.

In a competitive electricity market—that is, the real-world application under consideration—the user may possess the necessary technical expertise to perform his/her tasks effectively, although he/she may need to track significant volumes of new information that could affect his/her objectives and productivity. The net result may be a high level of information overload—and also task overload—that may lead to performance degradation. Furthermore, the user may be an untrained person, who wants to make effective use of the agent-based system. Accordingly, this work considers the second style of assistance mentioned above (and the underlying human-computer interaction model).

Within this delegative style, the user decides what needs to be done to reach his/her goals and which tasks need to be allocated to assistant agents. The agents operate in a fairly autonomous manner within bounds set by the user—they work on behalf of the user by executing tasks that have been assigned to them. In addition, the agents may assist the user by observing the input he/she presents to the interface and commenting his/her actions, by retrieving information (e.g., from the web) and/or computing information that can improve his/her decision making process, and by making suggestions and recommendations. Furthermore, the agents may interact with the user to solicit necessary information and, more importantly, to confirm important decisions.

Accordingly, the following desiderata were identified:

- The assistant agents should be able to operate in an autonomous way, although they must accept explicit directions from the user on what to do (or not) and how to do it.
- The agents should be capable of observing the input of the user, monitoring and keeping track of his/her activities, and mainly providing appropriate comments and criticisms to his/her actions.
- The agents should be able to communicate adequately what they are doing and why, providing the user with a clear understanding of the status (and strategy) of their actions.

Such desiderata are, we believe, very important in order for the assistant agents to be useful for both untrained and "busy" users.

8.3.2.3 The Interaction Paradigm

The paradigm for human-computer interaction adopted in this work is illustrated in Fig. 8.3. The interface is characterized by a collection of widgets (as shown by the snapshot of the simulation tool in the figure). The autonomous software agents are denoted by smiley faces with bow ties—specifically, the assistant agents are denoted by both light-green and light-blue faces and the market agents by light-orange faces. The user is denoted by the left-hand face with a blue tie.

The light-green interface agents are responsible for managing the graphical user interface—they operate in the middle ground between the user and the market agents. The light-blue assistant agents act mainly as personal assistants (or helpers). The market agents represent both the entities responsible for the various markets supported by the system (e.g., the operator that runs the day-ahead market or the operator that runs the futures market) and the entities participating in these markets (e.g., generating companies, retailers, aggregators and consumers). The interaction between the user and the graphical interface as well as the communication between the different agents are denoted by double solid arrows. The personal assistants can monitor and keep track of the user's actions (denoted by dashed arrows), and interact with programs external to the simulation tool.

The simulation tool is oriented towards the user in its support for human needs and interests, responsiveness to human inputs, and adaptivity to the user working style and preferences, although it provides a number of automated functions. To fully

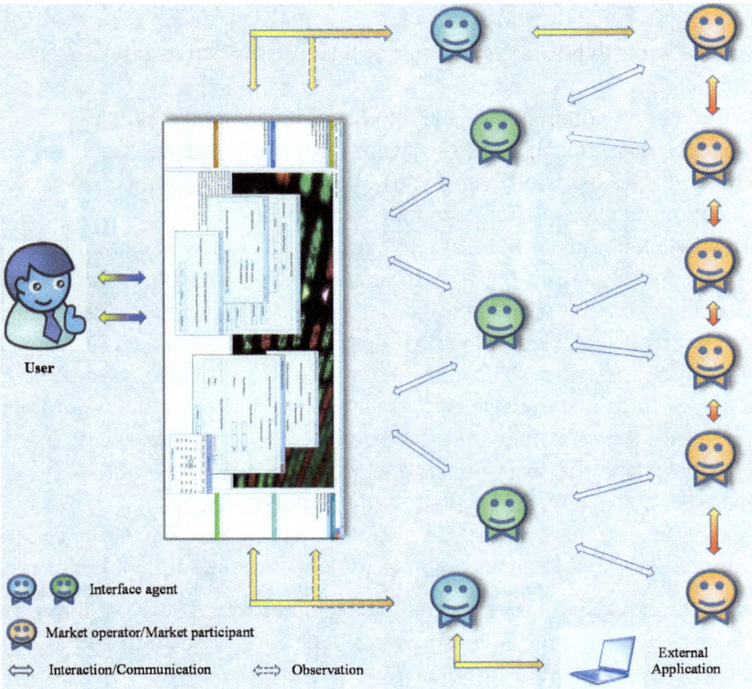

Fig. 8.3 Interface design: the human-computer interaction paradigm

accomplish this, and as noted earlier, we have made relevant efforts to consider a creative integration of direct manipulation interface techniques with automated services provided by autonomous interface agents. Also, we have accommodated, as much as possible, what is inescapably different about each particular market supported by the tool (the constructs needed to simulate a day-ahead market are substantially different from those needed to simulate, say, a forward market), while preserving consistency across markets. Accordingly, the tool considers different (light-green) interface agents for managing the specialized interfaces of the various simulated markets.[24] These interfaces share a common screen layout which, in turn, is operated by another (light-green) interface agent in an autonomous fashion.

Also, the tool is (highly) user centric in its support for human control, allowing the user to take responsibility for several key decisions to be made during each simulation, although he/she can turn over control of specific tasks to agents. Accordingly, the (light-green) interface agents are (roughly) located in the middle of Fig. 8.3—as we move towards the user, these agents are usefully made "visible" by operating in the interactive interface (e.g., by reading the user input or by displaying suggestions made

[24]Currently, the simulation tool includes a specialized interface for each simulated market. An earlier design placed the interactions related to all markets within a common (general-purpose) interface, but it was proven to be not adequate nor effective, causing (test) users to be confused about tasks associated with different markets and/or pricing mechanisms (e.g., the bid submission process involving either the system marginal pricing or the locational marginal pricing).

by other agents), and as we move in the other direction, these agents increasingly communicate with (light-orange) market agents that can actively handle operational tasks without direct user's guidance. The (light-green) interface agents can mediate, at least in part, the communications between the agents that take part in a given market (market participants) and the agent responsible for operating that market (market operator). Also, the (light-green) interface agents can mediate, in part, the communications between different market participants.[25]

To illustrate the control of the user over key decisions, a particularly important example follows. Both the bids to supply energy and the offers to purchase energy are central aspects of the day-ahead market. GenCo agents compute these bids autonomously by using, e.g., profit maximization models. Also, RetailCo agents prepare these offers autonomously by considering, e.g., data from demand/consumption forecast files. Typically, these agents present their bids/offers to the user for analysis and explicit confirmation (although the user is able to by-pass some or all confirmation steps if he/she wants to do that, i.e., the GenCo/RetailCo agents may autonomously submit bids/offers directly to the market operator).

More specifically, for each GenCo agent participating in the day-ahead market, the process of bid preparation and submission involves the following main tasks[26]:

- the GenCo agent generates a potential bid to sell energy. The agent may represent a thermal generating unit, a hydro power plant or a wind farm and is equipped with specific profit maximization models;
- the agent sends the potential bid to a particular (light-green) interface agent, who presents it to the user (i.e., displays the bid in a table and/or graph);
- the user examines the bid and can either: (i) accept it, (ii) reject it, or (iii) alter one or more numerical values, for some reason, and then accept the modified bid;
- in case of acceptance, the interface agent sends the bid to the market operator.

Apart from giving the user the feeling of being in control, we believe that this dissociation between normal market operation and active power industry agents is essential to the flexibility and extensibility of the tool. The underlying design principle or guideline—that is, the definition of autonomous interface agents that can mediate, in part, the communications between different market entities—emerged from practical experience and was adopted for all markets supported by the tool.

The development of the tool was also motivated by several complementary objectives, including (1) providing support to the user (in making strategic decisions), and to some extent, (2) relieving the user of routine tasks, and (3) intervening in situations where information and/or task overloads may lead to oversights or mistakes by the user. To accomplish the first objective, the assistant agents are able to retrieve information from the web, compute (extra) information that the user may find

[25]The simulation tool has been designed to provide the user with a high degree of positive control over system behavior, although it retains a strong measure of autonomy. Subsequently to the assignment of tasks by the user, he/she and the agents address their individual responsibilities in a fairly independent manner, initiating interactions with one another as needed.

[26]For retailer agents participating in the day-ahead market, the offer submission process is essentially identical to that of GenCo agents, and is therefore omitted.

helpful, display suggestions and make recommendations. In particular, some (light-blue) assistant agents can act as "information assistants", interacting autonomously with external programs (or agents), acquiring both relatively general information and problem-specific information, processing such information, and displaying it on the screen to support a (hopefully) more efficient decision making.[27]

Also, several (light-blue) assistant agents can (indirectly and in part) assist the user in dealing with the complexity of difficult decisions, particularly when the system confronts the user with important technical suggestions, and he/she is reluctant to take action. To illustrate this point, we return to the day-ahead market example, specifically to the part when the system displays a potential bid to sell energy, generated by a GenCo agent using a particular profit maximization model, and the user needs to make a decision about its acceptance, rejection, or modification (and then acceptance). In such a situation, the system is able to provide some guidance or assistance to the user, namely by looking for alternative strategic options (i.e., new bids), taking an active role in evaluating these options (i.e., rating these options using utility functions), and recommending some of them (referred to as agent-recommended solutions).

Furthermore, and more importantly, the system can suggest new types of business strategies to being pursued by the user. Bilateral contracting of electricity is a particularly relevant example. As we stated in the previous section, market agents playing the role of either "buyers" or "sellers" can negotiate the terms and conditions of long-term bilateral contracts. The agents exchange offers and counter-offers until they reach an agreement (or one of the parties decides to opt out of the negotiation). Each offer is prepared using a specific strategy and includes a set of energy prices (as well as other issues that are deemed appropriate). Typically, one of the negotiating parties represents the user, and therefore he/she is confronted with specific offers during the course of negotiation—that is, the user needs to make decisions about the acceptance/modification/rejection of specific offers. To aid the user in developing a better understanding of how a particular offer was prepared, the system may display information about the negotiation process and the disputants. Also, market agents are equipped with an expert system allowing them to recommend (or not) a change in strategy as negotiation unfolds. In all of these cases, market agents can initiate interactions with (light-blue) assistant agents as needed.

These various forms of guidance (or assistance) involve mainly market agents acting in a fairly autonomous manner, who may need to interact with (light-blue) assistant agents to solicit user-information that could significantly contribute to proper reasoning and improved performance. To this end, some (light-blue) assistant agents are endowed with both application-independent and application-dependent information about the user (called a user model or user profile). Application-independent information includes personal information, such as name, address, job and hobbies. Application-dependent information includes the user's goals, interests,

[27]To date, the simulation tool can "monitor" the Iberian Electricity Market (MIBEL). In particular, the (light-blue) assistant agents can interact with MIBEL (www.omie.es) to get the daily market-clearing prices of both Portugal and Spain. Future work will focus on monitoring other markets, notably the Nordic power market.

and preferences regarding the application under consideration, and may differ significantly from one user to another (e.g., some users can play the role of generating companies, others may act on behalf of large consumers, etc.).[28]

Now, considering the second and third objectives mentioned above, and as we have noted before, we chose to adopt a delegation style of interaction between the user and the system. This style enables the user to decide which tasks he/she feels comfortable to allocate to agents. In other words, it enables the user to delegate responsibility for routine tasks to agents, thus increasing the amount of time that he/she can dedicate to more challenging activities. Accordingly, and although not present within agents currently, future incorporation of a delegation capability will be useful to relieve the user of several less important tasks, enabling he/she to focus on power market activities.

8.4 Autonomous Software Agents

This section is organized as follows. Section 8.4.1 goes into slightly more depth on the two main agent types (i.e., market agents and assistant agents), as well as the various agent sub-types (e.g., market participants, market operators, assistant agents and interface managers), supported by MATREM. In particular, it presents a Unified Modelling Language (UML) class diagram showing the relationships among the different agent types and sub-types. Section 8.4.2 delves into the technical details of the conceptual (or abstract) model that underpins MATREM agents. First, the model initially adopted for agents—a "traditional" deliberative model—is presented and then the model currently being adopted—a simplified belief-desire-intention (BDI) model—is described in detail. Section 8.4.3 is devoted to the implementation of MATREM agents. It presents several key features of both JADE [21] and Jadex [53] and, more importantly, describes how the conceptual BDI model is being realized in a practical BDI architecture.

8.4.1 Agent Types: Market Agents and Assistant Agents

As mentioned earlier, the tool supports two centralized markets—a day-ahead market and an intra-day market—as well a bilateral market for trading standardized future

[28]To date, our focus for personalization has been on endowing (light-blue) assistant agents with knowledge about different users playing the roles of typical market participants (e.g., generating companies, retailers and consumers). One important area for future work is to consider machine learning techniques to allow the agents to learn the users' goals and preferences regarding the application domain. Furthermore, other area for future work is to analyze personalization from the point of view of the interaction between the user and the simulation tool (i.e., the assistant agents): discovering how the user wants to be assisted, learning when (and if) to interrupt the user, and learning his/her reactions towards different assistance actions (such as suggestions and warnings).

contracts—a futures market. Also, the tool supports a marketplace for negotiating the details of tailored (or customized) long-term bilateral contracts. In this way, market agents are the agents that act as market participants—such as, generating companies, retailers, aggregators, large consumers and small consumers—and the agents that are responsible for the simulated markets—that is, the various market operators and the system operator.

Assistant agents are categorized into interface managers and intelligent assistants.[29] Interface manager agents are the agents responsible for managing the mouse-and-keyboard based interfaces of the various simulated markets. Also, as stated earlier, these agents provide the facilities for specific "dialogues" between the user and each active market agent as well as "dialogues" between the different market agents. Intelligent assistant agents are the agents that provide support to the user in making strategic decisions. Some agents act as "information assistants", i.e., are able to interact autonomously with external programs or agents to acquire relevant information to the user. Other agents act as "trading assistants" and are capable of (indirectly and in part) assisting the user in dealing with the complexity of difficult decisions.

The different types of agents and their inheritance relationships are depicted in the UML class diagram of Fig. 8.4.[30] All agents inherit basic variables and methods from the `MatremAgent` class (root class). The classes `MarketAgent` and `AssistantAgent` are subclasses of `MatremAgent`. Market agents are specialized into market participants, system operators, and market operators. Accordingly, the various market participants supported by the tool—that is, generating companies, retailers, aggregators and consumers—inherit from the `MarketParticipant` class, which is a subclass of `MarketAgent`. Likewise, the system operator and the various market operators supported by the tool—that is, the operators of the two centralized markets as well as the operator of the futures market—inherit from the `SystemOperator` and the `MarketOperator` classes respectively, which are both subclasses of `MarketAgent`.

The classes `IntelligentAssistant` and `InterfaceManager` are subclasses of `AssistantAgent`. The two types of intelligent assistants supported by the tool—that is, information assistants and trading assistants—inherit from the `IntelligentAssistant` class. Interface manager agents are specialized into four different types of agents, responsible for managing the interfaces of the two

[29]For convenience, and also simplicity in exposition, the previous section considered the terms "light-green interface agents" and "light-blue assistant agents" to refer to interface managers and intelligent assistants, respectively.

[30]In this work, we adopt the graphical representation from the Unified Modelling Language [54]. Accordingly, classes are represented as rectangles with three compartments. The top compartment indicates the name of a class. The second and the third compartments list the variables and the methods of a class, respectively. For the sake of clarity and simplicity, the variables and methods of the various classes are not shown in Fig. 8.4. The generalization relationship is denoted by a solid directed line with a large open arrowhead, pointing to a superclass (or parent). More-specialized classes (subclasses) inherit the variables and methods of their parents, albeit they may have their own variables and methods.

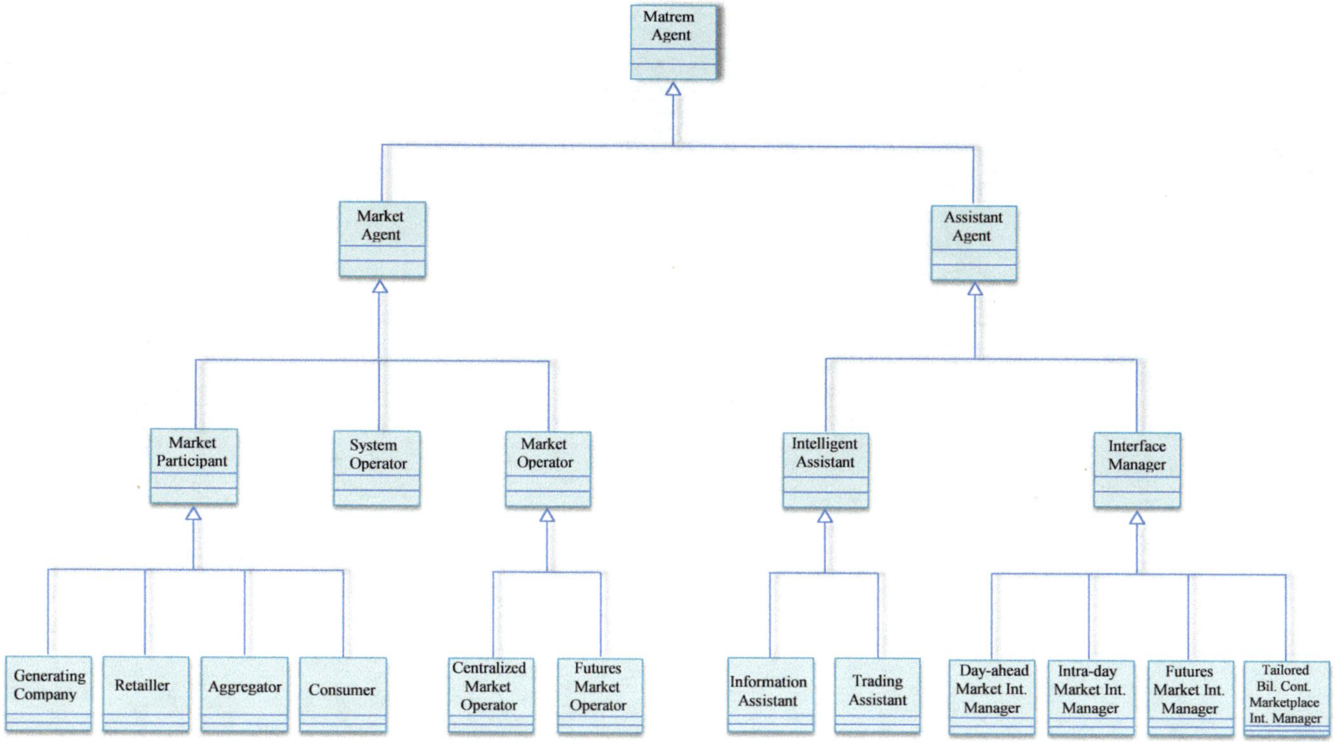

Fig. 8.4 UML class diagram of software agents in the simulation tool

centralized markets, the futures market, and the bilateral contracting marketplace. These four types of agents inherit from the `InterfaceManager` class.

8.4.2 *The Conceptual Agent Model: BDI Model*

The development of a comprehensive, high-fidelity, agent-based simulation tool for liberalized energy markets is naturally a long-term vision. Accordingly, this objective was pursued incrementally, with increasingly greater detail and complexity incorporated in the different versions of the tool. In particular, some initial design choices were made in the earlier versions of the tool to explore the flexibility of different agent types. These versions simplified several aspects of the operation of software agents—essentially, a "traditional" deliberative model (or abstract architecture) was initially adopted for agents.

In the earlier versions of the tool, all agents—either market agents or assistant agents—are computer systems capable of flexible action and able to interact, when appropriate, with other agents to meet their design objectives. They take sensory input from the environment—that is, a particular electricity market—and execute actions that affect it. Their effectoric capabilities are represented by a set of possible actions. In particular, two types of actions are considered: cognitive actions (e.g., performing a numerical computation) and communicative actions (typically, sending a message). The distinguishing feature of cognitive actions is that they are private, in the sense that they cannot be "observed" by other agents. Also, the agents performing a cognitive action have control over its completion. On the other hand, the effects of communicative actions are not under the complete control of the agents performing them, but also depend on the recipients of such actions.[31]

In addition to a repertoire of possible actions, the agents have internal data structures (information stores) to record the beliefs about themselves, the beliefs about the environment state and history, and the beliefs about the other agents in the environment. They can access the contents of the beliefs data store and change that contents (i.e., add new beliefs and revise the current beliefs). But, although the operations of adding sentences, revising sentences, and querying what is known often involve inference—that is, deriving new sentences from old ones—the agents can exhibit only very restricted inferential capabilities.

Also, the agents have internal data stores to record their top-level achievement goals. Typical goals include: "maximize-profit", "calculate-market-clearing-price", and "minimize-procurement-cost". They are able to perform a simple form of goal-based action selection—specifically, to combine information about their goals with the contents of their beliefs data store to select actions that further specific goals.

[31] Researchers working in some areas (e.g., philosophy, cognitive psychology and linguistics) may find the distinction between cognitive and communicative actions dubious. After all, an axiom of speech act theory is that agents requesting or informing are performing actions just like any other actions. This distinction is, however, rather natural and we believe suitable for the purposes of this work.

Furthermore, the agents can make use of utility functions to complement goal-based action selection—that is, they seek either to maximize their utility or to choose actions that will lead to good (rather than optimal) solutions (e.g., by adopting heuristic strategies during bilateral contracting of electricity). In this way, the agents can exhibit some degree of control both over their own internal state and over their actions, i.e., they are, at least in part, autonomous.[32]

Now, the simulation tool has been designed to be extensible, and extensions that are contemplated include improvements to its basic infrastructure and enhancements to the realism of the simulated power markets. A natural and rather straightforward extension is the refinement of the conceptual agent model. Hence, although the "traditional" deliberative model (or abstract architecture) can be considered simple and intuitive, it has been progressively "replaced" by a simple belief-desire-intention (BDI) model. The rationale for this decision is as follows. Existing BDI models have come to be possibly the best known and best studied models of practical reasoning agents.[33] In fact, most models are based on a respectable philosophical theory of human practical reasoning [55]. Also, their key components have an elegant abstract logical semantics, which have been taken up and elaborated upon widely within the agent-based community [56].

Furthermore, a whole range of practical development efforts relating to BDI systems have been undertaken with considerable success. For example, the Procedural Reasoning System (PRS) has been deployed in several major industrial applications (e.g., fault diagnosis on the space shuttle [57]). PRS has progressed from an experimental version to a fully fledged C^{++} implementation known as the distributed Multi-Agent Reasoning System (dMARS), which has been applied in perhaps the most significant multi-agent applications to date (e.g., air traffic management and business process control [58]). Other elegant and well-known BDI systems include COSY and GRATE* (see [59] for a review). Overall, although the question of exactly which combination of attitudes is most appropriate to characterize an agent has been the subject of some debate (see, e.g., [60, 61], and also Chap. 3), the BDI model is arguably the dominant force in the foundations of rational agency.

Before proceeding with the technical details of the agent model that underpins MATREM, we hasten to add two explanatory and cautionary notes. First, the model is neither a canonical nor a complete model of BDI agents. The aim is to present a plausible model that, we believe, captures some important features of various models of practical reasoning that employ the mentalistic notions of belief, desire

[32] The earlier versions of the tool incorporated a number of software agents equipped with an operational model resulting from the implementation of the "traditional" deliberative model (or abstract architecture). Although limited in reasoning and decision-making, such agents proved to be relatively satisfactory, since the tool suited the interests and needs of several untrained (test) users.

[33] The term "BDI model" has been coined by researchers working in closely related areas to describe slightly different types of models. Here, we use the term to describe any model of practical reasoning that makes use of the folk-psychology concepts of belief, desire and intention. In this way, a BDI model may or may not center on claims originally propounded by Bratman [55] about the role of intentions in focusing practical reasoning.

and intention. Second, the model is generic and, of necessity, fairly coarse grained. However, it may be readily adapted, extended and refined, to support the features of specific BDI agents—that is, it can be taken as a starting point from which to develop finer grained BDI agents.

Figure 8.5 presents the conceptual agent model. The agents have four major components: beliefs, desires, intentions and plans (or, more precisely, plans-as-recipes). *Beliefs* represent (possibly imperfect) information about the environment, the agents themselves and the other agents interacting with them. Beliefs are essential because the environment is dynamic (some past events need therefore to be remembered). Moreover, as the agents are resource bounded, it is desirable to cache important information rather than to recompute it from perceptual data. *Desires* represent preferences over well-defined future states of the environment. The agents will not, in general, be able to achieve all their desires, even if these desires are consistent. Accordingly, they typically select a consistent subset of desires to pursue—that is, given the situation represented by their beliefs, they fix upon a set of goals to accomplish. Hence, *goals* are desires that should be pursued by the agents.[34]

Since the agents are resource-bounded, they cannot pursue all their goals at once. Even if their set of goals is consistent, it is often necessary to select some goals to commit to. *Intentions* represent the chosen or committed goals and the agents will typically continue to try to achieve them until either they believe the intentions are satisfied, or they believe the intentions are no longer achievable. The notion of *commitment* characterizes the transition from goals to intentions. The agents also have a *plan library* containing a set of plan templates, or recipes, specifying particular courses of action that must be undertaken by them in order to achieve their intentions—that is, the agents cache various plans-as-recipes rather than try to recreate every new plan from scratch. *Plan templates* are essentially generic sequences of actions for use in future situations. Each plan template contains, at least, two components: a name and a body. The name of a plan template facilitates its retrieval from the library. The body defines a potentially complex course of action, which may consist of both goals (and subgoals) or primitive actions. Two types of plan templates are distinguished: composite and primitive. The body of a composite plan template specifies the decomposition of a goal into a set of subgoals. On the other hand, the body of a primitive plan template specifies an action or a sequence of actions that can achieve a goal.

In the following, the four major components of the agents are described more formally. Let $A = \{a_1, \ldots\}$ be the set of autonomous agents. Let L be a logical language—the precise nature of L is not relevant to this work and no assumptions are made about it, other than that it is at least a logical language. Let V_L and P_L be

[34] Practical reasoning involves two main processes [1]: deliberation and means-end reasoning. *Deliberation* is a complex process and consists mainly in defining a consistent set of goals from a (possibly inconsistent) set of desires and selecting (some) goals to commit to. For the sake of simplicity, the ongoing developments in the simulation tool consider that the agents have a consistent set of goals to achieve, but not a set of desires nor a deliberation process. Future work will focus on designing and implementing a (simple) deliberation process.

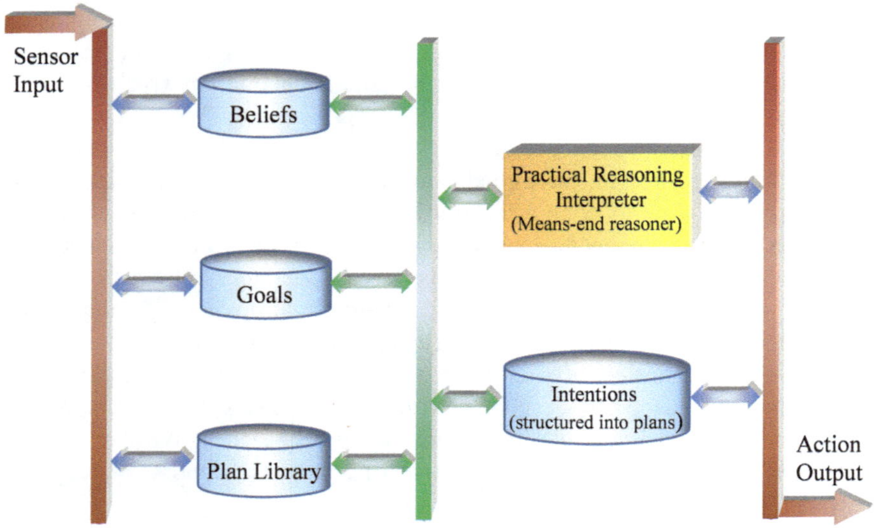

Fig. 8.5 The conceptual agent model: simplified BDI model

the sets of variable symbols and predicate symbols of L, respectively. Every agent $a_i \in A$ has the following key features:

- A set $B_i = \{b_1, \ldots\}$ of *beliefs* representing information about a_i, the world in which a_i resides, and the other agents in the world. Beliefs are formulae of L.
- A set $G_i = \{g_1, \ldots\}$ of *goals* representing world states to be achieved. Goals are also formulae of L.
- A library $PL_i = \{pt_1, \ldots\}$ of *plan templates* representing simple procedures for achieving goals. There are two types of plan templates: composite and primitive. Each plan template pt has a head and a body. The head is a 2-tuple: $hdr = \langle nm, lv \rangle$, where $nm \in P_L$ is the name of pt and $lv \in V_L$ is a list of variables (interpreted as universally quantified). The body of a composite plan template specifies the decomposition of a goal into more detailed subgoals. The body of a primitive plan template specifies a low-level action command.
- A set $P_i = \{p_1, \ldots\}$ of *plans* for execution, either immediately or in the near future. A plan p is a collection of plan templates structured into a hierarchical and temporally constrained tree. The nodes of the tree are instantiated plan templates retrieved from the library. The head of each plan template is referred to as *intention* and denoted by int. Intentions are formulae of L.

The four major components of the agents are controlled and managed by a practical reasoning interpreter. As noted earlier, practical reasoning involves deliberation and means-end reasoning. *Deliberation* consists mainly in defining a consistent set of goals and selecting some goals to commit to—that is, to decide what states of affairs to achieve. *Means-end reasoning* consists mainly in generating plans for achieving the committed goals—that is, to decide how to bring about the chosen states of affairs by using the available resources. Means-end reasoning is perhaps better known in the agent-based community as planning and will receive the preponderance of our attention in the remainder of this subsection (as we have already pointed out, the development of a deliberation process has been deferred to future work).

The main activity of the practical reasoning interpreter is to operationalize a means-end reasoner to compute sequences of actions whose execution leads to the achievement of the committed goals. Plan generation is based on a plan library and consists mainly in adapting plans-as-recipes to specific situations. This approach to planning is called planning from second principles and is often adopted by researchers working on practical reasoning agents and research fields like agent theories and agent architectures, of which planning is part, but not the main focus.

Conceptually, planning is a recursive process of plan expansion. Goals are incrementally refined into subgoals until a sufficiently fine-grained level of abstraction is reached, which is suitable for execution. More formally, the generation of a plan p from the plan templates stored in the library is performed through an iterative procedure involving the following four main tasks:

- *plan retrieval*: searching the library for any plan template whose head matches the description of a goal; when a suitable match is found, the chosen plan template is selected and its variables unified with corresponding values from the goal description; the retrieved plan templates are called applicable plan templates.
- *plan selection*: selecting the preferred plan template pt from the set of applicable plan templates (as all applicable plan templates suit the goal description, this selection is not critical for achieving the goal; however, a selection based on a specific notion of utility is always more suitable, since a plan that achieves a goal with some effort should be preferred to another plan achieving the same goal with higher effort);
- *plan addition or placement*: adding the preferred plan template pt to p; that is, placing pt at an appropriate point in the hierarchical subgoaling structure (or plan structure);
- *plan interpretation*: selecting a composite plan template from p, say pt, establishing a temporal ordering for the elements of its body, and picking the first ordered element (which is interpreted as a new goal).

Thus plans have a hierarchical structure that is embedded in the plan library. They are represented by a hierarchical and temporally constrained And-tree. The nodes of the tree are instantiated plan templates. Arcs form a hierarchy between pairs of nodes. Also, arcs represent ordering constraints.

8.4.3 Agent Implementation: Jadex

MATREM agents are currently being developed using the JADE (Java Agent DEvelopment) framework, probably the most widespread agent-oriented middleware currently in use [21]. JADE is a software platform that provides middleware-layer functionalities, independent of specific domains, allowing the realization of different types of distributed systems. It is built on top of the Java object-oriented programming language and benefits from the huge set of Java features and third-party libraries. Also, it has a flexible infrastructure facilitating the development of agent-based applications by means of a run-time environment implementing the life-cycle support features required by agents, the core logic of agents themselves, and a rich set of graphical tools.

Jade provides agent developers with a number of ready-to-use and easy-to-customize core functionalities, notably [21]:

- A fully distributed system inhabited by agents, each running as a separate thread, potentially on different remote machines.
- An effective agent life-cycle management. Agents can be created, suspended, resumed, migrated, cloned and killed.
- Efficient transport of asynchronous messages via a location-transparent application programming interface (API).
- Full compliance with the foundation for intelligent physical agents (FIPA) specifications (probably, the most widespread and accepted set of standards for multi-agent platforms and applications).
- A library of interaction protocols.
- Support for ontologies and content languages. Developers can also implement new content languages to fulfill specific application requirements.
- Implementations of both white pages and yellow pages.
- Support for agent mobility. Both agent code and, under certain restrictions, agent state can migrate between processes and machines.
- A set of graphical tools to support debugging and monitoring.
- An extensible kernel designed to allow developers to extend the functionality of the platform through the addition of kernel-level distributed services.

MATREM agents have their own Java threads, using them to control their life cycles and decide autonomously when to perform specific actions. The path of execution of each Java thread involves three main tasks:

1. *agent creation*: initialization operations and addition of initial behaviours;
2. *agent "life"*: execution of behaviours from the pool of active behaviours;
3. *agent termination*: clean-up operations.

The different activities that each agent has to perform are carried out within JADE behaviours. Behaviours represent tasks and can be added at any time to agents. Three types of behaviours are of particular importance: "one-shot" behaviours (complete in one execution phase), "cyclic" behaviours (never complete), and generic behaviours (complete when a given condition is met).

Agent communication is implemented in strict accordance with the FIPA specifications and consists mainly in sending and receiving messages. Each message includes several fields, notably: (i) the sender, (ii) the list of receivers, (iii) the communicative act or performative, which indicates what the sender intends to achieve by sending the message, (iv) the content, which contains the information to be exchanged by the message, and (v) some additional fields used to control concurrent conversations and to specify timeouts. The communication paradigm is based on asynchronous broadcast message passing and involves the following main tasks:

- a sender agent (S) prepares and sends a message to a receiver agent (R);
- the JADE run-time posts the message in the message queue of agent R;
- the agent R is notified about the receipt of the new message;
- the agent R gets the new message from the message queue and processes it.

Thus, each MATREM agent has a message queue—or mailbox—where the messages sent by other agents are placed and receives a notification whenever any new message is posted. To send a message, a MATREM agent just needs to know the identities of the receiver agent(s). There is no need to obtain the object reference of the receiver agent(s). There is also no temporal dependency between a sender agent and the receiver agent(s)—that is, the sender does not need to wait for a message response to continue processing. Furthermore, when, or even if, a specific receiver agent picks up a message from the mailbox for processing is a design choice of agent developers.

Now, Jadex [21, 53] is a BDI reasoning engine that runs over JADE, extending it with a number of BDI features, enabling the development of BDI agents. Jadex relies on established software engineering techniques, such as Java and XML, and includes a rich set of run-time tools. Also, Jadex fully supports the two main processes of practical reasoning—deliberation and means-end reasoning. This means that Jadex allows the construction of agents with explicit representations of mental attitudes and that automatically can deliberate about their goals and subsequently can pursue them by applying appropriate plans. Despite these and other virtues, however, the focus of the ongoing work is on a means-end reasoning process to generate plans for achieving specific states of affairs.

Jadex agents, and consequently MATREM agents, consist of two parts: a structural or declarative part and a behavioural or procedural part. The structural part comprises the beliefs, goals and plan heads of agents, which are represented using the extensible markup language (XML), and specified in agent definition files (ADFs). The behavioural part comprises the procedural knowledge contained in the plans of agents—the plan bodies—and is represented using plain Java. The connection between the two parts is established by an application programming interface enabling Java classes to access the beliefs, goals and plan heads of each agent.

Beliefs are the informational attitude of MATREM agents. The "beliefs data store" may include both named facts or named sets of facts. Goals are the motivational attitude of MATREM agents. Jadex supports four different types of goals [53]: perform, achieve, query and maintain goals. Currently, this work focuses on achieve (or achievement) goals: goals associated with desired world states, without specifying how to reach them.

Plans are the means by which MATREM agents achieve their goals. Each plan consists of two parts: a plan head and a plan body. The plan head contains information about the circumstances under which the plan will be used. The plan body represents the course of action that will be performed if the plan is chosen for execution.

Depending on the purpose of each plan, the degree of abstractness varies between fully abstract and very concrete. Abstract plans are specified in terms of goals and subgoals, whereas concrete plans consist of directly executable actions.

8.5 Conclusion

This chapter described the most important aspects of the work in which the author has been involved, realized in an agent-based simulation tool, called MATREM (for Multi-Agent TRading in Electricity Markets). MATREM allows the user to conduct a wide range of simulations regarding the behavior and outcomes of EMs, including markets with increasing penetrations of variable generation. The current version of the tool supports the following exchanges and marketplaces:

1. A power exchange: comprises a day-ahead market and a shorter-term market, known as intra-day market. Two pricing mechanisms are considered: system marginal pricing and locational marginal pricing.
2. A derivatives exchange: comprises a futures market, where private parties can trade standardized bilateral contracts, notably financial and physical future contracts. An electronic trading system automatically matches the bids and offers from different market participants.
3. A bilateral marketplace: enables market participants to privately negotiate the terms and conditions of two types of tailored (or customized) long-term contracts, namely forward contracts and contracts for differences. The negotiation process involves an iterative exchange of offers and counter-offers.

Also, the tool currently supports the following six different types of market entities: generating companies, retailers, aggregators, consumers, market operators and system operators.

MATREM relies on multiple software agents capable of flexible, autonomous action. Two key types of agents are currently being implemented:

1. Market agents: represent, in computational terms, the entities responsible for the various markets (e.g., the operator that runs the day-ahead market), as well as the entities participating in these markets (e.g., a generating company or a retailer).
2. Assistant agents: are further categorized into interface managers and intelligent assistants. Interface manager agents are responsible for managing the mouse-and-keyboard based interfaces of the various simulated markets, and also provide the facilities for specific "dialogues" between the user and the market agents. Intelligent assistant agents provide support to the user in making strategic decisions and can act as "information assistants", "trading assistants", etc.

The conceptual agent model is a belief-desire-intention (BDI) model. The four major components of the agents (i.e., the beliefs, goals, intentions and plan templates) are controlled and managed by a practical reasoning interpreter. The main activity of the interpreter is to operationalize a means-end reasoner to compute sequences of actions whose execution leads to the achievement of specific goals. Planning is based on a plan library and involves mainly a recursive process—goals are incrementally refined into subgoals until a sufficiently fine-grained level of abstraction is reached, which is suitable for execution.

The agents are currently being developed using both JADE (an agent-oriented middleware, fully integrated with the JAVA programming language) and Jadex (a BDI reasoning engine that runs over JADE, enabling the development of BDI agents). A graphical interface handles all interactions with the user and incorporates the following key functions:

1. Agent management: allows the user to manage all software agents (e.g., create new agents or kill active agents).
2. Scenario construction: allows the user to select a specific market to simulate (e.g., the day-ahead market) and to specify their participants (e.g., the generating companies and the retailers).
3. Simulation management: allows the user to specify, monitor and steer all simulations.
4. Report analysis: allows the user to view and analyze the simulation results.

The human-computer interaction paradigm is based on a creative integration of direct manipulation interface techniques with intelligent assistant agents, providing adequate control to the user, but allowing the agents to initiate interactions with one another as needed. The target platform for the system is a 32/64-bit computer running Microsoft Windows.

Now, some notes on the research areas and aims of this chapter. Noticeably, the work described here involves several research areas, notably energy markets and software agents, and draws upon various computational resources (e.g., JADE and Jadex). Accordingly, the author wishes to reiterate that both market design and agent technology are, at the time of writing, active areas of research. Thus, the development of the agent-based simulation tool is very much ongoing. Some important tasks that are currently being performed include:

1. Completing the implementation of the derivatives exchange allowing private parties to trade two different types of standardized bilateral contracts: future contracts and option contracts.
2. Improving the marketplace for negotiating customized bilateral contracts, specifically by considering new types of contracts tailored to the needs of renewable power producers.
3. Developing a market to match the imbalances caused by the variability and uncertain present in power systems.
4. Equipping autonomous software agents with a simplified BDI model and analyzing their behavior in different market situations.

Finally, the current version of MATREM has several shortcomings that will need to be overcome in subsequent prototypes to simulate energy markets in a more realistic way. Some desirable extensions and fruitful areas for future work are as follows:

1. Market entities: to consider new types of traditional power industry agents, notably transmission company agents (TransCos) and distribution company agents (DistCos).
2. Auction trading: to implement specific auction types allowing private parties to trade standardized bilateral contracts in both continuous and auction modes.
3. Negotiation strategies: to develop new negotiation strategies related to risk management in bilateral contracting, allowing private parties to negotiate the details of customized contracts more effectively.
4. Learning capabilities: to equip agents with machine learning techniques enabling them to acquire new knowledge and to use that knowledge to improve performance in different market situations.
5. User interface realism: to develop a graphical editor to construct and/or modify electric power industry scenarios.
6. Human-computer interaction: to analyze personalization from the point of view of the interaction between the user and the assistant agents, discovering how the user wants to be assisted and learning his/her reactions towards different assistance actions (such as suggestions and warnings).
7. Market monitoring: to develop assistant agents able to monitor other important markets (e.g., the Nordic power market).

Acknowledgements For the most part, the work described in this chapter was performed under the project MAN-REM (FCOMP-01-0124-FEDER-020397), supported by both FEDER Funds, through the program COMPETE ("Programa Operacional Temático Factores de Competividade"), and National Funds, through FCT ("Fundação para a Ciência e a Tecnologia"). Some parts of the work, notably the improvements in the day-ahead market and the current developments in the real-time market, were performed under the project IRPWind: Integrated Research Programme on Wind Energy, funded by the European Union's seventh programme for research, technological development and demonstration, under grant agreement 609795. The author also wishes to acknowledge the significant contributions made by a number of Ph.D. and M.Sc. students to the agent-based simulation tool, notably students from the NOVA University of Lisbon, University of Lisbon, University Institute of Lisbon (ISCTE) and Polytechnic Institute of Lisbon (ISEL).

References

1. Wooldridge, M.: An Introduction to Multi-agent Systems. Wiley, Chichester (2009)
2. Macal, C., North, M.: Tutorial on agent-based modelling and simulation. J. Simul. **4**, 151–162 (2010)
3. Russell, S., Norvig, P.: Artificial Intelligence: A Modern Approach. Pearson Education Inc, New Jersey (2010)
4. Wood, A., Wollenberg, B., Sheblé, G.: Power Generation, Operation, and Control. Wiley, New Jersey (2014)
5. Sioshansi, F.P. (ed.): Distributed Generation and its Implications for the Utility Industry. Academic Press, Oxford (2014)
6. Harp, S.A., Brignone, S., Wollenberg, B.F., Samad, T.: SEPIA: a simulator for the electric power industry agents. IEEE Control Syst. Mag. **20**(4), 53–69 (2000)
7. Koritarov, V.: Real-world market representation with agents: modeling the electricity market as a complex adaptive system with an agent-based approach. IEEE Power Energy Mag. **2**(4), 39–46 (2004)

8. Batten, D., George Grozev, G.: NEMSIM: finding ways to reduce greenhouse gas emissions using multi-agent electricity modelling. In: Perez, P., Batten, D. (eds.) Complex Science for a Complex World Exploring Human Ecosystems with Agents, pp. 227–252. Australian National University Press, Canberra (2006)
9. Sun, J., Tesfatsion, L.: Dynamic testing of wholesale power market designs: an open-source agent-based framework. Comput. Econ. **30**, 291–327 (2007)
10. Sensfuß, F.: Assessment of the impact of renewable electricity generation on the german electricity sector: an agent-based simulation approach. Ph.D. Dissertation, Karlsruhe University (2007)
11. Praça, I., Ramos, C., Vale, Z.: MASCEM: a multiagent system that simulates competitive electricity markets. IEEE Intell. Syst. **18**(6), 54–60 (2003)
12. Vale, Z., Pinto, T., Praça, I., Morais, H.: MASCEM - electricity markets simulation with strategically acting players. IEEE Intell. Syst. **26**(2), 9–17 (2011)
13. Cincotti, S., Santana, Gallo, G.: The Genoa artificial power-exchange. In: Filipe, J., Fred, A. (eds.) Agents and Artificial Intelligence (ICAART 2012), pp. 348–363. Springer, Berlin (2013)
14. Zhou, Z.Z., Chan, W.K., Chow, J.H.: Agent-based simulation of electricity markets: a survey of tools. Artif. Intell. Rev. **28**, 305–342 (2007)
15. Sensfuß, F., Genoese, M., Ragwitz, M., Möst, D.: Agent-based simulation of electricity markets - a literature review. Energy Stud. Rev. **15**(2), 1–29 (2007)
16. Weidlich, A., Veit, D.: A critical survey of agent-based wholesale electricity market models. Energy Econ. **30**, 1728–1759 (2008)
17. Guerci, E., Rastegar, M., Cincotti, S.: Agent-based modeling and simulation of competitive wholesale electricity markets. In: Rebennack, S., Pardalos, P., Pereira, M., Iliadis, N. (eds.) Handbook of Power Systems II, pp. 241–286. Springer, Heidelberg (2010)
18. European Union: Renewable Energy Road Map Renewable Energies in the 21st Century: Building a more Sustainable Future. Communication from the Commission to the Council and the European Parliament, COM(2006) 848 (10 January 2007) http://eur-lex.europa.eu/legal-content/EN/TXT/PDF/?uri=CELEX:52006DC0848&from=EN (Cited 11 Feb 2017)
19. European Union: Directive 2009/28/EC of the European Parliament and of the Council on the promotion of the use of energy from renewable sources and amending and subsequently repealing Directives 2001/77/EC and 2003/30/EC (23 April 2009) http://eur-lex.europa.eu/legal-content/EN/TXT/?uri=CELEX:32009L0028 (Cited 11 Feb 2017)
20. Jones, E.L. (ed.): Renewable Energy Integration: Practical Management of Variability, Uncertainty, and Flexibility in Power Grids. Academic Press, London (2014)
21. Bellifemine, F., Caire, G., Greenwood, D.: Developing Multi-agent Systems with JADE. Wiley, Chichester (2007)
22. Algarvio, H.F., Lopes, F., Santana, J.: Multi-agent retail energy markets: bilateral contracting and coalitions of end-use customers. In: 12th International Conference on the European Energy Market (EEM 2015), pp. 1–5. IEEE (2015)
23. Algarvio, H., Lopes, F., Santana, J.: Multi-agent retail energy markets: contract negotiation, customer coalitions and a real-world case study. In: Demazeau, Y., Takayuki, I., Bajo, J., Escalona M. (eds.) Advances in Practical Applications of Scalable Multi-agent Systems: The PAAMS Collection, pp. 13–23. Springer International Publishing (2016)
24. Vidigal, D., Lopes, F., Pronto, A., Santana, J.: Agent-based simulation of wholesale energy markets: a case study on renewable generation. In: Spies, M., Wagner, R., Tjoa, A. (eds.) 26th Database and Expert Systems Applications (DEXA 2015), pp. 81–85. IEEE (2015)
25. Algarvio, H., Couto, A., Lopes, F., Estanqueiro, A., Santana, J.: Multi-agent energy markets with high levels of renewable generation: a case-study on forecast uncertainty and market closing time. In: Omatu, S., et al. (eds.) 13th International Conference on Distributed Computing and Artificial Intelligence, pp. 339–347. Springer International Publishing (2016)
26. Kirschen, D., Strbac, G.: Fundamentals of Power System Economics. Wiley, Chichester (2004)
27. OMIP: Trading RuleBook. MIBEL Derivatives Market, Lisbon, Portugal (2016)
28. OMIP: General Contractual Terms: MIBEL PTEL Base Load Financial Futures Contracts. MIBEL Derivatives Market, Lisbon, Portugal (2016)

29. OMIP: Rules for Determining Electricity Indexes. MIBEL Derivatives Market, Notice 6/2006, Lisbon, Portugal (2013)
30. Lopes, F., Rodrigues, T., Sousa, J.: Negotiating bilateral contracts in a multi-agent electricity market: a case study. In: Hameurlain, A., Tjoa, A., Wagner, R. (eds.) 23rd Database and Expert Systems Applications (DEXA 2012), pp. 326–330. IEEE (2012)
31. Algarvio, H., Lopes, F., Santana, J.: Bilateral contracting in multi-agent energy markets: forward contracts and risk management. In: Bajo, J., et al. (eds.) Highlights of Practical Applications of Agents, Multi-Agent Systems, and Sustainability: The PAAMS Collection (PAAMS 2015), pp. 260–269. Springer International Publishing (2015)
32. Sousa, F., Lopes, F., Santana, J.: Contracts for difference and risk management in multi-agent energy markets. In: Demazeau, Y., Decker, K., Pérez, J., De la Prieta, F. (eds.) Advances in Practical Applications of Agents, Multi-Agent Systems, and Sustainability: The PAAMS Collection (PAAMS 2015), pp. 339–347. Springer International Publishing (2015)
33. Sousa, F., Lopes, F., Santana, J.: Multi-agent electricity markets: a case study on contracts for difference. In: Spies, M., Wagner, R., Tjoa, A. (eds.) 26th Database and Expert Systems Applications (DEXA 2015), pp. 88–90. IEEE (2015)
34. Department for Energy and Climate Change: Implementing Electricity Market Reform (June 2014) https://www.gov.uk/government/publications/implementing-electricity-market-reform-emr (Cited 11 Feb 2017)
35. Lopes, F., Mamede, N., Novais, A.Q., Coelho, H.: Negotiation in a multi-agent supply chain system. In: Third Int. Workshop of the IFIP WG 5.7 Special Interest Group on Advanced Techniques in Production Planning and Control, Firenze University Press, pp. 153–168 (2002)
36. Lopes, F., Coelho, H.: Concession strategies for negotiating bilateral contracts in multi-agent electricity markets. In: 23rd Database and Expert Systems Applications (DEXA 2012), pp. 321–325. IEEE (2012)
37. Lopes, F., Algarvio, H., Coelho, H.: Bilateral contracting in multi-agent electricity markets: negotiation strategies and a case study. In: International Conference on the European Energy Market (EEM-13), pp. 1–8. IEEE (2013)
38. Jaquero, V., Montero, F., Molina, J., González, P.: Intelligent user interfaces: past, present and future. In: Redondo, M., Bravo, C., Ortega, M. (eds.) Engineering the User Interface: From Research to Practice, pp. 259–270. Springer, London (2009)
39. Shneiderman, B., Plaisant, C.: Designing the User Interface: Strategies for Effective Human-Computer Interaction. Pearson Education Inc, Boston (2005)
40. Shneiderman, B.: The future of interactive systems and the emergence of direct manipulation. Behav. Inf. Technol. **1**(3), 237–256 (1982)
41. Shneiderman, B.: Direct manipulation: a step beyond programming languages. IEEE Comput. **16**(8), 57–69 (1983)
42. Kay, A.: User interface: a personal view. In: Packer, R., Jordan, K. (eds.) Multi-Media: From Wagner to Virtual Reality, pp. 121–131. W. W. Norton & Company Inc, New York (2001)
43. Maes, P.: Agents that reduce work and information overload. Commun. ACM **37**(7), 31–40 (1994)
44. Maes, P., Kozierok, R.: Learning: interface agents. In: 11th National Conference on Artificial Intelligence (AAAI-93), pp. 459–465. AAAI Press, California (1993)
45. Lieberman, H.: Autonomous interface agents. In: Pemberton, S. (ed.) ACM SIGCHI Conference on Human Factors in Computing Systems (CHI-97), pp. 67–74. ACM Press, New York (1997)
46. Rich, C., Sidner, C.: DiamondHelp: a generic collaborative task guidance system. AI Mag. **28**(2), 33–46 (2007)
47. Myers, K., Berry, P., Blyth, J., Conley, K., Gervasio, M., McGuinness, D., Morley, D., Pfeffer, A., Pollack, M., Tambe, M.: An intelligent personal assistant for task and time management. AI Mag. **28**(2), 47–61 (2007)
48. Birnbaum, L., Horvitz, E., Kurlander, D., Lieberman, H., Marks, J., Roth, S.: Compelling intelligent user interfaces: how much AI? In: Moore, J., Edmonds, E., Puerta, A. (eds.) 2nd International Conference on Intelligent User Interfaces (IUI-97), pp. 173–175. ACM Press, New York (1997)

49. Horvitz, E.: Principles of mixed-initiative user interfaces. In: Williams, M., Altom, M. (eds.) SIGCHI Conference on Human Factors in Computing Systems (CHI-99), pp. 159–166. ACM Press, New York (1999)
50. Schiaffino, S., Amandi, A.: User-interface agent interaction: personalization issues. Int. J. Hum.-Comput. Stud. **60**, 129–148 (2004)
51. Shneiderman, B., Maes, P.: Direct manipulation vs. interface agents. Interactions **6**(6), 41–61 (1997)
52. Armentano, M., Daniela, G., Amandi, A.: Personal assistants: direct manipulation vs. mixed initiative interfaces. Int. J. Hum.-Comput. Stud. **64**, 27–35 (2006)
53. Braubach, L., Pokahr, A., Lamersdorf, W.: Jadex: a BDI-agent system combining middleware and reasoning. In: Unland, R., Klusch, M., Calisti, M. (eds.) Software Agent-Based Applications, Platforms and Development Kits, pp. 143–168. Birkhäuser Verlag, part of Springer Science+Business Media, Basel, Switzerland (2005)
54. Booch, G., Rumbaugh, J., Jacobson, I.: The Unified Modeling Language User Guide. Addison Wesley Longman Inc, Reading (1999)
55. Bratman, M.: Intentions, Plans, and Practical Reason. Harvard University Press, Cambridge (1987)
56. Rao, A., Georgeff, M.: Decision procedures of BDI logics. J. Log. Comput. **8**(3), 293–344 (1998)
57. Ingrand, F., Georgeff, M., Rao, A.: An architecture for real-time reasoning and system control. IEEE Expert **7**(6), 34–44 (1992)
58. Georgeff, M., Rao, A.: A profile of the Australian AI institute. IEEE Expert **11**(6), 89–92 (1996)
59. Haddadi, A.: Belief-desire-intention agent architectures. In: O'Hare, G., Jennings, N. (eds.) Foundations of Distributed Artificial Intelligence, pp. 169–185. Wiley, New York (1996)
60. Shoham, Y.: Agent-oriented programming. Artif. Intell. **60**, 51–92 (1993)
61. Broersen, J., Dastani, M., Hulstijn, J., van der Torre, L.: Goal generation in the BOID architecture. Cogn. Sci. Q. **2**(3–4), 428–447 (2002)

Chapter 9
Renewable Generation, Support Policies and the Merit Order Effect: A Comprehensive Overview and the Case of Wind Power in Portugal

Fernando Lopes, João Sá and João Santana

Abstract The growth of wind power generation over the past decade has surpassed all expectations. The cost of the wind energy support policy was, however, quite significant and to a large extent has led to somewhat intensive debates. The merit order effect (MOE) is an important aspect to be considered in all debates, albeit sometimes oversimplified or even ignored. Accordingly, the central goal of this chapter is to analyze and quantify the reduction in the Portuguese day-ahead market prices achieved by wind power as a result of the MOE in the first half of 2016. The results generated by an agent-based simulation tool, called MATREM, indicate a price reduction of about 17 €/MWh for the entire study period. The (total) financial volume of the MOE reached the considerable value of 391.055 million €. Especially noteworthy is the net cost of the wind energy support policy, which takes into account the feed-in tariff, the market value of the wind electricity, and the financial volume of the MOE. This cost reached the value of −8.248 million € in January 2016, a negative value, indicating that a net profit has occurred in the month. The (total) net cost was 69.011 million € during the study period. Although considerable, this cost should be interpreted carefully, since it did not take into account the interaction of wind generation with the climate policy and the EU emission trading system (i.e., the carbon price effect on the electricity market).

F. Lopes (✉)
LNEG–National Laboratory of Energy and Geology,
Estrada do Paço do Lumiar 22, 1649-038 Lisbon, Portugal
e-mail: fernando.lopes@lneg.pt

J. Sá · J. Santana
Instituto Superior Técnico, Av. Rovisco Pais 1, 1049-001 Lisbon, Portugal
e-mail: joao.de.sa@tecnico.ulisboa.pt

J. Santana
e-mail: jsantana@ist.utl.pt

F. Lopes and H. Coelho (eds.), *Electricity Markets with Increasing Levels of Renewable Generation: Structure, Operation, Agent-based Simulation, and Emerging Designs*, Studies in Systems, Decision and Control 144, https://doi.org/10.1007/978-3-319-74263-2_9

9.1 Introduction

The evolution of renewable energy (RE) has increased substantially over the past decade, supplying approximately 19.2% of the world's final energy consumption at year-end 2014 [1]. The total renewable power capacity—excluding large hydro—has risen from 85 GW in 2004 to 785 GW by the end of 2015. The cumulative global wind capacity was 433 GW in 2015, an increase of 385 GW since 2004. The total global operating capacity of solar photovoltaic (PV) crossed the 100 GW milestone in 2012, reaching a total capacity of 227 GW in 2015 [2]. Renewable energy technologies are currently viewed as tools for improving energy security and mitigating and adapting to climate change. Also, they are recognized as investments that can provide direct and indirect economic advantages by reducing dependence on imported fuels, improving local air quality and safety, advancing energy access and security, propelling economic development, and creating jobs.

Renewable energy support policies have been the primary drivers of the expansion of renewable energy technologies by attracting investment and supporting technology advances. The vast majority of countries worldwide had support policies in place by the end of 2015—specifically, at least 173 countries had renewable energy targets and 146 countries had renewable energy support policies [2]. Feed-in policies—feed-in tariffs (FITs) and feed-in premiums—have been the most popular policy instruments, in place in 75 countries at the national level and in 35 states/provinces/territories. However, renewable energy tendering—also referred to as competitive bidding or auctioning—has gained significant momentum in recent years and is preferred to feed-in policies in a growing number of countries. As of late 2015, at least 64 countries had held renewable energy tenders, with record bids in terms of both low price and high volume. Also, net metering/net billing policies have been adopted in conjunction with other policy mechanisms (e.g., FITs and auctions) and were in force in 52 countries as of year-end 2015 [2].

Europe has been at the forefront of the renewable energy policy design and deployment, developing a strong and vibrant renewable energy industry. As early as 1997, the European Union (EU) published the White Paper for a Community Strategy and Action Plan [3], calling for the community to source 12% of its total energy, including 22% of electricity from renewables by 2010, an ambitious but realistic objective. In 2007, the EU published the Renewable Energy Roadmap [4], further developed and integrated in the Energy and Climate package. The particular EU Renewable Energy Directive 2009/28/EC [5] laid the ground for a policy framework on renewable energy sources (RES) until 2020. The often cited 20–20–20 targets form the core of the Directive and consist of three main pillars: (i) a binding target to increase the amount of energy consumption originating from renewable sources to 20% by 2020, (ii) a binding target to reduce greenhouse gas emission by 20% by 2020, and (iii) a nonbinding target to improve energy efficiency by 20% in relation to projections for 2020. In late 2014, the EU adopted a new regionally binding target, calling for a minimum of 27% renewable energy in final energy consumption by 2030 [2]. At the time of writing, discussions on a 2030/2050 renewable energy policy framework are ongoing.

Portugal has been one of the most enthusiastic countries in Europe in terms of renewable generation. The country adopted the target of achieving 31% of energy consumption from renewables by 2020 (and 40% by 2030) under its Commitment to Green Growth. Wind is one of the most developed renewable energy sources. After 15 years of intense deployment, Portugal reached 5270 MW of cumulative installed wind power capacity in 2016. Also, the annual wind power production was 12480 GWh in 2016 [6]. This production was influenced by favorable wind conditions over the central and northern regions of mainland Portugal, corresponding to the largest concentration of installed wind capacity. Furthermore, boosting an annual average of 2200–3000 h of sun in the mainland, and between 1700 and 2200 h in the Azores and Madeira islands, respectively, Portugal has a strong potential for solar energy [7].

The country has followed a strategy aimed at conciliating market mechanisms and the promotion of values of environmental preservation, sustainability and technological innovation. The adopted policy framework has largely driven the expansion of renewable energy technologies and created markets that support technical advances [8, 9]. Two remuneration systems for the sale of electrical energy generated from renewable sources are in place: a special regime and a general regime. The former is essentially a system of guaranteed remuneration involving a subsidized tariff and the acquisition in full of the energy generated by a last resort trader (LRT). The applicable tariff is determined by adding up several components, notably [7]: (i) the avoided investment costs on new power plants, (ii) the avoided costs of transmission, operation and maintenance, including fuel costs, and (iii) the environmental benefits arising from the use of renewable sources. In 2016, the solar photovoltaic tariff was very generous, with an average value of 295.67 €/MWh, and also remained relatively stable throughout the year, ranging from 290.34 €/MWh to 299.79 €/MWh. The wind power tariff was slightly variable and exhibited a somewhat inconsistent pattern, ranging from 86.37 €/MWh to 105.96 €/MWh, with an average value of 96.28 €/MWh [10]. The general regime consists basically in a generic remuneration scheme allowing generators to carry out their activities and sell their electricity under a market system—through organized markets or bilateral contracts. A particular market entity, referred to as "market facilitator", is responsible for the mandatory acquisition of energy generated from renewable sources and for placing it on the market, mitigating to some extent the risk inherent to the variability and uncertainty of renewable generation [9].

To date, Portugal has achieved a position of reference in respect of the generation of energy from renewable sources and the latest technologies in the electricity generation sector. However, the costs associated with pursuing a strategy that has actively promoted renewable generation, sustainability, and technical innovation, are significant and, to some extent, not compatible with the tariff deficit of the national electricity sector. Undoubtedly, a new remuneration model in the context of a free market will form the backbone of the next legislative changes to be introduced in the energy sector [9]. Until then and for now, there were only timid steps towards establishing a new remuneration regime for generators of energy from renewable sources, particularly for the wind power plants.

The growing importance of renewable energy in the power sector of Portugal has been largely driven by environmental concerns, a dedicated policy initiative, and the improving cost-competitiveness of renewable technologies. The continuous support triggered a rapid expansion of renewable energy deployment and helped renewables to be established as a mainstream source of energy, with more than a twofold increase over the past 10 years, from an annual production of 16188 GWh in 2006 to 33347 GWh in 2016 [6, 11]. As a remarkable consequence, the support payments reached a total of more than 1.75 trillion Euro in 2016 [10]. Accordingly, and in line with several political debates on the targets for renewable electricity generation, some somewhat intensive debates have taken place on the efficiency and the cost of the renewable energy support policy.

Despite the intense interest—and some controversy—it seems essential, however, to obtain a clear and complete picture of the potential interactions between the considerable volume of energy generated from renewable sources and the electricity sector, not only in a qualitative way, but also quantitatively, in monetary terms. Interestingly, renewable electricity generation interacts with the wholesale spot market, typically leading to a reduction of the market-clearing price, thereby affecting to some extent all players in the electricity sector. In fact, the supply curve—also called the merit order curve—goes from the least expensive to the most expensive power technologies and considers the costs and capacities of the different generating plants. Since the electricity generated from renewable sources has normally a very low marginal cost, it enters near the bottom of the supply curve. Practically speaking, this has the effect of pushing the most expensive sources of generation to the right, resulting in a lower wholesale electricity price, depending on the price elasticity of the power demand. This effect is called the merit order effect (MOE) [2, 12]. In addition to it, renewable electricity generation also interacts with the climate policy and the EU emission trading system (ETS), particularly by reducing the market price of allowances (or carbon price).

At present, strategic concerns over the cost of supporting renewables have cast doubt over the future deployment of RE technologies. Whilst the actual cost is "clear", many of the benefits of renewables are difficult to quantify directly. Energy independence, improved air quality and green jobs, whilst no doubt valuable, are difficult to attach an objective monetary value to. However, the MOE is a clear and quantifiable benefit that can ultimately impact consumer bills in a positive way. Accordingly, the central goal of this chapter is to analyze and quantify the reduction in the Portuguese wholesale prices achieved by wind power as a result of the MOE in the first half of 2016. The remainder of the chapter is structured as follows. Section 9.2 analyzes the rapid, albeit sustainable, growth of renewable generation and discusses the evolving policy landscape. Section 9.3 describes in detail the key principles underlying the merit order effect. Section 9.4 presents a study to investigate the MOE of the sustained—and quite significant—deployment of wind power in Portugal. Section 9.5 presents a detailed survey of the literature on the MOE of renewable electricity generation. Finally, Sect. 9.6 states the conclusions and outlines some avenues for future work.

9.2 Renewable Energy Technologies and Support Policies

The global perception of renewable energy has shifted considerably over the past decade. Early in the century, people widely acknowledged the potential of renewable energy, but large-scale deployment still had to be demonstrated. Today, continuing technology advances and rapid deployment of many renewable energy technologies—particularly in the electricity sector—have amply demonstrated their environmental benefits. Renewable technologies are also an economic driver, creating jobs, helping to diversify revenue streams, and stimulating new technological developments. Declining costs together with a global policy landscape have largely driven the expansion of renewable technologies. A handful of countries (e.g., Germany, Denmark, Spain and Portugal) have led the way, developing innovative policies that have driven much of the change witnessed in recent years.

9.2.1 Growth of Renewable Energy Generation

The evolution of renewable energy over the past decade has surpassed all expectations. The global installed capacity and production from all renewable technologies have increased substantially, the costs for most technologies have decreased significantly, and the supporting policies have continued to spread throughout the world. Out of the three end-user sectors—electricity, transportation solutions, and heating and cooling—renewables' share grew fastest in the electricity sector [1]. This growth is shown graphically in Fig. 9.1. Although hydro power has not shown a substantial variation of the total installed capacity (from 920 GW in 2007 to 1064 GW by the end of 2015), wind power and solar PV saw significant increases, with the former rising from 94 GW of installed capacity in 2007 to 433 GW in 2015, and the latter moving from 7.6 GW to 227 GW [2].

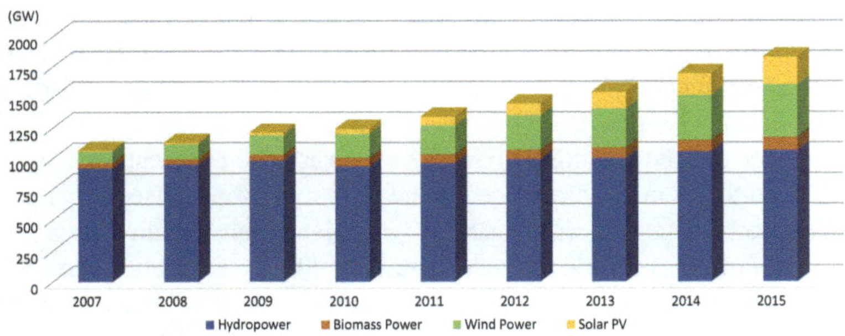

Fig. 9.1 Global installed capacity of renewable technologies in the world over the period 2007–2015 (based on data from [1, 2, 13, 14])

The first years of the century have witnessed the deployment and manufacturing of renewable energy concentrated essentially in Europe, United States of America, and Japan. Since then, deployment and manufacturing have expanded to other regions, and China has become a world leader, having increased the investment in the sector nearly every year over the past decade (the country accounted for nearly one-third of the global capacity added, i.e., more than 200 GW in the period 2004–2014). In 2012, the Middle East and Africa became important markets for the renewable industry. By the end of 2013, investment in renewable energy was also on the rise in Latin America as well as South-East Asia and Oceania [1]. Large amounts of money have flowed to developing and emerging countries across Africa, Asia, Latin America, and the Middle East, in response to a growing interest in renewables.

The year 2015 was a remarkable year for renewable energy, with the largest global capacity additions seen to date. The world saw several developments that all have a bearing on renewable energy, including a significant decline in global fossil fuel prices, various announcements regarding the lowest-ever prices for renewable power long-term contracts, and a historic climate agreement that brought together the global community. Renewables became cost-competitive with fossil fuels in many markets and were established around the world as mainstream sources of energy. The rapid growth in the electricity power sector was driven by several factors, notably cost-competitiveness of renewable technologies, dedicated policy initiatives, better access to financing, environmental concerns, growing demand for energy in developing and emerging economies, and the need for access to modern energy [2].

For the particular case of wind power, the global installed capacity has increased steadily over the past 20 years, with an increase of 270 GW in the period 2004–2014. The growth rate declined, however, in 2013 (down 10 GW to 35.5 GW), due primarily to the steep drop in US installations, from 13 GW in 2012, to just over 1 GW in 2013. The United States—which was the largest global market from 2006 to 2008 and in 2012—fell to sixth place behind Canada [1]. In 2014, the global wind power market resumed its advance, adding a record 51 GW—the most of any renewable technology—for a year-end total of 370 GW. An estimated 1.7 GW of grid-connected capacity was added offshore for a world total exceeding 8.5 GW [14]. In 2015, a new record 63 GW was added for a total of about 433 GW (wind had a record addition for the second consecutive year). The offshore sector had a strong year with an estimated 3.4 GW connected to grids, mostly in Europe, for a world total exceeding 12 GW [2].

Wind power was the leading source of new power generating capacity in Europe and the United States in 2015, and the second largest in China. Also, Canada, Brazil and India have become important markets with Mexico and South Africa growing rapidly. Furthermore, wind power met more than 20% of electricity demand in an increasing number of countries, including Denmark (42% of demand in 2015), Germany (more than 60% in four states), Uruguay (15.5%), Portugal, and Spain. Falling prices due to high competition and technology improvements made wind power an economically feasible power generation technology competing directly with heavily subsidized fossil fuels in an increasing number of markets.

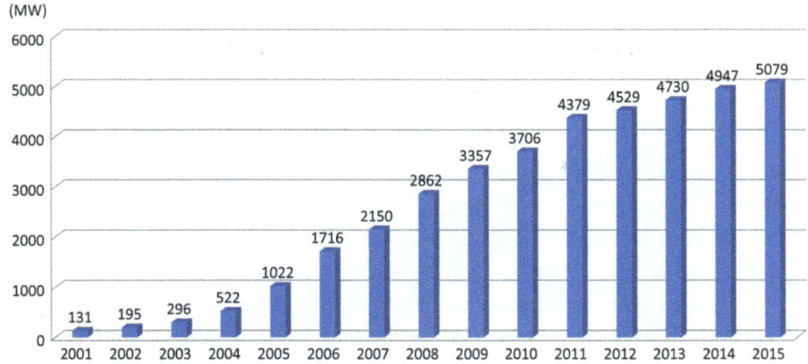

Fig. 9.2 Cumulative installed wind power capacity of Portugal in the period 2001–2015 (based on data from [8, 15–18])

In Portugal, the Government established a "National Renewable Energy Action Plan" in 2013, and laid down the energy sector objectives in a further "Commitment to Green Growth", in 2015. Some of the ambitious goals and policy incentives to the renewable energy investment are as follows [7]: (i) reaching 31% of renewable energy in gross end-consumer power consumption by 2020 and 40% in 2030, (ii) reducing the renewable energy price by 30–40%, and (iii) promoting the export of renewable energy to other European countries.

Figure 9.2 shows the evolution of the cumulative installed wind power capacity of Portugal over the period 2001–2015.[1] Since 2001, wind technology saw a continuous growth, motivated by the Government strategy in endogenous and renewable resources. Moreover, in the period between 2004 and 2011, the country experienced the same rapid growth of installed wind capacity that occurred in many other parts of the world. The global installed capacity crossed the 5 GW milestone in 2015, reaching a total capacity of 5.27 GW by the end of 2016 [6].

In the first three months of 2014, for the first time, the contribution of wind energy exceeded the global thermal generation (conventional thermal and fossil combined heat and power), allowing renewable energy sources, together with large hydro, to attend 90% of the electricity consumption [19]. Also, in 2014, the wind generated electricity contributed to suppress 24.2% of the total demand, in a renewable energy total of 63%, and where the hydro share was 32%. In 2015, the wind generated electricity was equivalent to 23.1% of the total demand [20].

The Portuguese wind market has three key players (EDP Renováveis, Iberwind and Generg), controlling roughly 47% share of the market, seven companies (ENEOP2, EEVM, Enel, Trustenergy, EDF EN, Enersis and Tecneira), comprising 32% share of the market, 7%, and the remaining 21% share of the market being distributed between several players having market shares below 2.5% each [7].

[1]The cumulative wind power capacity is based on data provided by Global Wind Energy Council (GWEC). The reason for choosing GWEC's figures is to present the evolution of the installed capacity using a consistent database based on a standardized methodology for different countries.

9.2.2 Renewable Energy Support Policies

The global policy landscape has largely driven the expansion of renewable energy technologies by attracting investment and creating markets that have supported technology advances. Since 2004, the number of countries promoting renewable energy with direct policy support has nearly tripled, from 48 to over 140, and an ever-increasing number of developing and emerging countries are setting renewable energy targets and enacting support policies [1]. In particular, as of year-end 2015, at least 173 countries had renewable energy targets and an estimated 146 countries had renewable energy support policies, at the national or state/provincial level. Policy mechanisms have received increased interest during that year, due in large part to the 21st Conference of the Parties (COP21) in Paris, where renewable energy technologies were highlighted as a means to mitigate emissions and to adapt to the impacts of the global climate change. By early 2016, the vast majority of countries worldwide had renewable energy support policies in place [2].

Feed-in policies (feed-in tariffs and feed-in premiums) have been the most popular renewable power support and also the primary driver of the renewable energy market growth so far.[2] As of year-end 2015, 110 jurisdictions at the national or state/provincial level had enacted feed-in policies [2]. However, several countries—particularly countries with mature renewable energy markets—have begun to implement important shifts in recent years. In Europe—the birthplace of the modern feed-in policy mechanisms—significant changes were made to a number of national FIT frameworks. For example, many countries at the national level (e.g., Italy, Spain and Greece) made FIT rate cuts, with most reductions focused primarily on solar PV and wind power [14]. Also, Germany removed FITs for solar PV projects of 0.5–10 MW in size in favour of new tender schemes. France and Poland also have used tendering to allocate large-scale renewable energy projects [2].

Globally, tendering (also referred to as competitive bidding or auctioning) has gained significant momentum in recent years and is preferred to feed-in policies in a growing number of countries. By the end of 2015, at least 64 countries had used renewable energy tenders, with record bids in terms of both low price and high volume. For example, in South America, an early adopter of renewable energy tenders, Brazil held several auctions throughout the year, with solar PV and wind power accounting for the majority of project allocations. Also, Peru held its fourth round of auctions, offering 1,300 GWh of biomass, wind and solar PV power [2]. And

[2] *Feed-in policies* guarantee to renewable generators a specified payment over a fixed period. Numerous options exist for defining the level of financial incentive, notably *feed-in tariffs* (the payment is structured as a guaranteed minimum price) and *feed-in premiums* (the payment typically involves the addition of a price premium to the spot price, capped at a maximum amount).

as noted, several European countries have started a transition to tendering, mainly in response to the European Commission's new State Aid guidelines.[3]

In addition to tendering, net metering/net billing has been used to support the deployment of small-scale, distributed renewable energy systems by enabling generators to receive payments for excess on-site generation. Also, in some cases, net metering/net billing policies were adopted in conjunction with other policy mechanisms—such as feed-in tariffs or auctions—that support larger-scale projects. The pace of adoption of such policies slowed in recent years, although this trend reversed in 2015, with four new policies announced at the national level and five added at the state/provincial level. Specifically, net metering/net billing policies were in force in 52 countries as of year-end 2015 [2].

To date, in numerous countries, particularly in European countries, renewables have achieved high shares of penetration in the electricity sector. Given that existing power systems have not been designed to cope with high levels of variable generation, numerous experts point that policy makers will continue to revise existing policy mechanisms to keep pace with changing market conditions, creating new policies that respond to the technical and non-technical challenges of higher renewable energy shares, and expanding renewable energy in the heating, cooling and transport sectors. Overall, experts point that future policies will evolve over time, but will remain an essential part of the renewable energy future [1, 2].

The Policy Framework for Renewable Energy in Portugal. The first Portuguese law guaranteeing grid access to independent power producers using renewable energy sources came into force in 1988.[4] It has also set a feed-in tariff scheme for the first time (with prices in the range 40–50 €/MWh). The legislative framework was revised in 1995 to take into account several different renewable energy sources, including wind power.[5] Also, the feed-in tariff scheme was revised in 1999,[6] and a more complex formula was introduced, taking into account the avoided costs of investing in conventional power plants, the avoided costs of operating and maintaining a conventional power plant, the avoided environmental costs in terms of CO_2 emissions, and the inflation rate [21].

In 2001, consistent with the European Directive on renewable electricity [22], Portugal launched the E4 Programme (Energy Efficiency and Endogenous Energies), setting ambitious objectives related to the amount of energy consumption originated from renewable energy sources (including hydropower). In the same year, the feed-in tariff formula was updated, with the introduction of a new factor, to differentiate between technologies [8]. Specifically, the formula was adjusted by introducing a coefficient Z that affects the environmental savings differently for each technology. The corresponding feed-in tariff increased to an average value of 80 €/MWh [7].

[3]The European Commission State Aid guidelines, issued in 2014, instructed EU countries to begin using tendering to allocate support to new renewable energy projects in 2015–2016, and also required a shift to renewable energy tenders for the majority of projects in 2017.

[4]Decree-Law no. 189/88, enacted on 27 May, 1988.

[5]Decree-Law no. 313/95, of 24 December, 1995.

[6]Decree-Law no. 168/99, passed on 18 May, 1999.

Also in 2001, the new legislation gave a further boost to the wind energy sector by clarifying the license-granting process for grid access and simplifying the administrative procedures. Along with these measures, a special tax payable to the local municipalities, of 2.5% of the total revenue from wind projects, was introduced [21].

Between 2001 and 2005, a major source of investment support was the Incentive Scheme for Rational Use of Energy, which provided capital grants for different types of renewable installations [8]. In 2005, revisions to the previous feed-in tariff legislation limited the power purchase agreements to only the first 33 GWh produced per each MW installed, or 15 years, whatever is reached first, and also decreased the feed-in tariff.[7] A tender for 1800 MW of wind power was also released in three phases. Bid winners gave several discounts, which ranged between 5% for phases A and B, to a maximum of 23% for one of the projects in phase C, meaning that new wind projects received less than 80 €/MWh. Specifically, the reference tariff was around 73 €/MWh, but the lowest bid was only 56 €/MWh [21].

The National Action Plan for Renewable Energy (NAPRE) was presented to the European Commission in 2010. It included a target of 6875 MW of wind power by 2020, of which only 75 MW would be of offshore wind. In 2011, the Portuguese economy was under scrutiny by the International Monetary Fund (IMF), the European Central Bank (ECB), and the European Commission.[8] As a result, a Memorandum of Understanding (MoU) was prepared and several measures were defined for the energy sector, including the renegotiation of existing contracts in renewables and, for new contracts, a revision of FITs ensuring that they do not overcompensate producers for their costs. In other words, efficiency in the promotion of renewable energies was one of the touchstones of the MoU, which imposed a number of restrictive measures, with special emphasis on the revision of the support schemes, to be achieved through: (i) a reduction of the feed-in tariff applicable to contracts currently in force and future contracts, and (ii) the use of less mature technologies. In compliance with the MoU and the Portuguese macroeconomic situation, the NAPRE was revised in 2012.[9] The revision led to a decrease of the wind power capacity to 5300 MW in 2020, corresponding to the installation of the remaining power granted in the 2005 tender, and only a few other equipment projects [8].

Following a free market logic, the Portuguese Government introduced a general remuneration system (or general regime) for generators of energy from renewable sources.[10] Generators were allowed to carry out their activities and sell their electricity under a market system—through organized markets or bilateral contracts. The inherent principle was in fact a completely different principle, contrary to what had happened to date, when generators were, mandatorily and directly, part of an electricity production regime—aptly called special regime—involving a subsidized

[7]Decree-Law no. 33-A/2005, passed on 16 February, 2005.

[8]Statement by the IMF, the ECB and the EC on the first review mission to Portugal. Press release 11/307 (August 12, 2011).

[9]Decree-Law no. 51/2012, enacted on 20 May, 2012. It allowed for the installation of 20% more power than the power stated in the grid connection allowance, in return for a discount on the FIT.

[10]Decree-Law 215-B/2012, enacted on 8 October, 2012.

tariff, and the acquisition in full of the energy generated was made by a last resort trader (LRT). Generators had the possibility to be assisted by a market facilitator, entity responsible for the mandatory acquisition of energy generated from renewable sources and for placing it on the market [9].

In 2013, new remuneration rules for (non-hydro) electricity generators from renewable sources were approved.[11] The rules provided that the guaranteed feed-in tariff will be maintained for an additional period of 5–7 years, after the end of the initial 15-year period (for such tariff). For example, wind power producers were allowed to choose, at the end of the period of 15 years from the respective start of operations, a guaranteed tariff in the range between 74 €/MWh and 98 €/MWh, for an additional period of 5 years (see, e.g., [7] for details of other options). Also in 2013, a Plan for the period 2013–2020 was presented by the Portuguese Government.[12] The strategic reconfiguration was based on reaching an appropriate level of national generation capacity by applying a logic of economic rationality and freedom of initiative of the promoters, without depending on subsidies and guaranteed remuneration. The path to be followed was that of a free market logic [9].

Over the recent years, Portugal has followed a strategy aimed at conciliating market mechanisms and the promotion of values of environmental preservation, sustainability and technological innovation. Both the general and the special regimes are currently still in place. For the specific case of a guaranteed remuneration (the special regime), the electricity generated is delivered to the LRT against payment of the remuneration attributed to the electricity generating plant, mainly in accordance with tariffs following the new remuneration rules. The tariffs take into account the following three key components [7]:

- Avoided investment costs on new power plants.
- Avoided costs of transmission, operation and maintenance, including fuel costs.
- Environmental benefits arising from the use of renewable sources.

The Portuguese Energy Services Regulatory Authority (ERSE) monitors and publishes monthly information on the feed-in tariffs and the special regime production generally. In 2016, the solar photovoltaic tariff was very generous, with an average value of 295.67 €/MWh, and also remained relatively stable throughout the year, ranging from 290.34 €/MWh to 299.79 €/MWh. The wind power tariff was slightly variable and exhibited a somewhat inconsistent pattern, ranging from 86.37 €/MWh to 105.96 €/MWh, with an average value of 96.28 €/MWh [10].

Overall, Portugal has achieved a position of reference in relation to the generation of energy from renewable sources and the latest technologies in the electricity sector. However, the costs associated with pursuing the selected strategy—that is, one that has actively promoted renewable generation and technical innovation—are significant and, to some extent, not compatible with the tariff deficit of the national electricity sector. The recent legislative changes were already in line with a more

[11] Decree-Law 35/2013, enacted on 28 February, 2013.

[12] Council of Ministers Resolution 20/2013, of 10 April, 2013.

flexible strategy, appropriate to the need to reduce the costs of productively carrying it out. Yet, for now, there were only timid steps towards establishing a new remuneration regime for generators of energy from renewable sources. Undoubtedly, a new remuneration model in the context of a free market will form the backbone of the next legislative changes.

9.3 Merit Order Effect

Strategic concerns over the cost of renewable energy support policies have led to significant changes in policy mechanisms and cast doubt over the future deployment of renewables. Whilst these mechanisms may increase consumer bills in the short term, renewables also act in various ways that ultimately can reduce them. In particular, the *merit order effect* (MOE) is a shift of market prices along the supply curve due to the market entry of power stations with low production costs, typically zero or near-zero, displacing the power stations with the highest costs from the market (see Sect. 9.3.1). The literature on this very subject often discusses two different effects: a price effect and a volume effect. The *price effect* is simply the value of the MOE per megawatt hour. The *volume effect* refers to the total savings brought about by VG penetration during a particular year (see Sect. 9.3.2).[13]

9.3.1 Influence of Renewable Generation on Market Prices

In a wholesale spot market, generators compete to supply demand by submitting bids to sell energy at some price for every hour of the day of operation.[14] Retailers and possibly other market participants submit offers to purchase energy at some price for every hour of the day under consideration. A market operator collects the bids to sell energy and sorts them according to the price, building a supply curve—that is, a curve showing the price as a function of the cumulative quantity.[15] Also, the market

[13]Both the price effect and the volume effect are often associated with the *direct effect* of variable generation on spot market prices. Additionally, an *indirect effect* is sometimes mentioned in the literature, related to the climate policy and the EU emission trading system (ETS). The rationale for this effect is as follows. Increasing levels of renewable generation reduces the demand for electricity generated by fossil fired power plants, thereby reducing the demand on the emissions trading market (or carbon market). This, in turn, leads to a reduction of the market price of allowances (or carbon price), as long as the supply curve has a positive slope, thus creating savings for the different entities that take part in the ETS, which may ultimately be reflected in the cost of electricity production (see, e.g., [12, 23]).

[14]A spot market is a market where delivery is immediate [24]. Typically, notation is somewhat abused, and a day-ahead market is often considered a spot market.

[15]The supply curve is also referred to as the *merit order curve*.

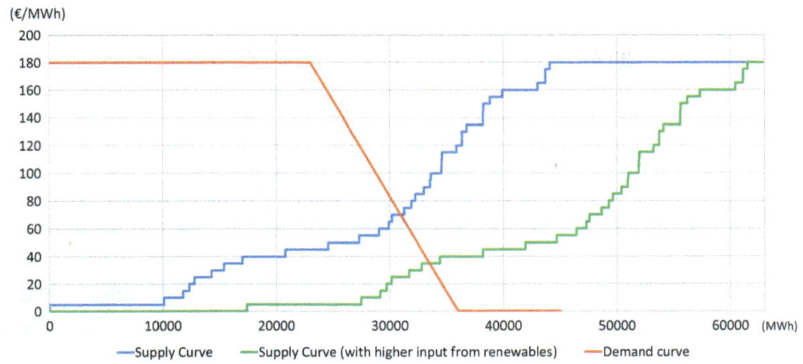

Fig. 9.3 Effect of renewable electricity generation on the spot market price

operator collects the offers to buy energy, ranks them from the highest energy price to the lowest energy price, and builds a demand curve.[16] In a next step, the marker operator clears the market and determines the market prices and the traded volumes. For each hour, the market-clearing price is defined by the intersection of the supply and demand curves. All bids to sell energy with prices lower than or equal to the market-clearing price are accepted. Likewise, all offers to buy energy with prices greater than or equal to the market-clearing price are accepted.

Figure 9.3 shows two typical supply curves (blue and green curves) and a demand curve (orange curve) for an electricity spot market, as well as the points where they meet, which determine the market-clearing price. The power portfolio is made up of a range of electric power generation technologies—from low-carbon, renewable energy technologies (e.g., wind and solar PV) to traditional, fossil fuel technologies (e.g., coal and natural gas combined cycle). Each technology has a *marginal cost* associated with it—that is, the cost of producing an additional unit of electricity at any moment in time [25]. For renewable generators, which have no fuel costs and low maintenance costs, the marginal cost is near-zero. For fossil-fuel-fired power plants, the marginal cost is predominately the cost of fuel.

The left-hand supply curve (blue curve) represents the aggregated bids of the various generating companies to provide energy and goes from the least expensive to the most expensive power technologies. Generators are called upon according to the merit order to meet demand, such that those with the lowest marginal costs are the first ones brought online, and those with the highest are brought on last.[17] Accordingly, the bids from renewable generators enter the supply curve at the very

[16]The demand curve may simply be a vertical line defined by considering the value of the load forecast, since the demand for electricity may be, and typically is, considered (highly) inelastic and set according to a forecast of the load.

[17]A list of generators in ascending order of marginal cost is known as a *merit order* [26].

bottom, followed by coal and gas units, while condensing plants are those with the highest marginal costs of power production.[18]

The demand curve represents mainly the demand of retailers and large direct consumers. This curve is relatively steep, meaning that the demand for electricity is considered inelastic—that is, it remains almost unchanged in spite of a given percentage change in the price. Since demand is near independent of the spot price in the short term, minor changes in the supply may result in major price changes (see Fig. 9.3).

The way in which supported renewable electricity generation influences the spot market prices can be described as follows. Since renewable generators have low marginal costs (and therefore enter near the bottom of the supply curve), when the supply of power from renewable energy sources increases, it has the effect of shifting the supply curve to the right. At a given demand, this results in a lower wholesale price, as illustrated by the right-hand supply curve (green curve) of Fig. 9.3. Power stations with high generation costs may be vulnerable to being pushed out of the market, since they may find it hard to compete at a lower price.

Alternatively, the influence of supported renewable electricity generation on spot market prices can be interpreted as a "negative" demand impact. In fact, the electricity generated by renewable energy sources is privileged in a way that it is bought by grid operators to pass it on to specific electricity traders. Accordingly, the demand that needs to be purchased on the market is reduced correspondingly, thereby shifting the demand curve to the left ("negative" shift). As long as the supply curve has a positive slope, the reduced demand leads to a lower wholesale electricity price.[19]

Now, it is worth noting that electricity is bought in wholesale spot markets as well as contracted months or years in advance via bilateral contracts. As such, it is not immediately obvious how potential savings brought about by reduced wholesale prices will be passed through to end users. Nevertheless, since the spot market price is typically the leading price indicator for all electricity trades, the price paid through contracts will be based, to some extent, upon that market price. Accordingly, a reduction in the spot price will ultimately impact the price of new contracts.

9.3.2 Volume Effect

The volume effect is expressed as the difference between the market price excluding variable generation and the market price including VG times the total demand for

[18]Notice that large hydropower stations are not mentioned explicitly, since hydro bids are normally considered strategic, depending on precipitation and the level of water in reservoirs.

[19]Notice that this particular way to interpret the influence of RES on market prices, the so-called "negative" demand impact, is mentioned in this section for reasons associated with completeness, but is not illustrated in Fig. 9.3.

electrical energy [23]:[20]

$$V = \sum_{h=1}^{t} (x_h - p_h) \times d_h \tag{9.1}$$

where V represents the financial volume of the MOE (in €), t is the time period under consideration (in hours, i.e., $t = 8760\,\text{h}$ for a one-year period), x_h is the hourly spot market price excluding renewable energy generation (€/MWh), p_h is the hourly spot market price when renewable energy is part of the generation mix (€/MWh), and d_h is the total electricity demand (MWh).[21]

From the consumer perspective, an interesting indicator for the discussion of the actual cost of renewable electricity support is the specific value of the merit order effect [23]:

$$S = \frac{V}{R} \tag{9.2}$$

where V is the financial volume of the MOE (€), R is the electricity generated by renewable energy sources (MWh), and S is the specific value of the MOE (€/MWh).

9.4 Wind Generation and the Portuguese Electricity Prices

This section investigates the merit order effect of the intensive deployment of wind power in Portugal and analyzes the potential consumer's cost of the wind feed-in tariff when including the MOE. It presents a careful study on the Iberian electricity market (MIBEL), making use of data extracted from the managing entity of the spot market (OMIE), as well as data from the Portuguese (REN) and Spanish (REE) electrical grids.

The section is organized into three major parts. The first part, called "preliminary analysis", consists in analyzing the market in days with different levels of wind power production. This simple, albeit interesting and relatively important, analysis is helpful in identifying the influence of wind generation on the spot market prices (see Sect. 9.4.1). The second part, referred to as "simulation-based study", involves the simulation of the Iberian market prices in the period between January 1 and June 30 of 2016. The analysis is carried out using the agent-based system called MATREM (for Multi-Agent TRading in Electricity Markets). Two scenarios are considered:

- Scenario A: the supply and demand curves are built from specific bids and offers submitted to MIBEL.
- Scenario B: the supply curve is built as in scenario A. However, to simulate what would have been the market prices in the absence of wind generation, the value of the electricity demand is changed correspondingly.

[20]The volume effect is also known as the financial volume of the merit order effect.

[21]Equation (9.1) considers that the total energy demand is traded at the spot market.

The effect on the market price is estimated as the difference between the prices of electricity in scenarios A and B (see Sect. 9.4.2). The last part of the section summarizes the results and discusses the conclusions reached (see Sect. 9.4.3). As noted earlier, Portugal's target of a 31% share of total final energy from renewables by 2020 (and 40% by 2030), coupled with a strong support policy predating 1988, have been instrumental in making the country a reference in terms of renewable electricity generation. The wind energy sector achieved a maturity status within the national power system. Accordingly, the conclusions obtained from the study may be considered relevant to other countries that are promoting renewables with direct policy support.

9.4.1 Preliminary Analysis

A simple approach to analyze the influence of wind power on spot power prices is to investigate the interaction between the level of wind production and the market price. In this way, we examine the market in two working days from the fourth week of April 2016, namely April 18, 2016 (a Monday) and April 22, 2016 (a Friday). The level of electricity demand is similar in both days. Although rather *ad hoc*, this approach has the advantage of isolating the impact of wind generation from other factors that may affect the market price, notably the level of demand, the evolution of fuel prices, the differences in the levels of electricity generation with hydro sources, and the unexpected unavailability of thermal plants.

Figure 9.4 depicts the market price for each hour of the two weekdays.[22] The rates of wind power penetration differ substantially from one day to another[23]:

- On 22 April, the level of wind generation was low during most of the day, while the level of thermal-based generation was quite significant. The estimated rate of wind power penetration was 2%. The market price stood at a considerable average value (specifically, at 34.32 €/MWh).
- On 18 April, a significant level of wind generation was observed during several hours. The estimated rate of wind power penetration was 32%. The market price decreased substantially (the average value was 22.33 €/MWh).

Thus, there was a significant impact on the power price, which might increase in the long term if even larger shares of wind power are fed into the system.

Certainly, factors other than the wind power production may influence the price on the day-ahead (spot) market. But the close interaction between wind power and the

---

[22]The following sources of data were used for the analysis: hourly generation from wind [27, 28] and hourly day-ahead (spot) prices published by OMIE [29].

[23]Wind energy penetration was expressed as the ratio of wind power generation to electricity demand. Accordingly, it represents the approximate share of consumption met by real wind energy production. Albeit on an annual basis, similar definitions of wind energy penetration appear in the technical literature, as well as definitions of wind power capacity penetration (see, e.g., [30, 31]).

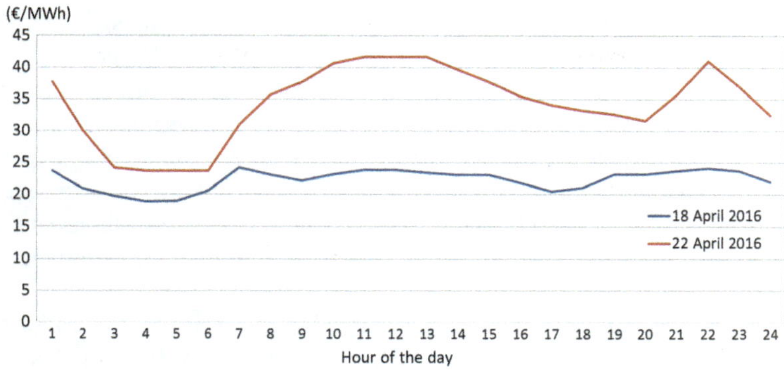

Fig. 9.4 Hourly spot prices from OMIE on 18 April 2016 (32% of wind power penetration) and 22 April 2016 (2% of wind power penetration)

market price is clearly verified—the price was reduced from around 34.32 €/MWh (on 22 April) to 22.33 €/MWh (on 18 April). And generally speaking, when the market price is lowered, such a reduction is ultimately beneficial to all power consumers, since it applies to all electricity traded in the market.

9.4.2 Simulation-Based Study

The literature on renewable electricity generation and spot market prices is rather considerable. The existing studies may be broadly classified into the following three groups [32]:

1. *Simulation-based studies*: rely on simulation models and in-house software based on realistic representations of the electricity market and the clearing process. Typically, use real data (although hypothetical data may be used as well).
2. *Empirical studies*: aim at estimating the relationship between market prices and renewable generation. Typically, use econometric models and real market data, which reflect the true conditions observed "after-the-fact".
3. *Limited information studies*: provide some information on the price effect of renewable production, but typically do not yield a precise quantification.

The studies of each group have specific strengths and weaknesses. However, as noted, the simulation-based studies rely on a realistic representation of the electricity market and a clear understanding of the complexities inherent to the price formation process. Accordingly, the remainder of this section presents an insightful simulation-based study conducted with the help of the agent-based simulation tool called MATREM.

The tool simulates a power exchange where supply bids and demand offers are aggregated to find a clearing price at which supply and demand are equal (see, e.g., [33–35]). MATREM also simulates a derivatives exchange where private parties can trade standardized bilateral contracts. This exchange uses an electronic trading system that automatically matches the bids and offers from various market participants. Furthermore, MATREM simulates a marketplace for negotiating the details of tailored (or customized) long-term bilateral contracts, specifically contracts designed to cover the delivery of large amounts of energy over long periods of time (see, e.g., [36–38]). To this end, buyer and seller agents are equipped with a negotiation model that handles two-party and multi-issue negotiation (see, e.g., [39, 40]).[24]

The study makes use of data published by OMIE, the managing entity of the daily Iberian electricity market, as well as data extracted from REN (the Portuguese electrical grid) and REE (the Spanish electrical grid). Specifically, the following sources of data are considered:

- Hourly energy prices and quantities submitted to the daily Iberian market (data published by OMIE [29]).
- Hourly generation from wind for both Portugal and Spain (data reported by REN [27] and REE [28]).
- Hourly day-ahead (spot) prices and traded energy quantities (data published by OMIE [29] an also reported by REN [41]).

The time period of the study has the duration of six months: from January 1, 2016 to June 30, 2016 (a total of 4368 h).

Now, to simulate the Iberian market prices for this period, there is a need to define the software agents that participate in the simulated day-ahead (spot) market—that is, the agents that submit bids to sell energy (the suppliers or sellers) as well as the agents that submit offers to buy energy (the demanders or buyers). A detailed examination of MIBEL reveals a number of bids and offers on the order of thousands (for a particular hour of operation). Thus, to perform computer simulations as close to the reality as possible, while overcoming the added computational complexity of considering a very large number of software agents, we make the following two simplifying assumptions:

- The electricity supply industry is represented by 39 software agents. The agents submit bids to sell energy at prices in the range between 0 €/MWh and 180 €/MWh.
- The demand for electrical energy is perfectly inelastic. A single agent submits an offer to buy the entire electricity demand at 180 €/MWh.

Table 9.1 presents the 39 supplier agents and summarizes their energy bids. Each bid is expressed as a quantity and price pair. The table also presents the demander agent and the corresponding offer to buy electrical energy.

[24] *Chapter* 8 is entirely devoted to the agent-based simulation tool and the interested reader is referred to it for further technical details of the power and derivatives exchanges, the bilateral marketplace, the user interface and the human-computer interaction paradigm, as well as the various types of software agents.

Table 9.1 Software agents and their energy bids/offers for a particular hour of the market horizon

Agent	Agent type	Energy quantity (MWh)	Energy price (€/MWh)
1	Supplier	Wind power production in Portugal	0
2	Supplier	Wind power production in Spain	0
3	Supplier	Quantity of the remaining bids at 0 €/MWh	0
4	Supplier	Quantity of all bids in the range]0, 5] €/MWh	5
5	Supplier	Quantity of all bids in the range]5, 10] €/MWh	10
. . .	. . .	. . .	. . .
38	Supplier	Quantity of all bids in the range]170, 175] €/MWh	175
39	Supplier	Quantity of all bids in the range]175, 180] €/MWh	180
40	Demander	Quantity associated with the market price	180

The rationale for defining 39 agents to represent the electricity supply industry is as follows. Consider first the bids to buy energy at 0 €/MWh submitted to the Iberian electricity market, at any given hour of the market horizon. The decision to associate a software agent to the Portuguese gross generation from wind power, another agent to the Spanish gross generation from wind power, and a third one to all the remaining bids at 0 €/MWh, seems to be intuitive and natural.

Now consider the bids to buy energy at a price greater than 0 €/MWh (the price of the bids submitted to MIBEL ranges between 0 €/MWh and 180.30 €/MWh). The decision to decompose the interval]0, 180] into 36 sub-intervals, based on a somewhat arbitrary increment of 5 €/MWh (i.e., to define the sub-intervals]0, 5],]5, 10], . . . ,]175, 180]), and to associate a software agent to each sub-interval, more specifically to the bid to supply energy at the price corresponding to the upper bound of each sub-interval, seems to be simple, satisfactory and rather elegant (yet somewhat *ad hoc*). Accordingly, a software agent playing the role of a supplier submits a bid to sell the quantity of energy corresponding to the sum of the quantities of all bids (submitted to MIBEL) in the range]0, 5] at 5 €/MWh. Another agent submits a bid to supply the sum of the quantities of all bids in the range]5, 10] at 10 €/MWh, and so on. Notice that the generation technologies associated with the bids submitted to MIBEL are not publicly available and, as a result, no attempt was made to associate each software agent with only a single technology.

Figure 9.5 shows the supply and demand curves published by OMIE [29] on Friday 5 February 2016 at 9 a.m. (a typical day of operation). Specifically, the figure shows the "initial" supply and demand curves (light orange and light blue curves, respectively). Also, the figure shows the actual supply curve (red curve), obtained by considering the generation restrictions of complex sale bids [42], and the actual

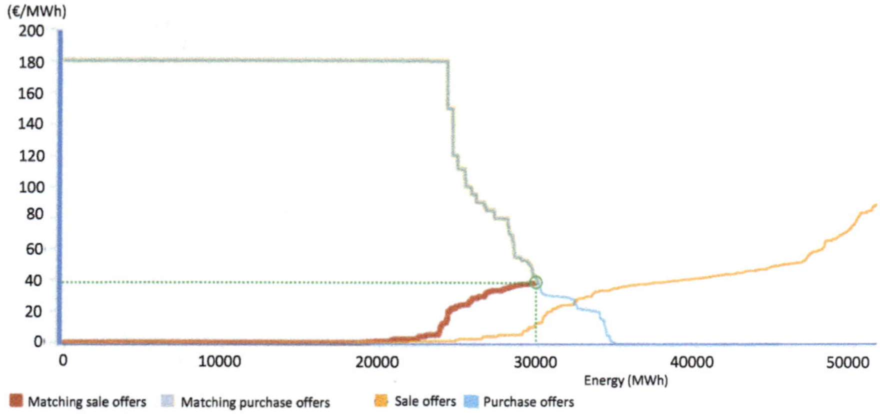

Fig. 9.5 Supply and demand curves published by OMIE on 5 February 2016 at 9 a.m. (adapted from [29])

demand curve (grey blue curve). The intersection of theses curves (green point) determines the market-clearing price (39 €/MWh) and the equilibrium energy quantity (29959 MWh).

As noted earlier, the study involves two energy scenarios related to the Iberian electricity market (i.e., the Portugal-Spain region):

- *Scenario A*: the supply curve is based on the "initial" supply curve (see Fig. 9.5). More specifically, the supply curve is built from the bids of 39 supplier agents, defined by considering the bids submitted to MIBEL and increments of 5 €/MWh. The demand curve is built from the offer of a single demander agent (assuming that the demand for electricity is perfectly inelastic). The offer of this agent involves an energy quantity determined by considering the market-clearing price (39 €/MWh, in Fig. 9.5) and the simple sale bids submitted to MIBEL.
- *Scenario B*: the supply and demand curves are built as in scenario A. However, to simulate what would have been the market prices in the absence of wind generation, the value of the electricity demand is changed correspondingly.

Also, to simulate the Iberian market in a correct and realistic fashion, the study accounts for the price differences between Portugal and Spain that result from market splitting in the daily horizon.[25] Market splitting involves basically the segmentation of the Iberian market into two independent markets due to congestion in the Portugal-Spain interconnection, typically leading to different prices for the Portuguese and Spanish areas, yet making possible to exhaust the available capacity safely. In the interests of completeness, and also for the sake of clarity, we present next a brief description of the operation of the Iberian market, placing emphasis on the supply and demand curves and, mainly, the price formation process, which may involve or

[25] *Market splitting* in the daily horizon is the main mechanism to jointly manage the Portugal-Spain interconnection, following a proposal made by the Regulatory Council [44] (see also [45]).

not involve market splitting, depending on the equilibrium quantity (see below for a summary, and [43] for full details).

Supply bids made by generating companies, which may include complex conditions, are submitted per production unit, and specify independent quantities and prices for each hour of the market horizon. These bids are sorted by ascending price and a market supply curve is built for each hour. Adjustable hydroelectric power stations tend to appear in the upper part of the curve, since their opportunity cost is high. To the contrary, run-of-river plants usually appear on the lower part of the curve, as they cannot store water for long periods of time. Also, nuclear power plants usually appear in the lower part of the curve in the Spanish zone, as their opportunity cost is low. The middle section of the curve includes coal plants. The highest end of the curve has combined cycle plants and, as noted, the part of hydraulic power with scant reserves.

Demand offers in the day-ahead market cannot include complex conditions. These offers are ranked in order of decreasing price and a demand curve is built for each hour. The highest part of the curve corresponds to the demand associated with distributors (regulated supplies), typically involving the instrumental price of 180.30 €/MWh. The middle and lower parts of the curve include consumption corresponding to pumping stations and to providers for their supply in the free market, who present offers specifying energy prices different from the instrumental price.

The intersection of the supply and demand curves determines the market equilibrium.[26] Figure 9.5 represents a situation in which there is a single price for the Iberian Peninsula (i.e., there is no congestion in the Portugal-Spain interconnection, and thus the Portuguese and the Spanish regions are treated together, resulting in a single Iberian market). However, for all hours in which the equilibrium corresponds to a level of use of the lines that join the two countries greater than the available capacity—that is, the maximum net transfer capacity—market splitting will occur, and two price areas will be considered.

More specifically, the following two situations may occur when matching the supply and the demand for electrical energy [43]:

- If the traffic in the Portugal-Spain interconnection is less than or equal to the available capacity in a given direction, the equilibrium price is the same for Portugal and Spain, since there is economic viability (guaranteed by the matching of the supply and the demand), and also technical viability (guaranteed by the existence of capacity in the networks to realize the economic dispatch). This situation is referred to as market integration.
- If the traffic in the interconnection is greater than the available capacity in a given direction, the actual equilibrium situation—and the corresponding economic dispatch—cannot be realized. The Portuguese and Spanish regions are treated separately, with particular supply and demand curves for each region. An amount corresponding to the commercial capacity in the interconnection in the exporting direction is placed in the demand curve for the exporting system and an equivalent

[26] Strictly speaking, and as depicted in Fig. 9.5, the market-clearing price results from the intersection of the so-called actual supply and demand curves.

amount is placed in the supply curve for the importing system (i.e., the supply and demand curves are changed by the value of the interconnection capacity). The intersection of the resulting supply and demand curves for each of the two regions is used to determine the market-clearing prices for Portugal and Spain. This situation is referred to as *market splitting*.

The occurrence of market splitting generates a price differential, since the supply of the exporting market that ensures maximum traffic in the interconnection is paid at the equilibrium price of that market, while the corresponding demand is paid at the equilibrium price of the importing market.[27]

Market splitting is used to handle structural congestion known before scheduling (i.e., when allocations are not final). Congestion emerging after scheduling, when allocations are final, is solved using coordinated balancing activities (CBAs) between the Portuguese and the Spanish system operators. Such CBAs consist mainly in counter-trading measures—that is, energy transactions induced by the system operators in real time which superimpose the pre-existing cross-border trading schedule, making it possible even if congestion arises. The idea underlying the method is based on principles of transparency, although it gives rise to some additional costs (but see [43]).

Returning to the simulation-based study, the following two situations are considered to account for market splitting:

- For all hours in which there is no congestion in the Portugal-Spain interconnection, and thus there is a single price for Portugal and Spain, the agent-based simulations involve both the Portuguese and the Spanish regions.
- For the remaining hours, i.e., for all hours in which market splitting occurs, and thus different price areas are considered for Portugal and Spain, with particular supply and demand curves for each region, the agent-based simulations involve the Portuguese region only.

Also, it is especially noteworthy at this stage that the study involves the simulation of the Iberian market prices using the agent-based tool called MATREM. The capability of MATREM to produce realistic day-ahead (spot) market prices is, therefore, very important and should be rigorously and thoroughly analyzed (e.g., in terms of value, reliability and statistical consistency). Hence, in a calibration (and benchmarking) procedure, real hourly electricity prices published by OMIE are compared with simulated hourly prices generated by the simulation tool for scenario A. More specifically, the following two comparisons are considered:

- A first comparison between the time series of hourly prices from OMIE and the simulated prices on Friday 5 February 2016.
- A second comparison, similar to the first one, but involving the time series of hourly prices from OMIE and the simulated prices for the time period of the study (i.e., from January 1, 2016 to June 30, 2016).

---

[27] This price differential multiplied by the traffic in the interconnection corresponds to the *congestion income*.

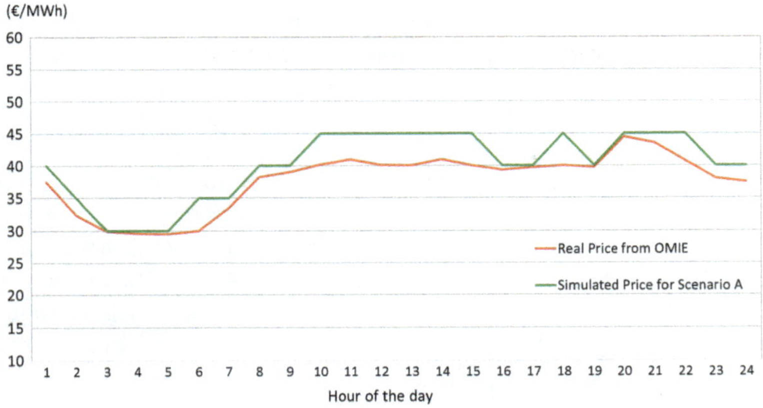

Fig. 9.6 Comparison of hourly spot prices from OMIE and simulated prices on 5 February 2016

The results of the first comparison—a somewhat preliminary comparison—are shown in Fig. 9.6. The high correlation between the real and the simulated prices indicates that MATREM is a promising system for simulating the spot market prices (despite the slight overestimation of the real hourly prices).

Table 9.2 summarizes the results of the second comparison. The first lines of the table show the average real electricity prices published by OMIE, the average simulated prices for scenario A, and the mean absolute error in the simulated prices, for each month of the time period of the study. The bottom line shows the average results for the entire 6-month period. The simulated prices correctly fit those reported by OMIE. The average of the mean absolute errors is less than 2.50 €/MWh, a rather low value in relative terms, which may be considered not significant.

Technically, the comparison of the real and the simulated prices shows that MATREM is a reliable system for simulating the Iberian day-ahead (spot) market.

Table 9.2 Comparison of real and simulated market prices for the period between January and June of 2016

Month/Period (of 2016)	Average real electricity price (€/MWh)	Average simulated price for scenario A (€/MWh)	Mean absolute error (€/MWh)
January	35.71	38.01	2.43
February	26.48	28.88	2.52
March	27.49	29.76	2.43
April	23.35	25.86	2.50
May	24.84	27.25	2.43
June	38.30	40.70	2.44
January to June	29.39	31.76	2.46

Notice, however, that the simulated prices generally tend to be slightly higher than the real prices. A possible explanation for this slight increase can be the fact that the supplier agents prepare bids to purchase energy based on a somewhat arbitrary increment of 5 €/MWh. A reduction of this increment could be an interesting issue for further investigation.

Overall, as pointed out throughout this section, the study involves 40 software agents (30 suppliers and a demander agent), two energy scenarios related to the Iberian electricity market (scenarios A and B), and a 6-month trading period (a total of 4368 h). Accordingly, there is a need to perform 8736 simulation runs (i.e., 4368 simulation runs for scenario A and 4368 for scenario B). Each simulation run corresponds to a given hour of operation and involves basically the following:

1. Set simulation- and agent-specific parameters.
2. Obtain the hourly generation from wind for both Portugal and Spain.
3. Obtain the hourly energy prices and quantities submitted to the daily Iberian market.
4. Analyze the occurrence of market splitting and consider either the Portuguese and the Spanish regions or the Portuguese region only.
5. Prepare the bids of all supplier agents (perform the procedures outlined above).
6. Prepare the offer of the demander agent.
7. Submit the bids and the offer to the day-ahead market (incorporated in the MATREM system).
8. Determine the market-clearing price and announce it to market participants.
9. Prepare a simulation report.

9.4.3 Results and Discussion

Tables 9.3, 9.4 and 9.5 summarize the results of the simulation-based study. Table 9.3 shows the average monthly rate of wind power penetration, the evolution of the average monthly prices actually observed in OMIE, the average monthly prices corresponding to the simulated dispatch taking into account the real wind generation (scenario A), and the average monthly prices corresponding to the simulated dispatch in the absence of wind generation (scenario B). The simulated prices for scenario A vary between 25.86 €/MWh (in April) and 40.70 €/MWh (in June). These prices are considerably lower than the prices computed for scenario B, which range between 39.89 €/MWh (in May) to a maximum of 62.56 €/MWh (in January). The bottom line of the table presents the average simulated prices for the entire period of the study, showing that the intensive deployment of wind power in Portugal has a substantial impact on market prices.[28]

[28]Notice that several factors other than wind generation may influence the market price (e.g., the evolution of fuel prices, the level of electricity generation with hydro sources, or maintenance activities). In fact, the month with the highest average wind power penetration rate was not the one with the lowest average real electricity price. Specifically, in February, the average penetration

Table 9.3 Simulation results: average market prices for scenarios A and B

Month/Period (of 2016)	Average wind power penetration (%)	Average real electricity price (€/MWh)	Average simulated price for scenario A (€/MWh)	Average simulated price for scenario B (€/MWh)
January	33.45	35.71	38.01	62.56
February	35.98	26.48	28.88	53.49
March	27.34	27.49	29.76	46.86
April	27.97	23.35	25.86	41.21
May	23.10	24.84	27.25	39.89
June	20.63	38.30	40.70	48.51
January to June	28.11	29.39	31.76	48.56

Table 9.4 shows the key indicators related to the merit order effect (MOE). As noted earlier, the MOE is a shift of market prices along the merit-order (or supply) curve due to the market entry of power stations with low production costs. The price effect is the value of the MOE per megawatt hour. This effect is estimated directly from the absolute value of the difference between the simulated prices for scenarios A and B. The financial volume of the MOE refers to the total savings brought about by wind power penetration during the time period of the study (see Eq. 9.1). The specific value of the merit order effect is expressed as the ratio of the financial volume of the MOE to the electricity generated by wind power (see Eq. 9.2).

As shown in the second column of Table 9.4, the reduction of the average market price during the study period varies between 8 €/MWh and 25 €/MWh, reaching the highest value in January and February, and the lowest value in June. As expected,

Table 9.4 Simulation results: key indicators related to the merit order effect

Month/Period (of 2016)	Average price reduction (€/MWh)	Financial volume of the merit order effect (million €)	Specific value of the merit order effect (€/MWh)	Average feed-in tariff (€/MWh)
January	25	96.171	71	99
February	25	90.988	66	98
March	17	70.668	62	96
April	15	56.671	54	98
May	13	47.008	53	97
June	8	29.549	37	97
January to June	17	391.055	59	97

rate reached a maximum of 35.98% and the average market price was 26.48 €/MWh. In April, the average penetration rate was (only) 27.97%, but the average market price fell to a minimum of 23.35 €/MWh. A possible explanation for this effect can be the evolution of the reference price of natural gas during the period under consideration—that is, slightly above 15 €/MWh in January, falling to nearly 12 €/MWh in April, and raising to nearly 15 €/MWh in June [46].

the price effect is more significant in the months with the highest levels of wind power penetration, indicating that wind generation has a (relatively large) decreasing impact on the Iberian market prices (although, as already noted, other factors may also influence the market prices). On average, a price reduction of about 17 €/MWh is estimated for the entire 6-month period (see the bottom line of Table 9.4).

The financial volume of the MOE ranges between 29.549 million € (in June) to a maximum of 96.171 million € (in January). As expected, June is the month of the study period with the lowest financial volume (since it is actually the month with the lowest wind penetration rate). For the entire study period, the (total) volume of the MOE reaches 391.055 million €. The specific value of the MOE varies between 37 €/MWh (in June) and 71 €/MWh (in January). To some extent, this indicator is important for the discussion of the efficiency and actual cost of the wind energy support policy. Specifically, it allows a (preliminary) comparison to the average feed-in tariff (FIT) for wind energy, which varies between 96 €/MWh (in March) and 99 €/MWh (in January). For the entire 6-month period, the specific value of the MOE is 59 €/MWh and the average FIT reaches the value of 97 €/MWh.

Table 9.5 shows various different costs related to the wind energy support policy. The market value of the electricity produced from wind turbines during the period of the study is undoubtedly a very important aspect for the discussion of the actual cost of the support policy. This value may be roughly estimated by multiplying the hourly day-ahead (spot) prices observed in OMIE by the corresponding hourly generation from wind in Portugal.[29] Leaving minor aspects aside, the direct cost of the support policy—also referred to as the effective cost or additional cost—is obtained from the feed-in tariff and the market value of the wind electricity.

Specifically, the direct cost of support for wind generation is estimated by the following equation:

$$DC = \sum_{h=1}^{t} (\mathit{fit} - p_h) \times w_h \tag{9.3}$$

where t is the time period under consideration (in hours, i.e., $t = 4368\,\text{h}$), *fit* is the feed-in tariff for wind energy (€/MWh), p_h is the hourly spot market price when renewable energy is part of the generation mix (€/MWh), w_h is the hourly generation from wind power (MWh), and DC represents the cost of the support policy when considering the market value of wind electricity (€).

The direct cost ranges between 46.244 million € (in June) to a maximum of 100.929 million € (in February). As expected, this cost is more significant in the months with the highest levels of wind power penetration (i.e., January and February). For the study period, it reaches the value of 460.066 million €. The financial volume of the merit order effect takes into account the reduction in the electricity prices as a result of wind generation, and therefore should be interpreted as a saving (from

[29]Notice that a considerable volume of electricity is traded via bilateral contracts, whose price may deviate from the spot market price (e.g., the notifications of physical delivery of energy from forward contracting conducted in OMIP are, for all purposes, considered supply bids at a specific instrumental price [43]). Nevertheless, the price paid through bilateral contracts is based, to some extent, upon the spot market price, since it is the leading price indicator for all electricity trades.

Table 9.5 Simulation results: key costs related to the wind energy support policy

Month/Period (of 2016)	Average feed-in tariff (€/MWh)	Direct cost of support policy (million €)	Financial volume of the merit order effect (million €)	Net cost support policy (million €)
January	99	87.923	96.171	−8.248
February	98	100.929	90.988	9.941
March	96	80.475	70.668	9.807
April	98	78.663	56.671	21.992
May	97	65.832	47.008	18.824
June	97	46.244	29.549	16.695
January to June	97	460.066	391.055	69.011

the consumer perspective). As indicated previously, the (total) volume of the MOE reaches 391.055 million €. This considerable value shows that the actual cost of the wind energy support policy is dramatically reduced once the merit order effect is taken into account.

To this end, the net cost of the support policy is estimated as the difference between the direct cost of the support policy and the financial volume of the MOE. It has the value of −8.248 million € in January, a negative value, indicating that a net profit has occurred in this month. In other words, the wind electricity promotion did not entail an additional cost for consumers. On the contrary, there has been a significant saving of more than 8 million €. Whether this saving (created on the wholesale market) is passed on to consumers heavily depends on the competitiveness of the electricity supply system, especially the retail market. For the other five months of the study period, the net cost is positive, meaning a financial burden for consumers (as a result of the wind energy promotion). Specifically, the net cost varies between 9.807 million € (in March) and 21.992 million € (in April).

The (total) net cost reaches the considerable value of 69.011 million € during the 6-month period of the study. However, in addition to the market value of the wind electricity and the reduction in the market prices achieved by wind power as a result of the MOE, another aspect should be taken into account, namely the interaction of wind electricity generation with the climate policy and the EU emission trading system (ETS). Indeed, wind generation reduces the demand for electricity generated by fossil fuel fired power plants, thereby reducing the demand on the emissions trading market (or carbon market). This, in turn, leads to a reduction of the market price of allowances (or carbon price), typically creating savings for the different entities that take part in the ETS.

The carbon price is normally part of the variable cost of fossil fired power plants, and thus a lower carbon price should lead to a lower variable cost of conventional power plants. This reduction in cost results in a shift of the supply curve of the spot market downward and, as long as this curve has a positive slope, creates a price effect—often referred to as the indirect effect—similar to the merit order effect. Although very interesting, the quantification of this effect is a very complex task and, therefore, deferred to future work. Accordingly, the net cost reported above should be interpreted carefully, as as preliminary cost of the wind energy support policy, and definitely not as a result that reflects the actual cost of the support policy.

9.5 Renewable Generation and Spot Prices: Literature Review

The impact of increasing levels of renewable generation on electricity markets is, at the time of writing, a contemporary topic, and the literature on this very important subject is rather substantial. Studies tend to focus on specific countries or regions, since the mix of generation technologies and the renewable energy support policies vary considerably. Furthermore, most studies involve slightly different sets of assumptions and somewhat complex methodologies and algorithms. Accordingly, this section presents a detailed technical review of nine representative articles that exist in the literature.[30] The key aspects covered in the review are as follows:

1. Country or region, time period and renewable energy technology.
2. Type of study (i.e., simulation-based or empirical) and the corresponding model.
3. Key assumptions.
4. Results and main conclusions.

Some articles deal essentially with the price effect of renewable electricity generation on market prices, whereas others focus on related topics such as the impact on subsidy schemes and/or the market design. Hence, we analyze mainly the following: (i) the influence of VG on market prices, producing a merit order effect, and (ii) the potential cost and efficiency of the renewable energy support policy.

Sensfuß et al. [47] analyze the impact of feed-in supported electricity generation on the German market. The time period has the duration of six years (2001–2006). The key assumptions are as follows: (i) electricity demand is inelastic in the short-term perspective of the day-ahead market, and (ii) all energy is traded in the simulated day-ahead market. The authors use the PowerACE cluster system [23], an agent-based simulation tool. The results indicate that the price reduction due to renewables is considerable. The largest reduction in the average market price was 7.83 €/MWh and occurred in the year 2006. The total volume of the MOE was 4.98 billion € in 2006. Also, the market value of the generated renewable electricity was estimated as 2.5 billion € in 2006, which is almost 45% of the support payments (5.6 billion €). The authors verify that the difference between the support payments and the sum of the market value of RE and the volume of the MOE (4.98 billion €) leads to a net profit for consumers.

Miera et al. [48] analyze the impact of increasing levels of wind generation on the spot market prices in Spain. The authors consider a 29-month period (2005, 2006

[30] A good review of work on the impact of wind power generation on spot market prices up to 2009 is presented in [12]. See [32] for a general survey of subsequent work on renewable electricity generation and power markets.

and the first five months of 2007). The key assumptions are as follows: (i) electricity demand is totally inelastic, (ii) imports and exports are those actually observed in the period under consideration, (iii) the dispatch of hydro plants, which was assumed, is the one actually observed during that period, and (iv) restrictions due to ramps and those associated to the number of stops/starts that may reasonably occur in a plant during a year are not considered. The reduction in market prices was 7.08 €/MWh in 2005, 4.75 €/MWh in 2006 and 12.44 €/MWh between 1 January and 31 May 2007 (i.e., a reduction of 11.7, 8.6 and 25.1%, respectively). The total savings brought about by wind power penetration during a particular year (i.e., the volume effect) were estimated as 1746 M€ in 2005, 1200 M€ in 2006, and 1348 M€ in 2007 (Jan–May). The net cost savings for consumers were 942 M€ in 2005, 306 M€ in 2006, and 898 M€ in 2007 (Jan–May). The authors conclude that the policy implications of these results are highly relevant, since they provide an additional argument for RE support and contradicts one of the usual arguments against RE deployment: the excessive burden on the consumer.

Munksgaard and Morthorst [49] describe the redesigned wind power policy measures following the liberalization of the Danish electricity market, estimate the impact of the new tariffs on the market prices for the period 2004–2006, and assess whether such tariffs make an incentive to invest in wind power. The results indicate that the prices would have been 1 €/MWh higher in 2004, 4 €/MWh higher in 2005, and 2.5 €/MWh higher in 2006, if wind-power production had been absent. Accordingly, the authors point out that consumers "have to pay" a subsidy to wind-power producers, but the subsidy is to some extent compensated by the reduced prices. Also, the authors compare the revenue from wind power production with the costs of production to assess whether an incentive exists to invest in wind power. The results indicate that producer subsidies included in the new tariffs give incentives to invest in new wind-power plants, although risk-averse investors could be reluctant to invest, especially in wind power on land, as those investments are exposed to a return below the return of financial assets when electricity prices are low.

Hannes Weigt [50] analyzes the extent to which wind turbines can replace fossil fuel capacity and studies the merit order effect of wind energy on German wholesale prices during the period from January 2006 till June 2008. The author develops an optimization model to estimate the differences in the production costs and market prices caused by wind penetration. The results indicate that the wind potential in Germany will not allow a significant reduction of installed conventional capacities. However, with regard to the cost-saving potential of wind energy, the author found that wind generation has a downward impact on both prices and generation costs. On average, a price reduction of about 10 €/MWh was obtained during the study period, going up from 6.26 €/MWh (in 2006) to 13.13 €/MWh (in the first half of 2008). Also, a total saving of 4.1 billion € was obtained during the observation period (1.3 billion in 2006, 1.5 billion in 2007, and 1.3 billion in the first half of 2008). Adding the possible savings from reductions in emission allowance prices leads the author to state that the overall impact of wind energy on consumer prices is positive.

Gil et al. [51] investigate the impact of large-scale wind power on the spot market prices in Spain for the period between April 1, 2007 and December 31, 2010 (a total of

32905 h). The authors adopt a so-called "ex-post" approach to model the relationship between the electricity prices and the wind power output "after-the-fact". The results of an econometric model indicate a price reduction of 9.72 €/MWh during the study period (i.e., a 18% drop in price with respect to a hypothetical "no-wind" scenario). The corresponding savings (from the consumer perspective) were estimated as 7.84 billion € and exceed the cost of the wind energy support policy when considering the market value of wind electricity (estimated as 5.72 billion €). The authors conclude that the quantified figures, particularly the net profit of 2.12 billion €, show a strong evidence of a positive economic benefit to all electricity consumers and the society in general.

Tveten et al. [52] investigate the merit order effect of the large scale deployment of solar power in Germany and analyze the potential consumer's cost of the solar feed-in tariffs when including the MOE. The authors point out that there is a substantial difference between solar power and wind power or run-of-the-river hydro power, since solar energy generation (SEG) normally reaches its maximum during the hours of peak electricity demand. The authors develop a quantitative model for the market price as a function of the electricity generation level. The model is based on the ordinary least squares regression technique and used to predict the electricity prices in Germany, from July 28, 2010 to July 27, 2011, with and without solar electricity generation. Additionally, the model is used to quantify the MOE from SEG and to determine the net consumer's cost of the solar FITs per unit of electricity consumption. The results indicate that the SEG has caused a 7% decrease (3.9 €/MWh reduction) in the average electricity prices. The average daily maximum price and daily price variation are also found to decrease, by 13 and 23%, respectively. Furthermore, when including the MOE of SEG, the net cost of solar FITs is found to be 23% lower than to the charge listed in the consumer's electricity bill. Accordingly, the authors stress the importance of including the MOE when evaluating the total costs and benefits of the FIT policy mechanism.

Cludius et al. [53] analyze the merit order effect of wind and photovoltaic (PV) electricity generation in Germany for the period 2008–2012. The authors mention that consumers are divided into a privileged group, who pay 0.05 ct/kWh for the German Renewable Energy Sources Act (EEG) surcharge, and a non-privileged group (mainly households), who pay a surcharge calculated on a yearly basis. They use regression analysis of historical time-series data to analyze the MOE of wind and PV. The estimated merit order effect ranges from 5.06 to 10.80 €/MWh, for wind, and from 0.98 to 4.56 €/MWh, for PV. The total effect of wind and PV lies between 6.04 €/MWh in 2010 and 10.13 €/MWh in 2012. The authors point out that these results highlight significant redistributive transfers under the current design of the EEG. Specifically, the estimated MOE for 2012 (10 €/MWh or 1 ct/kWh) likely overcompensates the group of privileged consumers for their contribution to the support scheme (0.05 ct/kWh). Accordingly, the authors state that the burden on non-privileged consumers could be reduced if the surcharge for the privileged consumers takes into account the MOE of renewable generation. In particular, for the year 2012 (around 150 TWh exempt consumption), such key consideration would have meant a decrease in the EEG surcharge for non-privileged consumers of roughly 0.4 ct/kWh.

Azofra et al. [54] examine the influence of different levels of wind power production on the spot market prices in Spain for a 12-month period (the year 2012). The authors develop a descriptive model of the market by means of an "ex-post" approach and use it to simulate the market prices. They consider 111 scenarios: a real scenario corresponding to the wind power generation of 2012 (referred to as "100% scenario"), and 110 hypothetical scenarios, going from 0 to 110% of wind generation. The two key assumptions are as follows: the demand for electricity is perfectly inelastic and the production of the hydro plants is the one actually observed in 2012. The results indicate a price reduction of 7.42 €/MWh, if 2012 was less windy, involving a wind power generation of 90%, and 10.94 €/MWh, if 2012 was windier, involving a generation of 110%. The net profit for consumers was 128.2 million € for the "90% scenario" and 697.8 million € for the "110% scenario". The authors conclude that wind power generation is beneficial to the Spanish electrical system, albeit stating that the results must be cautiously analyzed due to possible errors inherent to the descriptive model as well as the assumptions made.

Azofra et al. [55] extends the work presented in [54] by analyzing the individual impacts of wind and solar power generation on the spot market prices in Spain. The authors also use a model to simulate the market prices for the year 2012 and make the two aforementioned simplifying assumptions. They consider the following two hypothetical scenarios: scenario A, involving the actual mix of generation technologies but excluding the electricity produced by wind, and scenario B, similar to scenario A, but excluding the energy generated by solar PV (instead of wind). The results indicate a price reduction of 9.10 €/MWh for scenario A, corresponding to a saving of 2.4010 billion €, and a reduction of 2.18 €/MWh for scenario B, corresponding to a saving of 576.6 million €. The authors conclude that wind energy promotion resulted in a net profit of 364.0 million €. However, solar energy promotion resulted in a financial burden for consumers around 2034.1 million €.

Overall, the nine selected articles consider three key European countries (Spain, Germany and Denmark), a sufficiently long period of time (between 2001 and 2012), key generation technologies (wind, solar and renewables generally), two broad approaches to analyze the merit order effect of renewables (electricity market modeling and empirical analysis), and clear discussions about the cost and efficiency of the renewable energy support policies. They were chosen to provide a representative sample of the wealth of material on the merit order effect of renewable electricity generation on the spot market prices.[31]

The articles show that high levels of renewable electricity production is consistently associated with a reduction of the spot market prices. Also, they show that if both the market value of the wind electricity and the savings brought about by wind power penetration are taken into consideration, the cost of the wind energy support policy is substantially reduced (typically, this cost becomes negative, i.e., a net profit

[31] As noted earlier, there is currently a considerable literature on the impact of increasing levels of renewable generation on electricity markets. For further information about the Spanish and the Germany electricity markets, the interested reader is referred to [56–58] and [32, 59–61], respectively. See also [62] for further information on the Western Danish price area of the Nord Pool's Elspot market, [63] for the Italian power market, and [64] for the Belgium market.

occurs for consumers).[32] In accordance with these conclusions, the results of our simulation-based study indicate an average price reduction of about 17 €/MWh during the first half of 2016. Furthermore, the net cost of the wind energy support policy has the value of −8.248 million € in January 2017, indicating a net profit in this month. However, this cost reaches the value of 69.011 million € during the 6-month period of the study (from January 1, 2016 to June 30, 2016), meaning a financial burden for consumers as a result of wind energy promotion. As noted in the previous section, our study does not take into account the carbon price effect on the electricity market. Additionally, the typical seasonal traits characteristic of wind generation are not considered (since the time period has the duration of six months only).

9.6 Conclusion

This chapter has analyzed and quantified the reduction in the Portuguese day-ahead market prices achieved by wind power as a result of the merit order effect. It began by analyzing the growth of renewable energy generation over the past decade and discussing the evolving policy landscape, placing emphasis on the Portuguese policy framework.

Following this introductory material, the chapter described in detail the key principles underlying the merit order effect and presented a study to investigate the MOE of the significant deployment of wind power in Portugal during the first half of 2016 (i.e., from January 1, 2016 to June 30, 2016). The main results generated by an agent-based simulation tool, called MATREM, are as follows:

- The reduction of the average market price ranged between 8 and 25 €/MWh, reaching the highest value in January and February, and the lowest value in June.
- The price reduction was about 17 €/MWh during the entire study period.
- The financial volume of the MOE ranged between 29.549 million € (in June) to a maximum of 96.171 million € (in January).
- The (total) volume of the MOE reached the considerable value of 391.055 million € during the study period.
- The specific value of the MOE ranged between 37 €/MWh (in June) and 71 €/MWh (in January).
- For the entire 6-month period, the specific value of the MOE was 59 €/MWh and the average FIT reached the value of 97 €/MWh.
- The net cost of the wind energy support policy, computed by taking into consideration the feed-in tariff, the market value of the wind electricity, and the financial

[32] A cautionary and explanatory note is in order here. The nine selected studies—and other relevant pieces of work proposed in the literature—make use of a diverse range of models and consider different sets of simplifying assumptions. As a consequence, assessing and relating such individual research contributions to draw general conclusions is always a nontrivial (and daunting) task. Any comparative analysis should be carried out carefully and bear in mind the limitations associated with disparate research efforts. Nevertheless, the two conclusions presented here are quite general and, we believe, perfectly acceptable.

volume of the MOE, reached the value of −8.248 million € in January, a negative value, indicating that a net profit has occurred in this month. For the other five months of the study period, the net cost ranged between 9.807 million € (in March) and 21.992 million € (in April).
- The (total) net cost was 69.011 million € during the 6-month period of the study.

The chapter concluded by presenting a detailed technical review of nine representative articles that exist in the MOE literature. The following two general conclusions were drawn from the articles (albeit presenting a cautionary note stating the limitations to draw general conclusions from disparate research efforts):

- High levels of renewable electricity production may be consistently associated with a reduction of the spot market prices.
- If both the market value of the wind electricity and the savings brought about by wind power penetration are taken into consideration, the cost of the wind energy support policy may be substantially reduced (and may even become negative, indicating a net profit for consumers).

The results of our simulation-based study are in strict accordance with the first conclusion. Also, the net cost of the support policy was negative in January 2016, which is in accordance with the second conclusion. However, the (total) net cost for the 6-month period of the study was positive, indicating a financial burden for consumers as a result of the wind energy promotion. Although only partially consistent with the second conclusion, this result should be interpreted carefully, since it did not take into account the carbon price effect on the electricity market nor the typical seasonal traits characteristic of wind generation.

Finally, some notes on the scope of the chapter and the effectiveness of the simulation-based study. The work described here involves several different topics (e.g., renewable energy, feed-in policies and energy markets), draws upon various computational resources (e.g., MATREM), is based on a particular approach to analyze the merit order effect, and considers a number of simplifying assumptions. Accordingly, the simulation-based study to investigate the merit order effect of wind power in Portugal has several shortcomings. These shortcomings need to be overcome in subsequent studies in order to examine the impact of wind electricity generation on the Portuguese day-ahead market prices in a fully rigorous way. Some important aspects for future work are as follows:

1. Software agents and computational complexity: to consider a larger number of software agents to represent the electricity supply industry. This will allow us to decrease the (somewhat arbitrary) increment of 5 €/MWh and, hopefully, to obtain better correlations between the real and the simulated market prices (as the increment goes to 0 €/MWh).
2. Energy scenarios and experimental procedure: to define energy scenarios similar to the current ones, but considering a (slightly) different experimental procedure, involving a shift of the supply curve to simulate what would have been the market prices in the absence of wind generation.

3. Observation period: to extend the time period of the study (e.g., to consider a one-year period or even a longer period). This will allow us to analyze the typical seasonal traits characteristic of wind generation. Also, this is likely to improve the quality of the experimental results.
4. Carbon price effect: to consider the interaction of wind generation with the climate policy and the EU emission trading system. In particular, to quantify the carbon price effect on the electricity market and to compute the net cost of the wind energy support policy by taking into consideration this effect (in addition to the feed-in tariff, the market value of the wind electricity, and the financial volume of the MOE).
5. Renewable energy technology: to consider several different generation technologies, notably wind power and solar photovoltaic, with the main aim of analyzing the individual impact of each technology on the Portuguese day-ahead market prices.

Acknowledgements This work was performed under the project MAN-REM (FCOMP-01-0124-FEDER-020397), supported by FEDER funds, through the program COMPETE ("Programa Operacional Temático Factores de Competividade"), and also National funds, through FCT ("Fundação para a Ciência e a Tecnologia"). The authors wish to thank Rui Castro, from INESC-ID and also the Technical University of Lisbon (IST), and João Martins and Anabela Pronto, from the NOVA University of Lisbon, for their tireless ability to read the draft and the valuable comments and helpful suggestions to improve the chapter.

References

1. Lins, C., Williamson, L., Leitner, S., Teske, S.: The first decade: 2004–2014, 10 years of renewable energy progress. Technical Report, Renewable Energy Policy Network for the 21st Century (REN21), Paris, France (2014)
2. Sawin, J., Seyboth, K., Sverrisson, F.: Renewable 2016: global status report. Technical Report, Renewable Energy Policy Network for the 21st Century (REN21), Paris, France (2016)
3. European Union: Energy for the Future: Renewable Sources of Energy. White Paper for a Community Strategy and Action Plan, COM(97) 599 (26 November 1997). http://europa.eu/documents/comm/white_papers/pdf/com97_599_en.pdf (Cited on 22 April, 2017)
4. European Union: Renewable Energy Road Map Renewable Energies in the 21st Century: Building a more Sustainable Future. Communication from the Commission to the Council and the European Parliament, COM(2006) 848 (10 January 2007). http://eur-lex.europa.eu/legal-content/EN/TXT/PDF/?uri=CELEX:52006DC0848&from=EN (Cited on 22 April, 2017)
5. European Union: Directive 2009/28/EC of the European Parliament and of the Council on the promotion of the use of energy from renewable sources and amending and subsequently repealing Directives 2001/77/EC and 2003/30/EC (23 April 2009). http://eur-lex.europa.eu/legal-content/EN/TXT/?uri=CELEX:32009L0028 (Cited on 22 April, 2017)
6. Direção Geral de Energia e Geologia: Renováveis: Estatísticas Rápidas. Technical Report 146, DGEG, Lisboa, Portugal (2016)
7. Macedo, V.: Portuguese renewable energy sources: overview. Technical Report, Macedo Vitorino & Associados, Lisboa, Portugal (2015)
8. Shukla, S., Sawyer, S.: 30 years of policies for wind energy: lessons from 12 wind energy markets. Technical Report, International Renewable Energy Agency (IRENA), Abu Dhabi, United Arab Emirates (2012)

9. PLMJ-International Legal Network: Renewable Energies: 2013 Legislative Changes in the Sector. Informative Note, PLMJ, Lisbon, Portugal (2014)
10. ERSE - Portuguese Energy Services Regulatory Authority: Special Regime Production. http://www.erse.pt/pt/desempenhoambiental/prodregesp/2016/Comunicados/PRE_2016.xls (downloaded on 22 April, 2017)
11. Direção Geral de Energia e Geologia: Renováveis: Estatísticas Rápidas. Technical Report 134, DGEG, Lisboa, Portugal (2015)
12. Bianchin, R., Ray, S., Munksgaard, J., Morthorst, P., Sinner, A.: Wind Energy and Electricity Prices: Exploring the "Merit order Effect". Technical Report (Literative Review), European Wind Energy Association (EWEA), Brussels, Belgium (2010)
13. Sawin, J., Sverrisson, F.: Renewables 2014: global status report. Technical Report, Renewable Energy Policy Network for the 21st Century (2014)
14. Sawin, J., Sverrisson, F., Rickerson, W.: Renewables 2015: global status report. Technical Report, Renewable Energy Policy Network for the 21st Century (REN21), Paris, France (2015)
15. Council, Global Wind Energy: Global Wind Statistics 2011. Technical Report, GWEC, Brussels, Belgium (2012)
16. Council, Global Wind Energy: Global Wind Statistics 2012. Technical Report, GWEC, Brussels, Belgium (2013)
17. Council, Global Wind Energy: Global Wind Statistics 2013. Technical Report, GWEC, Brussels, Belgium (2014)
18. Council, Global Wind Energy: Global Wind Statistics 2014. Technical Report, GWEC, Brussels, Belgium (2015)
19. INEGI and APREN: Wind Farms in Portugal. Technical Report, Institute of Mechanical Engineering and Industrial Management (INEGI) and Portuguese Renewable Energy Association (APREN), Portugal (2014)
20. INEGI and APREN: Wind Farms in Portugal. Technical Report, Institute of Mechanical Engineering and Industrial Management (INEGI) and Portuguese Renewable Energy Association (APREN), Portugal (2015)
21. Council, Global Wind Energy: Annual Market Update 2010. Global Wind Report, GWEC, Brussels, Belgium (2010)
22. European Union: Directive 2001/77/EC of the European Parliament and of the Council on the promotion of electricity produced from renewable energy sources in the internal electricity market. (27 September 2001) http://eur-lex.europa.eu/legal-content/EN/TXT/PDF/?uri=CELEX:32001L0077&from=EN (Cited on 22 April, 2017)
23. Sensfuß, F.: Assessment of the impact of renewable electricity generation on the german electricity sector: an agent-based simulation approach. Ph.D. Dissertation, Karlsruhe University (2007)
24. Hunt, S.: Making Competition Work in Electricity. Wiley, Chichester (2002)
25. Kirschen, D., Strbac, G.: Fundamentals of Power System Economics. Wiley, Chichester (2004)
26. Stoft, S.: Power Systyem Economis: Designing Markets for Electricity. IEEE Press and Wiley Interscience, New York (2002)
27. REN: Redes Energéticas Nacionais, Daily Summary. http://www.centrodeinformacao.ren.pt/PT/InformacaoExploracao/Paginas/EstatisticaDiariaDiagrama.aspx (Cited on 22 April, 2017)
28. REE: Red Eléctrica de España, Península, Seguimiento de la demanda de energía eléctrica. https://demanda.ree.es/movil/peninsula/demanda/total (Cited on 22 April, 2017)
29. OMIEL: Operador del Mercado Ibérico de Energía (Spanish Electricity Market Operator). Market Results (online data). http://www.omie.es/files/flash/ResultadosMercado.swf (Cited on 22 April, 2017)
30. Wind Energy - The Facts: Grid Integration, Setting the Scene. https://www.wind-energy-the-facts.org/wind-energy-penetration-and-integration.html (Cited on 29 April, 2017)
31. Corbetta, G., Mbistrova, A., Ho, A., Pineda, I.: Wind in power: 2015 european statistics. Technical Report, European Wind Energy Association (EWEA), Brussels, Belgium (2016)
32. Würzburg, K., Labandeira, X., Linares, P.: Renewable generation and electricity prices: taking stock and new evidence for Germany and Austria. Energy Econ. **40**, S159–S171 (2013)

33. Vidigal, D., Lopes, F., Pronto, A., Santana, J.: Agent-based simulation of wholesale energy markets: a case study on renewable generation. In: Spies, M., Wagner, R., Tjoa, A. (eds.) 26th Database and Expert Systems Applications (DEXA 2015), pp. 81–85. IEEE (2015)
34. Algarvio, H., Couto, A., Lopes, F., Estanqueiro, A., Santana, J.: Multi-agent energy markets with high levels of renewable generation: a case-study on forecast uncertainty and market closing time. In: Omatu, S., et al. (eds.) 13th International Conference on Distributed Computing and Artificial Intelligence, pp. 339–347. Springer International Publishing (2016)
35. Algarvio, H., Couto, A., Lopes, F., Estanqueiro, A., Holttinen, H., Santana, J.: Agent-based simulation of day-ahead energy markets: impact of forecast uncertainty and market closing time on energy prices. In: Tjoa, A., Vale, Z., Wagner, R., (eds.) 27th Database and Expert Systems Applications (DEXA 2016), pp. 166–70. IEEE (2016)
36. Lopes, F., Rodrigues, T., Sousa, J.: Negotiating bilateral contracts in a multi-agent electricity market: a case study. In: Hameurlain, A., Tjoa, A., Wagner, R. (eds.) 23rd Database and Expert Systems Applications (DEXA 2012), pp. 326–330. IEEE (2012)
37. Sousa, F., Lopes, F., Santana, J.: Contracts for difference and risk management in multi-agent energy markets. In: Demazeau, Y., Decker, K., Pérez, J., De la Prieta, F.: (eds.) Advances in Practical Applications of Agents, Multi-Agent Systems, and Sustainability: The PAAMS Collection (PAAMS 2015), pp. 339–347. Springer International Publishing (2015)
38. Sousa, F., Lopes, F., Santana, J.: Multi-agent electricity markets: a case study on contracts for difference. In: Spies, M., Wagner, R., Tjoa, A. (eds.) 26th Database and Expert Systems Applications (DEXA 2015), pp. 88–90. IEEE (2015)
39. Lopes, F., Mamede, N., Novais, A.Q., Coelho, H.: Negotiation in a multi-agent supply chain system. In: Third International Workshop of the IFIP WG 5.7 Special Interest Group on Advanced Techniques in Production Planning and Control, pp. 153–168. Firenze University Press (2002)
40. Lopes, F., Coelho, H.: Strategic and tactical behaviour in automated negotiation. Int. J. Artif. Intell. **4**(S10), 35–63 (2010)
41. REN: Redes Energéticas Nacionais, Preos Mercado Spot, Portugal e Espanha. http://www.mercado.ren.pt/PT/Electr/InfoMercado/InfOp/MercOmel/Paginas/Precos.aspx (Cited on 29 April, 2017)
42. OMIE: Daily and Intraday Electricity Market Operating Rules (2014). http://www.omie.es/files/20140509_reglas_v11_ingles.pdf (Cited on 24 June, 2017)
43. MIBEL Regulatory Council: Description of the Operation of MIBEL (2009). http://www.erse.pt/eng/electricity/MIBEL/Documents/Description_Operation_MIBEL.pdf (Cited on 14 May, 2017)
44. ERSE and CMVM: Proposta de Mecanismo de Gestão conjunta da interligação Espanha-Portugal. Entidade Reguladora dos Serviços Energéticos and Comisssão do Mercado de Valores Mobiliários (2006). http://www.erse.pt/pt/mibel/compatibilizacaoregulatoria/Documents/PROPOSTADEMECANISMODEGEST%C3%83OCONJUNTADAINTERLIGA%C3%87%C3%83OESPANHAPORTUGAL.pdf (Cited on 14 May, 2017)
45. ERSE, CMVM, CNE and CNMV: Proposta do Conselho de Reguladores para a Repartição da Capacidade de interligação entre os Mecanismos de "Market Splitting" e Leilões Explícitos de Capacidade no Âmbito do MIBEL. Entidade Reguladora dos Serviços Energéticos, Comisssão do Mercado de Valores Mobiliários, Comísión Nacional de Energía and Comísión Nacional del Mercado de Valores (2007). http://www.erse.pt/pt/mibel/compatibilizacaoregulatoria/Documents/PropostaCRreparticaodacapacidadedeinterligacao.pdf (Cited on 14 May, 2017)
46. EEX: European Energy Exchange, Natural Gas Daily Reference Price. https://www.eex.com/en/market-data/natural-gas/spot-market/daily-reference-price (Cited on 14 May 2017)
47. Sensfuß, F., Ragwitz, M., Genoese, M.: The merit-order effect: a detailed analysis of the price effect of renewable electricity generation on spot market prices in Germany. Energy Policy **36**, 3086–3094 (2008)
48. Miera, G., González, P., Vizcaíno, I.: Analysing the impact of renewable electricity support schemes on power prices: the case of wind electricity in Spain. Energy Policy **36**, 3345–3359 (2008)

49. Munksgaard, J., Morthorst, P.: Wind power in the Danish liberalised power market-policy measures, price impact and investor incentives. Energy Policy **36**, 3940–3947 (2008)
50. Weigt, H.: Germany's wind energy: the potential for fossil capacity replacement and cost saving. Energy Policy **86**, 1857–1863 (2009)
51. Gil, H., Gomez-Quiles, C., Riquelme, J.: Large-scale wind power integration and wholesale electricity trading benefits: estimation via an ex post approach. Energy Policy **41**, 849–859 (2012)
52. Tveten, A., Bolkesjo, T., Martinsen, T., Hvarnes, H.: Solar feed-in tariffs and the merit order effect: a study of the German electricity market. Energy Policy **61**, 761–770 (2013)
53. Cludius, J., Hermann, H., Matthes, F., Graichen, V.: The merit order effect of wind and photovoltaic electricity generation in Germany 2008–2016: estimation and distributional implications. Energy Econ. **44**, 302–313 (2014)
54. Azofra, D., Jiménez, E., Martínez, E., Blanco, J., Saenz-Díez, J.: Wind power merit-order and feed-in-tariffs effect: a variability analysis of the Spanish electricity market. Energy Convers. Manag. **83**, 19–27 (2014)
55. Azofra, D., Martínez, E., Jiménez, E., Blanco, J., Azofra, F., Saenz-Díez, J.: Comparison of the influence of photovoltaic and wind power on the Spanish electricity prices by means of artificial intelligence techniques. Renew. Sustain. Energy Rev. **42**, 532–542 (2015)
56. Gelabert, L., Labandeira, X., Linares, P.: An ex-post analysis of the effect of renewables and cogeneration on spanish electricity prices. Energy Econ. **33**, S59–S65 (2011)
57. Linares, P., Santos, F., Ventosa, M.: Coordination of carbon reduction and renewable energy support policies. Clim. Policy **8**, 377–394 (2008)
58. Moreno, F., Martínez-Val, J.: Collateral effects of renewable energies deployment in Spain: impact on thermal power plants performance and management. Energy Policy **39**, 6561–6574 (2011)
59. Nicolosi, M., Fürsch, M.: The Impact of an Increasing share of RES-E on the conventional power market - the example of Germany. Zeitschrift füer Energiewirtschaft **33**(3), 246–254 (2009)
60. Traber, T., Kemfert, C.: Gone with the wind? electricity market prices and incentives to invest in thermal power plants under increasing wind energy supply. Energy Econ. **33**, 249–256 (2011)
61. Traber, T., Kemfert, C., Diekmann, J.: German electricity prices: only modest increase due to renewable energy expected. Weekly Report No. 6/2011, German Institute for Economic Research (DIW), Berlin, Germany (2011)
62. Jónsson, T., Pinson, P., Madsen, H.: On the market impact of wind energy forecasts. Energy Econ. **32**, 313–320 (2010)
63. Clò, S., Cataldi, A., Zoppoli, P.: The merit-order effect in the italian power market: the impact of solar and wind generation on national wholesale electricity prices. Energy Policy **77**, 79–88 (2015)
64. Delarue, E., Luickx, P., Dhaeseleer, W.: The actual effect of wind power on overall electricity generation costs and CO2 emissions. Energy Conv. Manag. **50**, 1450–1456 (2009)

Chapter 10
Demand Response in Electricity Markets: An Overview and a Study of the Price-Effect on the Iberian Daily Market

Fernando Lopes and Hugo Algarvio

Abstract The electricity industry is undergoing a deep transformation as Europe moves towards a greener, healthier future—the growth of renewable generation has surpassed all expectations and demand response (DR) has emerged as a key element of market design. Most European countries have already opened their markets to the participation of demand response and, over the long-term, DR will probably reach its full potential as the entire range of DR programs will be made available to retail customers. To date, however, progress has been only gradual. There is currently a need to understand and quantify the major impacts and benefits of DR, to facilitate an effective implementation of DR programs. Accordingly, this chapter investigates the impact of different levels of DR on the Iberian market prices, during the period 2014–2017, and analyzes the potential benefits for market participants and retail customers. The results generated by an agent-based simulation tool, called MATREM, are striking. In the year 2017, for instance, a modest load reduction of 5% when prices rose above 80 €/MWh yielded the (very large) benefit of 76.62 million €. Also, the same decrease in load when prices exceeded 90 €/MWh provided the (still large) benefit of 39.05 million €. The chapter concludes with specific recommendations—for consideration by state institutions, system operators, electric utilities and other market participants—to foster demand response in Portugal through both incentive-based and price-based programs.

F. Lopes (✉)
LNEG–National Laboratory of Energy and Geology,
Estrada do Paço do Lumiar 22, 1649-038 Lisbon, Portugal
e-mail: fernando.lopes@lneg.pt

H. Algarvio
Instituto Superior Técnico, Avenida Rovisco Pais 1, 1049-001 Lisbon, Portugal
e-mail: hugo.algarvio@tecnico.ulisboa.pt

F. Lopes and H. Coelho (eds.), *Electricity Markets with Increasing Levels of Renewable Generation: Structure, Operation, Agent-based Simulation, and Emerging Designs*, Studies in Systems, Decision and Control 144, https://doi.org/10.1007/978-3-319-74263-2_10

10.1 Introduction

A major structural problem has become apparent with the introduction of competition in the electricity supply industry—a disconnection between wholesale and retail markets. In fact, the cost of electric power varies on a very short time scale, but most retail customers face flat, average-cost based electric rates, which give them no indication that electricity prices change over time, nor any incentive to reduce usage during periods of high market prices. This lack of price-responsive demand, or demand response (DR), deprives the wholesale market of a natural mechanism for relieving temporary pressure on prices, gives generators the opportunity to exercise market power and, in the long-term, may lead to a need of investments in expensive generation capacity. Put differently, an active response of retail customers to changes in the price of electricity over time provides reliability benefits, disciplines market power, improves economic efficiency, and reduces the need to build new generation facilities [1].

Demand response refers broadly to actions by retail customers that change their consumption of electric power in response to price signals, incentives, or directions from system operators and market participants. The term is widely used, although it defies attempts to produce a single universally accepted definition. Nevertheless, the following definition is frequently adopted in both the academic and practitioner literature [2]:

> Demand response means "changes in electric usage by end-use customers from their normal consumption patterns in response to changes in the price of electricity over time, or to incentive payments designed to induce lower electricity use at times of high wholesale market prices or when system reliability is jeopardized".

Customers respond to demand response events typically by curtailing load—reducing electricity usage at times of high market prices without making it up later—or also by shifting load—rescheduling energy usage away from times of high market prices to other time periods.

Demand response is essential in competitive energy markets to assure an efficient interaction between supply and demand [3]. Additionally, in markets with high penetration of variable generation (VG), such as wind and photovoltaic solar power, demand response is crucial to provide increased levels of flexibility at relatively low cost. Indeed, such markets require greater flexibility to manage the increased variability and uncertainty introduced by VG, and DR can provide, at least to a certain extent, that flexibility. Load can be reduced when renewable generation is not abundant and, by adjusting the timing of power consumption, demand can be increased when supply from low-carbon resources is abundant. To be more concrete, load flexibility can be increased in three different ways [4]: (i) peak shaving, which involves a reduction of peak consumption during tight system conditions to release pressure on generation and grid capacity needs, (ii) valley filling, which consists in increasing or shifting consumption to time periods of ample renewable generation, and (iii) ramp reduction, which involves a reduction of steep ramping needs at peak hours by shifting load to times when system requirements are lower.

During the course of the decade, the European Union acknowledged the importance of demand response in the internal energy market, particularly as a central instrument for increasing the flexibility of the market and enabling optimal use of networks. The Directive 2012/27/EU [5], on energy efficiency, states that "Member States shall ensure that national energy regulatory authorities encourage demand side resources, such as demand response, to participate alongside supply in wholesale and retail markets". The communication from the Commission [6] points out that "The potential of the demand side response at the Union scale is enormous". The legislative proposal 2016/0379 (COD) provides a regulatory framework for the dispatch and curtailment of generation and demand response [7].

At the time of writing, most European countries have already opened their markets to the participation of demand response, but progress has been only gradual, and various fundamental problems are still waiting to be addressed more thoroughly. Over the long-term, demand response will probably reach its full potential as the entire range of DR programs will be made available to retail customers. In the meantime, however, there is a need to understand and quantify the major impacts and benefits of demand response, to facilitate an effective implementation of DR programs. Accordingly, this chapter presents an insightful study to investigate the impact of demand response on the Iberian electricity market prices and analyzes the potential benefits for market participants and retail customers. The aim is to quantify the impact of different levels of demand response—modeled as load reductions between 1 and 5% when prices rise above a threshold between 80 and 100 €/MWh—on the Iberian market prices at times of system constraints, and to verify the extent of financial benefits. The observation period has the duration of 42 months: from January 1, 2014 to June 30, 2017.[1]

The remainder of this chapter is structured as follows. Section 10.2 describes the most important DR programs and presents an overview of demand response in Spain and Portugal. Section 10.3 introduces the Iberian electricity market (MIBEL), particularly the day-ahead market, and describes the short-term market impact of DR, illustrating the description with a practical example using real data.[2] Section 10.4 presents a study to investigate the price effect of DR on MIBEL and discusses, in detail, the experimental results. Finally, Sect. 10.5 states the conclusions and makes specific recommendations to foster demand response in Portugal.

[1]The authors are aware of no other work to investigate the price effect of DR on the Iberian market at times of system constraints. Fernández et al. [8] analyze the economic impact of a DR program—the voluntary price for small consumers—on the Iberian market during the period from April 2014 till March 2015. However, the study considers load reductions of 1.5, 3 and 6% uniform for all hours of a 24-hour day. Also, there are various other studies that link specific levels of DR to decreases in market prices, mostly involving markets from the United States, indicating that the benefits may be quite significant (see, e.g., the three-percent solution [9]).

[2]Notice that demand response can perform three key functions [4, 10]: energy, capacity and ancillary services. Put another way, demand response can participate on wholesale energy markets (i.e., day-ahead and real-time markets), capacity markets, and balancing and ancillary services markets. Throughout this chapter, we focus primarily on day-ahead markets and the material that follows clearly reflects this bias.

10.2 Demand Response in Competitive Electricity Markets

Demand response programs offer the opportunity for electricity consumers to reduce or shift their load either in exchange for an incentive or in response to price signals.[3] The technical literature proposes several ways to classify existing DR programs. Hirst and Kirby [11], for example, group existing programs into three generic categories: dynamic pricing, voluntary load reductions at times of high prices, and customer sales of ancillary services to the system operator. Braithwait et al. [1] adopt the classification proposed by Hirst and Kirby [11]. The authors focus primarily on markets for energy, rather than ancillary services, and draw a distinction among three types of the (second) category of load reduction programs: traditional load management programs, energy buy-back programs, and demand bidding programs.

In an earlier book on demand response in liberalized electricity markets [12], the International Energy Agency (IEA) makes a distinction between system-led and market-led demand response. Hogan [13] considers three general types of demand response: real-time pricing, explicit contract and imputed demand response. The first two types correspond to a response to electricity prices, while the third type treats demand response as a generation resource. In a recent book on market design and regulation during the transition to low-carbon power systems [4], IEA adopts the three types of demand response proposed by Hogan [13], and discusses DR in the context of wholesale energy markets, capacity markets, and balancing and ancillary services markets.

The US action plan on demand response [14] considers both dispatchable DR (i.e., planned changes in consumption that customers agree to make in response to directions from someone other than them) and non-dispatchable DR (i.e., programs and products in which customers decide whether and when to reduce consumption based on retail rates that change over time). The annual reports on demand response and advance metering (e.g., [15–17]), published by the Federal Energy Regulatory Commission (FERC), make a distinction between incentive-based and time-based demand response programs. The former category includes direct load control, interruptible, demand bidding/buyback, emergency demand response, and capacity market and ancillary service market programs. The latter involves real-time pricing, critical peak pricing, variable peak pricing, and time-of-use rates.

The mapping of demand response in Europe (e.g., [18]), published by the Smart Energy Demand Coalition (SEDC), draws a distinction between two groups of demand response, namely explicit and implicit DR, highlighting that the former type is also referred to as "incentive-based" and the latter as "price-based", and also stressing that neither form of DR is a replacement for the other (e.g., customers may participate in explicit DR through aggregators, and at the same time, in implicit DR through more or less dynamic tariffs).

[3]A *demand response program* is a mechanism for communicating prices and willingness to pay between wholesale and retail power markets, with the immediate objective of achieving load changes, particularly at times of high wholesale prices [1].

Overall, even though there exist several ways to classify existing demand response programs, we adopt the classification proposed in an earlier report of the US Department of Energy [2], which categorize DR programs into two groups: incentive-based and price-based. The main reasons for adopting this classification are as follows. First, it involves two groups of programs very similar to the two groups considered in the annual reports published by FERC, and second, it seems to be more general than other existing classifications. The remainder of this section is structured so that the two groups of programs are presented in separate subsections. Specifically, Sect. 10.2.1 deals with incentive-based DR programs and Sect. 10.2.2 with price-based DR programs.

10.2.1 Incentive-Based Demand Response Programs

Incentive-based DR programs represent contractual arrangements to elicit demand reductions from customers either when the system operator believes reliability conditions are compromised or when market prices are high. These programs give participating customers load-reduction incentives that are separate from, or additional to, their retail electricity rates. Typically, customer enrollment and response are voluntary, although some programs penalize customers that enroll but fail to respond or fulfill their contractual commitments when events are declared. The most common programs include [2, 15]:

1. *Direct load control programs*: a program operator remotely shuts down or cycles customers' electrical equipment on short notice. Programs primarily offered to residential or small commercial customers.
2. *Interruptible/curtailable rates*: customers receive a rate discount or bill credit in exchange for agreeing to reduce load during system contingencies. Programs typically offered by system operators to the largest industrial (or commercial) customers.
3. *Demand bidding programs*: programs that encourage large customers to bid into the wholesale energy market and offer to provide load reductions at a price at which they are willing to be curtailed.
4. *Emergency demand response programs*: customers receive incentive payments for load reductions during reliability-triggered events, but curtailment is voluntary. The level of payments is typically specified beforehand.
5. *Capacity market programs*: customers commit to provide pre-specified load reductions when system contingencies arise, receiving guaranteed capacity payments. Programs typically offered by agents that operate installed capacity markets to customers that meet eligibility requirements.
6. *Ancillary services market programs*: customers bid load curtailments in organized wholesale markets as operating reserves.

Incentive-based programs may be introduced at virtually all timescales of the management of electric power systems (see [2] for details).

Incentive-based programs are related to various key submarkets of power markets, namely wholesale energy markets, capacity markets, and balancing and ancillary services markets. The particular category of demand bidding programs involves the submission of offers to curtail load into wholesale energy markets (i.e., day-ahead and real-time markets). Since this work focuses primarily on day-ahead markets, rather than capacity or ancillary services markets, demand bidding programs will receive the preponderance of our attention in the following paragraphs.

Demand Bidding Programs. Individual consumers, or demand response aggregators (DRAs) acting on behalf of many consumers, prepare load-reduction bids to submit to the day-ahead market (DAM).[4] The bids may vary due to a number of factors, but typically include a bid price (i.e., the price at which customers are willing to reduce load) and an amount of reduction (i.e., the quantity of demand customers are willing to reduce). Some programs also allow, for example, bids with a curtailment initiation cost (CIC), which places a floor on the total payment received if the bids are accepted, as well as bids for consecutive hours of load reduction, often referred to as strips (see, e.g., [19]).

The DAM for a given day d is normally cleared at 12 noon of the previous day $d-1$.[5] When a load-reduction bid is scheduled, the benefits from participating in a DR program are typically based on the bid information and the market-clearing price. For instance, customers may be paid the greater of the market-clearing price or the bid price for the demand-reduction amount scheduled. Or, alternatively, the greater of the market-clearing price or the CIC, which ensures a minimum level of benefit. Also, the incentive payment may be based on the reduction of the market price due to the participation of demand response on the DAM. In this case, calculating the size of the incentive payment involves determining the market-clearing price twice for each settlement interval where demand response clears: once including DR and once excluding DR (see, e.g., [20]).

The measurement of the load reductions for which customers are paid has proven to be a difficult and controversial problem. The key issue is that load reductions cannot be measured directly. Only energy consumption can be metered. Accordingly, load reductions are typically inferred by subtracting the actual consumption during periods in which customers are dispatched from a higher hypothetical baseline level, referred to as customer baseline load (CBL), that would have been consumed if customers had not been dispatched. A number of methods have been proposed to estimate the baseline level, typically involving an average of usage over a given period of time (see, e.g., [21] for a technical review).

---

[4]To be eligible for most demand bidding programs, customers must able to reduce load by a minimum of 100 kW. Small consumers are allowed to participate in some programs through demand response aggregators.

[5]*Chapter* 2 gives a brief overview of energy markets and *Chap.* 8 describes the DAM supported by the agent-based system called MATREM. The interested reader is therefore referred to them for further details of market operation and electricity pricing. See also Sect. 10.3.1 for a detailed overview of the Iberian Electricity Market.

Some methods can lead to a relatively accurate statistical estimate of the baseline level, but usually require a large quantity of data and are complex to implement. Other methods are relatively simple to implement, although may lead to over- and under-estimations. Hence, the choice of a specific method remains a trade-off between the availability of data, the accuracy of the inherent algorithm, and the simplicity of implementation. Furthermore, it is also fundamental to consider the risk of gaming and wrongful behavior. Indeed, one important problem with such methods consists in preventing gaming opportunities in which customers intentionally modify usage to artificially increase the amount of load response for which they are compensated. An example of gaming is provided by the case in which the base level is estimated by considering usage in the hour (or hours) immediately preceding a load reduction period. Customers may intentionally increase their load in the few hours prior to responding, establishing a baseline level above their normal usage, thus receiving an incentive payment that does not reflect their typical usage pattern [1]. Apparently, as reported in [4], the managers of the baseball stadium in Baltimore turned on the lights during daytime to artificially increase the baseline level.

Penalty rates are typically applied to the difference between the customer baseline load assigned to each hour of a load reduction period and the metered use in that hour. In other words, customers scheduled for curtailments that have been eligible for incentives payments, but that subsequently fail to curtail, are typically charged for the non-curtailed load.[6] Non-performance penalties serve to reinforce the obligation of customers to be available and deliver load reductions when called. However, establishing appropriate penalty levels may be challenging. Program designers should balance the attractiveness of DR program to customers against the potential consequences of increased penalty levels.

To conclude, we hasten to add an explanatory and cautionary note. Demand biding programs have been the subject of some controversy, particularly over the issue of defining an adequate incentive payment for successful bids. In the United States, FERC issued Order 745 [22] requiring that all systems operators pay the full market price, but only in a subset of hours passing a "net benefits test". The Electric Power Supply Association (EPSA) sued FERC over Order 745, and a large number of economists supported the position of EPSA on compensation. EPSA also challenged whether FERC had jurisdiction over demand response, which they claimed was a retail activity within state jurisdiction. In May 2014, the DC Circuit Court vacated Order 745 on jurisdictional grounds, indicating that FERC overstepped its authority by encroaching on states' jurisdiction of the retail electricity market, and also noting substantive errors with compensation rules. However, FERC appealed the decision to the US Supreme Court. The impacts of this process have yet to be realized and may result in substantial changes in how demand biding programs may participate on wholesale energy markets.

[6]For example, customers unable to reduce load by the bid amount during the scheduled time pay the higher of the day-ahead price or the real-time price for the amount of the incomplete scheduled load reduction [19].

Incentive-Based Demand Response in Spain and Portugal. Incentive-based DR in Spain consists basically in only one program [23]: interruptible contracts. In mainland Spain, the program was introduced in 2008, with a threshold of 5 MW to participate.[7] Its main purpose was to provide a flexible and rapid response to the needs of the system operator (SO), the Spanish electrical grid (REE), in situations of imbalances between generation and demand. Participation was opened to large industrial customers as well as any other customers able to reduce a minimum of 5 MW of load when required by the SO. In 2011, there were 152 contracts in force, with a total interruptible power of about 2200 MW [24]. In 2015, a new framework start to be applied to mainland Spain, allowing to bid 5 MW blocks or 90 MW blocks of curtailable load.[8]

Since 2014, the interruptible capacity was assigned by the SO through public auctions, where all customers satisfying pre-defined requirements could participate. The first auction, for 2015, took place in November–December 2014: a total of 3020 MW of interruptible load was assigned, with a cost of 508 million €. The auction for 2016 took place in September 2015, allocating 2890 MW of interruptible load, with a cost of 503 million €. In November 2016 another auction has been carried out, allocating 2975 MW for 2017, with a total cost of 525 million € [25].

Participants in the DR program include energy consumers from the construction industry (steel, concrete, glass, etc.), material factories (paper, chemistry, etc.) and desalinization plants (in the Canary Islands). There are five different types of contracts that differ in the notification time (from 0 to 2 h) and the duration of the interruption (from 1 to 12 h). For example, contracts of "type 1" involve a notification time of 2 h and a maximum interruption of 12 h, while contracts of "type 5" involve a notification time of 0 min and a maximum interruption of 1 h. Interruptions can take place for up to 240 h per year (120 h per month), with a maximum of one interruption per day (12 h maximum per day) and five interruptions per week (60 h per week).

Customers typically send their load forecasts to the SO monthly and the baseline levels are estimated individually based on the submitted data [23]. Payments are computed according to a formula involving the average energy price, the yearly energy consumption, a load modulation coefficient, annual discounts and equivalent hours of yearly usage. Penalties are applied when customers do not reduce their power by the agreed amount [24].

As in Spain, Portugal is limited to interruptible contracts for large industrial customers [24]. However, in 2015 and 2016, the country did not experience any relevant problems requiring the implementation of measures aimed at guaranteeing the coverage of peak demand, and thus no interruptible load was curtailed [26, 27]. Furthermore, the country may be considered "closed" to the participation of incentive-based DR in the market, largely due to a lack of regulatory structures defining roles and responsibilities, baselining, payments and all other technical aspects required for implementing demand response programs [23].

[7]Orden ITC/2370/2007, BOE-A-2007-14798, enacted on 4 August, 2007.

[8]Orden IET/2013/2013, BOE-A-2013-11461, passed on 1 November, 2013.

10.2.2 Price-Based Demand Response Programs

Price-based demand response refers to changes in usage by end-use customers in response to changes in the price of electricity over time. Customers adjust the timing of their electricity usage to take advantage of lower-priced periods and/or avoid consuming when prices are higher, thus reducing their electricity bills. Customer response is typically driven by an internal decision-making process and load modifications are entirely voluntary. The most common price-based DR programs—also referred to as time-varying retail tariffs—include [2, 15]:

1. *Time-of-use*: rates with different prices for usage during different periods of time, usually defined for a 24 hour day. The number of periods vary by time of day (e.g., peak and off-peak, or peak and mid-peak and off-peak).
2. *Real-time pricing*: rates in which the price for electricity vary continuously (i.e., fluctuates hourly) during a 24 hour day, reflecting to a certain extent changes in the wholesale price, as opposed to time-of-use rates, which are largely based on preset prices. There are several variants, notably one-part and two-part real-time pricing (but see below).
3. *Critical peak pricing*: rates that are typically a hybrid of time-of-use and real-time pricing. The basic structure is time-of-use, although provision is made for replacing the normal peak price with a much higher event price under specified trigger conditions (e.g., when supply prices are very high).

Price-based programs may be incorporated at different time scales of the management of electric power systems (see, e.g., [2]). In brief, time-of-use (TOU) rates are fixed months in advance, real-time pricing (RTP) provides hourly prices to customers with day-ahead or near-real-time notice, and most critical peak pricing (CPP) rates involve a critical peak price that is called on a day-of basis.

Now, energy suppliers face various sources of risk in deciding how to price the electricity to provide to customers, including wholesale price variability (i.e., uncertainty about future wholesale power prices) and load variability (i.e., uncertainty about future customer loads). By taking into account the exposure of suppliers to wholesale price risk, retail tariffs may be viewed as forming a spectrum in itself (see, e.g., [28] for a taxonomy of retail electricity products). At one extreme are flat tariffs—that is, customers may consume as much power as they want at a guaranteed price during certain periods of time. Energy suppliers face the entire price risk. At the other extreme are spot-price tariffs, in which suppliers offer to provide whatever amounts of electricity customers want to consume at an hourly price that is tied directly to the wholesale price of power. Customers bear all the risk associated with uncertain wholesale prices. Between the two extremes, there is a wide range of possible retail tariffs. Two broad categories are guaranteed prices, which are known in advance but may differ in specific time periods (e.g., time-of-use pricing), and variable or dynamic prices, that change on an hourly basis during at least some time periods to match changes in wholesale prices (e.g., real-time pricing and critical peak pricing).

Dynamic pricing—sometimes called time-based pricing—provides customers with time-varying prices that reflect wholesale energy prices, offering a benchmark for demand response.[9] Retail tariffs provide a natural link between wholesale and retail markets. Indeed, they link retail prices with wholesale prices and also provide a (price) signal to customers, which guide them in making electricity usage decisions. Real-time pricing is the most common tariff for large (and medium) industrial and commercial customers [15]. Residential (and small business) customers represent a special challenge for dynamic pricing, since most customers lack information on their electricity-using appliances and equipment and are not familiar with demand response enabling technologies that can facilitate effective energy management [2]. Real-time pricing is often considered a tariff too sophisticated and rather difficult to understand for customers in this group [4]. Yet RTP is the default tariff option for small consumers in some countries (e.g., Spain). Accordingly, it will receive the preponderance of our attention in the following paragraphs.

Real Time Pricing for Electricity. The concept of real-time pricing is relatively simple, but the rules, regulations, and procedures can sometimes be confusing. In practice, RTP programs work essentially as follows.[10] Pricing information is provided to customers via automated phone call, text message, email, etc. Typically, customers receive hourly, market-based electricity prices, known as real-time prices, that vary according to the actual prices of energy in the market.[11] Also, they may receive alerts when electricity prices are trending high or when electricity is expected to be in high demand (see, e.g., [29]). Such information provides a financial incentive for customers to avoid consuming when prices are high and/or to move consumption away from peak times (to lower-priced hours), thereby creating opportunities for substantial savings. Customers should monitor the information received and be flexible in the ways they choose to use electricity. To this end, they need to understand their electricity consumption patterns in substantial detail and also need to be aware of their capabilities to curtail or shift discretionary usage. All customers able to change their normal consumption patterns to take advantage of lower-priced periods will end up paying less for energy. Electricity bills are calculated by taking into account the hourly prices and the corresponding hourly usage, meaning that there is a need of a special meter, called smart meter, to record the energy usage at every hour of the billing period (see, e.g., [30]).

---

[9]Notice that the natural way to account for demand response in day-ahead markets, when some customers face dynamic retail prices, is for retailers to offer price-responsive loads. For instance, a retailer that expects to serve 5000 MW in a hot summer period, armed with the information that RTP customers will reduce load by 250 MW if prices reach 80 €/kWh, may submit offers to purchase 5000 MW if prices remain low or 4750 MW if prices are expected to rise above 80 €/kWh. Such price-responsive demand offers provide the mechanism for informing the market about the extent of demand response, and typically result in lower day-ahead prices.

[10]Generally speaking, *real-time pricing* means tariffed retail charges for delivered electric power that vary hour-to-hour, determined to some extent from wholesale market prices, using pre-specified methodologies.

[11]Recall that the price of electricity varies with the time of day, day of week, or season of the year, and typically peaks when load peaks.

One-Part Real-Time Pricing. Real-time pricing is typically divided into two very different rate design structures. One-part RTP rates are characterized by hourly prices based on (expected) wholesale prices plus a mark-up, which normally takes the form of a simple adder or multiplier [31]. For example, some rates charge hourly prices that include all or a substantial level of fixed costs, in addition to marginal energy costs [32]. Typically, one-part rates charge ex-post, based on actual spot prices, although some rates consider day-ahead forecasts of the expected wholesale prices to set hourly prices.

One-part RTP rates have several advantages, notably [32]: (i) hourly prices apply to all electric usage by customers, (ii) the need to use historical reference periods is avoided or minimized, and (iii) customers with the same load and service characteristics are charged similar amounts. One-part rates have, however, a few flaws that have limited their success, at least to some degree [28]: (i) they are not revenue neutral to customers, limiting their appeal to consumers whose load characteristics produce rate savings, (ii) they may lead to potentially large variations in bills, and (iii) hourly prices may be inappropriately distorted by adders or multipliers.

Two-Part Real-Time Pricing. Two-part RTP rates are relatively more complex than one-part RTP rates. In particular, two-part rates combine two financial building blocks—a forward contract and spot pricing—to form a retail product. The forward contract guarantees a price for a fixed amount of load in each hour (the forward price is typically equal to that of the otherwise applicable standard rate). Customers then balance their actual consumption against the forward contract by purchasing incremental load or selling decremental load at hourly RTP prices, which are based on spot prices, and typically adjusted by a retail mark-up or mark-down [28].

Most two-part RTP rates consider the historic electricity usage of retail customers—that is, the customer baseline load (CBL)—as the basis for the forward contract quantity. Hence, the two-part design normally begins with the definition of the customer baseline load. This historical load shape is priced using standard (non-RTP) rates, which may involve different unit prices for different blocks of time (such as TOU rates).[12] Any actual load changes from the baseline level are priced at hourly RTP prices. This means that both increments (i.e., increases in consumption above that of the CBL) and decrements (i.e., reductions in hourly load below that of the CBL) are priced at hourly prices [33].

The two-part design allows customers to achieve savings by curtailing usage at times when prices are higher and by using more energy during off-peak periods [15]. Customers who increase consumption above the CBL may reduce their electricity bills. Customers who decrease consumption during high prices may also benefit. In fact, decrements are priced at hourly RTP prices, which translates into suppliers crediting customers for reductions in usage below that of the CBL. This decremental crediting is an extremely powerful feature of two-part rates and should not be overlooked, since it encourages the significant price response attributes of RTP [33].

[12]Two-part RTP customers who do not deviate from the baseline level pay the same amount under the two-part rate or the standard (non-RTP) rate. Accordingly, the two-part RTP rate is sometimes called "revenue neutral at the CBL" [28].

Especially noteworthy is the fact that the forward contract provides risk management benefits to both customers and sellers (or suppliers). On the one hand, the contract allows customers to lock in a price for a large portion of their load—that is, they face price and quantity risks only on the incremental and decremental load. This represents a significant reduction in the risk exposure relative to one-part RTP products, as discussed above. On the other hand, the forward contract simplifies the risk management task of suppliers, since it considers a specified quantity, and thus there is no quantity risk on that portion of customers' load. Suppliers still face some price risk on the forward contract, but this risk is easily offset by either their own generation or by purchasing wholesale forward contracts [28].

Price-Based Demand Response in Spain and Portugal. Electricity bills in Spain are among the highest in Europe. In 2013, the Spanish Government initiated a reform of the electricity sector, focusing on the economic stability of the electric system.[13] In 2014, the Voluntary Price for Small Consumers (VPSC)—that is, a real-time pricing tariff for residential and small business customers, involving a specific methodology for price calculation—came into force.[14] Also in 2014, the European Directive 27/2014/EU, on energy saving, was transposed to the Spanish legislation.[15]

The VPSC tariff is applied to consumers who have smart meters integrated into the information and telecommunication (IT) system of a reference trading company (i.e., a supplier allowed to offer the tariff). Electricity prices are calculated for each day and hour by adding up the following components [34]: (i) hourly wholesale prices, (ii) regulated network charges, and (iii) a regulated retail margin. Pricing information is provided online to customers. More specifically, the Spanish electrical grid (REE) publishes the hourly prices for a given day (d) at 8:15 p.m. of the previous day ($d-1$), in strict accordance with Royal Decree 216/2014.[16]

The VPSC tariff is the default pricing option for nearly 28 million customers in Spain (although customers can subscribe to a different tariff). At the end of 2014, 11.91 million smart meters were installed, of which 10.19 million were successfully integrated into the IT system of a trading company, meaning that approximately 36% of the eligible customers have subscribed the VPSC tariff [35]. In June 2016, smart meter roll-out was at around 17.53 million (62%) of customers [36]. The full deployment of smart meters is expected to be completed by the end of 2018 [8].

Price-based demand response in Portugal is limited to different forms of time-of-use rates [23]. The country may be considered "passive" in relation to the implementation of dynamic pricing options for retail customers. Despite a long-standing interest, and some controversy, no rules are in place for a clear participation of price-based demand response in the market.

[13]Royal Decree-Law 9/2013, BOE-A-2013-7705, of 12 July, 2013.

[14]Royal Decree 216/2014, BOE-A-2014-3376, enacted on 28 March, 2014.

[15]Royal Decree-Law 8/2014, BOE-A-2014-7064, passed on 5 July, 2014. Also, Law 18/2014, BOE-A-2014-10517, of 17 October, 2014.

[16]https://www.esios.ree.es/en/pvpc (accessed on 15 May, 2017).

10.3 Demand Response and Daily Market Prices

This section provides a detailed discussion of how customer load reductions lower spot market prices. The section is organized as follows. Section 10.3.1 presents a detailed overview of the Iberian Electricity Market (MIBEL). Following this introductory material, Sect. 10.3.2 describes how demand response averts the need to use the most costly-to-run power plants during periods of high demand, driving wholesale market prices down. The description is illustrated with a practical example using data published by the managing entity of the Iberian market (OMIE).

10.3.1 Overview of the Iberian Electricity Market

The Royal Decree 2019/1997, passed on 26 December, establishes the basic structure of MIBEL, by distinguishing five units: the day-ahead market, the intra-day market, futures markets, non-organized markets and the system adjustment services market. The international agreement between the Kingdom of Spain and the Republic of Portugal, ratified in Santiago de Compostela on 1 October 2004, approves the current organizational structure, which involves the Spanish market operator (OMIE), operating the day-ahead and intra-day markets, and the Portuguese market operator (OMIP), operating the futures market. The market operating rules comply with the Royal Decree 2019/1997 and Act 24/2013, of 26 December (Electricity Sector Act).

The day-ahead market (DAM) has been a reliable and representative meeting point between supply and demand since 1 January 1998, for the Spanish region, and 1 July 2007, for the Portuguese region [37]. Sessions are structured in scheduling periods equivalent to a calendar hour, with a scheduling horizon divided into 24 consecutive schedule periods of the Central European Time (CET). Sale bids may be simple or complex, depending on their content. Simple sale bids are essentially bids to sell electricity specifying a price and an amount of power for each hourly period. Complex sale bids comply with the requirements governing simple bids and, in addition, include some or all of the following conditions [38]:

- Indivisibility: the indivisible block of a bid must be matched in its entirety.
- Minimum income: a bid is only considered submitted for matching purposes when the seller obtains a minimum income.
- Scheduled stop: if a bid is not matched due to the application of the minimum income condition, then it can be treated as a simple bid in the first block of the first three hourly periods of the daily scheduling horizon.
- Production capacity variation or load gradient: establishes, for each production unit, a maximum upward or downward difference in energy variation, between two consecutive hourly scheduling periods.

Purchase bids can only be simple—that is, the offers to buy energy cannot include complex conditions.

The supply bids made by generating companies, which may include complex conditions, are submitted per production unit, and specify independent quantities and prices for each hour of the market horizon. These bids are sorted by ascending price and a market supply curve is built for each hour. Plants under the special regime and nuclear power plants (in the Spanish region) usually appear in the lower part of the curve, as their opportunity cost is low. Also, run-of-river plants usually appear in this part of the curve, as they cannot store water for long periods of time. The middle section of the curve includes mainly coal plants. The highest end of the curve has combined cycle plants and the part of hydraulic power with scant reserves. Typically, supply bids from hydroelectric power plants do not include complex conditions. However, most supply bids from thermal power plants include complex conditions [37].

The demand offers in the day-ahead market, which cannot include complex conditions, are ranked in order of decreasing price, starting with the highest, until reaching the lowest, and a demand curve is built for each scheduling period in the daily horizon. The highest part of the curve corresponds to the demand associated with distributors (regulated supplies), typically involving the instrumental price of 180.30 €/MWh.[17] The middle and lower parts of the curve include consumption corresponding to pumping stations and to providers for their supply in the free market, who present offers specifying energy prices different from the instrumental price.

The market operator (MO) matches the bids for the sale and purchase of energy using the Euphemia algorithm [38]. The purpose of this algorithm is to optimize the welfare, which corresponds to the sum of the benefit from the sale bids (i.e., the difference between the marginal price received and the price of the matched sale bid), plus the benefit from the purchase bids (i.e., the difference between the price of the matched purchase bid and the marginal price received), plus the congestion charge. The algorithm performs the matching process with the accuracy of the price and energy values exceeding the ceiling of decimals specified for the submission of bids. Especially noteworthy are the two sale curves developed by the MO for each production zone[18]:

- A first curve containing all the blocks of all the simple bids and all the blocks of all the bids that have declared the condition of indivisibility. The energy tendered at the same price is aggregated to that price with no differentiation.
- A second curve containing all the blocks of the economic order of precedence that are not contained in the first curve, without aggregating the energy tendered at the same price, and featuring the identification of the production blocks that belong to the same tender.

Both curves do not feature any identification of the production units to which they correspond to.

[17]The instrumental sale price, or minimum price (price floor), is set at 0 €/MWh, and the instrumental purchase price, or maximum price (price cap), is set at 180.30 €/MWh [38].

[18]Here, for the sake of clarity in exposition, the two sale curves are referred to as the "first" or "initial" curve and the "second" or "actual" curve.

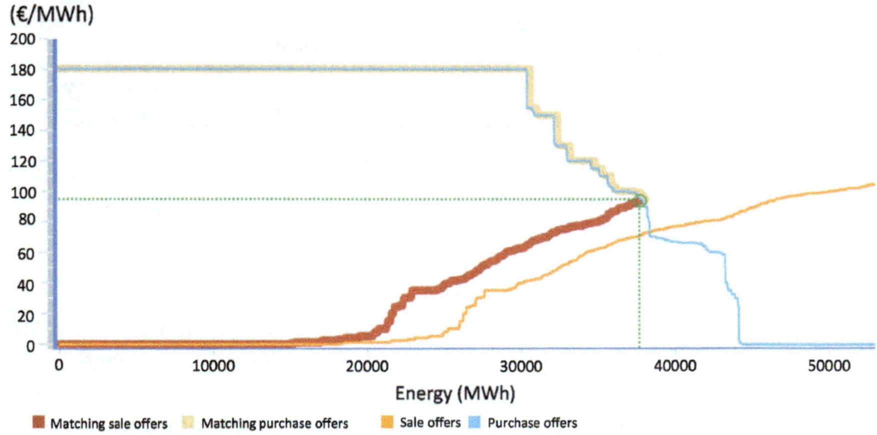

Fig. 10.1 Supply and demand curves published by OMIE on 20 January 2017 at 2 p.m. (adapted from [39])

For each market session, the market operator publishes and makes public the following information [38]:

- Hourly prices and total energy negotiated per hour on the day-ahead market.
- Supply and demand curves. Supply bids and demand offers, with an indication of the prices and quantities for each segment of energy offered.
- Business of each international interconnection per hour, indicating: (i) maximum import and export sales capacity for each interconnection, (ii) occupied capacity in each direction of the interconnection, and (iii) free capacity in each direction of the interconnection.

Figure 10.1 shows the supply and demand curves published by OMIE [39] on Friday 20 January 2017 at 2 p.m. (an hourly scheduling period when the energy price was extremely high). The intersection of the "first" supply curve (light orange curve) with the demand curve (light blue curve) determines the "initial" market-clearing price (nearly 73 €/MWh) and the equilibrium energy quantity. As noted above, the "actual" supply curve (red curve) is determined by considering the generation restrictions of the complex sale bids. The intersection of this curve with the demand curve (grey blue curve) determines the "final" market-clearing price (nearly 94.70 €/MWh) and the equilibrium quantity (around 38083 MWh).

Notice that the integration of complex supply bids in the market clearing optimization process has a remarkable influence on the energy price—that is, the price resulting from the intersection of the "initial" supply and demand curves (around 73 €/MWh) is considerably lower than the market-clearing price (nearly 94.70 €/MWh). In other words, the actual supply curve visibly differs from the initial supply curve, due to the effect of the complex generation bids. In contrast, since buyers are not allowed to make complex bids, the actual demand curve may be regarded as part of the initial demand curve (see Fig. 10.1).

Also, notice that a thorough analysis of the demand curve reveals the (typical) downward-sloping shape. Furthermore, the supply and demand curves exhibit the following characteristics:

- The supply curve starts with a flat section corresponding to the sale bids at (nearly) the minimum price (typically, bids from renewable power producers).
- From the initial flat section, the slope of the supply curve is positive, since the sale bids are ranked in ascending order of price.
- The demand curve starts with a flat section corresponding to the purchase offers at the maximum price (typically, offers from retailers of electrical energy).
- After the initial flat section, the slope of the demand curve is negative (and very steep), since the demand bids are ranked in descending order of price. Also, this slope tends to be more steeply than the slope of the supply curve, due to the characteristic inelasticity of demand.

10.3.2 Short-Term Market Impact of Demand Response

Demand response refers broadly to incentive payments designed to induce lower electricity use at times of high wholesale market prices, or to changes in electric use by end-use customers in response to changes in the price of electricity over time. In regions with wholesale markets, the curtailment of a given amount of load, especially at hours of high market prices, results in a variety of financial and operational benefits. These benefits fall into the following four groups [2]:

1. *Participant financial benefits*: DR involves explicit financial payments and, in addition, motivates lower electricity usage at times of high prices.
2. *Market-wide financial benefits*: DR averts the need to use the most costly-to-run power plants during periods of high demand, driving wholesale market prices down. Over the longer term, sustained DR also lowers aggregate system capacity requirements, reducing the need to build additional generation, transmission or distribution capacity infrastructure.
3. *Market performance benefits*: DR acts as a deterrent to the exercise of market power by generating companies.
4. *Reliability benefits*: DR reduces the likelihood and consequences of forced outages that often impose financial costs and inconvenience on customers.

Market-wide financial benefits are related to both short-term and long-term market impacts. Short-term market impacts are the most immediate and easily measured

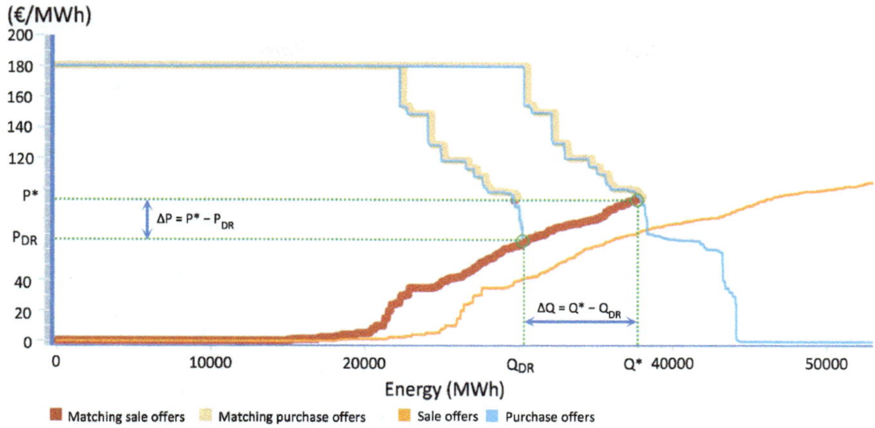

Fig. 10.2 Impact of demand response on the spot market price (adapted from [39])

source of financial benefits from demand response and, therefore, will receive the preponderance of our attention in the remainder of this subsection.[19]

Figure 10.2 illustrates the way in which demand response influences the spot market prices. The figure shows the supply and demand curves published by OMIE [39] on Friday 20 January 2017 at 2 p.m. (as first presented in Fig. 10.1). Let P^* be the market-clearing price and Q^* the equilibrium energy quantity. Since the retail suppliers of customers enrolled in demand response programs submit offers to buy energy at the instrumental price of 180.30 €/MWh (and therefore appear in the highest part of the demand curve), when consumption decreases (e.g., from Q^* to Q_{DR}), it has the effect of shifting the demand curve to the left ("negative" shift), thereby lowering the highest marginal cost. As long as the supply curve has a positive slope, the reduced demand leads to a lower wholesale electricity price, as illustrated by the left-hand demand curve of Fig. 10.2 (grey blue curve). The "new" market-clearing price P_{DR} is defined by the intersection of the "actual" supply curve (red curve) with the left-hand demand curve.[20]

Put simply, a (substantial amount of) demand reduction lowers the market price for all wholesale electricity purchasers, and also may push out of the market expensive (and probably high-polluting) power plants. This is the basis of the so-called merit order effect of energy saving [8], which can be considered similar to the well-known merit order effect of renewable electricity generation (see *Chap.* 9). Also, if, over time, customers routinely respond to high prices by reducing load, then additional

[19]A full description of the benefits of demand response in electricity markets is beyond the scope of this chapter. For further information, the interested reader is referred to [2, Sect. 3].

[20]Figure 10.2 presents a hypothetical situation for illustrative purposes only. For the sake of clarity, the load reduction from customers participating in DR programs—that is, the demand reduction that moves consumption from Q^* to Q_{DR}—was intentionally enlarged.

(bill) savings may result to such customers. Furthermore, the spot market price is typically the leading price indicator for all electricity trades, and therefore the price paid through bilateral contracts is based, to some extent, upon that market price. Accordingly, if demand response consistently reduces market prices and volatility, then bilateral contract prices may (ultimately) be pushed down.

10.4 Price-Effect of Demand Response on the Iberian Market

This section investigates the beneficial effects of different levels of DR on the day-ahead prices of MIBEL and analyzes the potential benefits that result to market participants and retail customers. The section is organized into three major parts. The first part analyzes, in detail, the hourly prices published by OMIE in the period 2014–2017. This part also investigates the interaction between a specific level of DR and the market prices on January 25, 2017 (a Wednesday), when the prices were at their highest. This simple, albeit relatively important, analysis is helpful in identifying the influence of DR on the spot power prices (see Sect. 10.4.1).

The second part presents an insightful study to quantify the impact of different levels of DR on the Iberian electricity market prices at times of system constraints (see Sect. 10.4.2). The study is conducted with the help of the agent-based simulation tool called MATREM (for Multi-Agent TRading in Electricity Markets). The time period of the study has the duration of 42 months: from January 1, 2014 to June 30, 2017. The following 10 scenarios are considered:

- Scenario A: the supply and demand curves are built from the bids and offers submitted to MIBEL (base-case scenario).
- Scenarios B1–B3, C1–C3, D1–D3: the supply curve is built as in scenario A. However, to simulate what would have been the market prices in the presence of certain levels of DR, the values of the electricity demand are changed correspondingly.

The last part of the section summarizes the results and discusses the conclusions reached (see Sect. 10.4.3).

10.4.1 Preliminary Analysis

Demand response in wholesale energy markets emphasizes reductions in electric usage by end-use customers when prices are high. As mentioned earlier, individual customers, or demand response aggregators acting on behalf of various customers, may participate directly in energy markets by submitting bids to curtail load. Also, large customers, who can be direct market participants, may reduce consumption when they observe market prices rising above the maximum benefit that they can

Table 10.1 Iberian market prices for the Portuguese area in the period between January 1, 2014 and June 30, 2017 (based on data from [40–44])

Month/Year	Average real electricity price (€/MWh)			
	2014	2015	2016	2017
Jan	31.47	51.82	36.39	71.52
Feb	15.39	42.57	27.35	51.39
Mar	26.20	43.22	27.70	43.95
Apr	26.36	45.49	23.50	44.18
May	42.47	45.18	24.93	47.12
Jun	51.19	54.74	38.28	50.22
Jul	48.27	59.61	40.36	–
Aug	49.91	55.59	41.14	–
Sep	58.91	51.92	43.61	–
Oct	55.39	49.89	52.78	–
Nov	46.96	51.46	56.25	–
Dec	47.69	52.92	60.27	–
Year	41.86	50.43	39.44	–

obtain from consuming electricity. Small customers may do the same to the extent they are exposed to wholesale spot prices through retail arrangements.

Energy prices drive, therefore, the potential revenues of demand response in electricity spot markets. Accordingly, we examine the prices published by OMIE in the period between January 1, 2014 and June 30, 2017. Table 10.1 presents the average monthly electricity prices for the Portuguese area and Fig. 10.3 shows graphically the highest price for each month of the observation period. For the sake of clarity, we present next a brief description of the main figures.

The year 2014 was a very important year for MIBEL, with the commissioning of the new interconnection between the Iberian Peninsula and France and, therefore, a greater convergence of prices and lower volatility between Portugal, Spain and the rest of the European Union. The Governments of Portugal and Spain set strong foundations for a sustainable growth of the economies of both countries and, within this economic scenario, the demand for electricity has fallen less than in previous years. The energy traded on the spot market was about 259 TWh. The average prices in Portugal and Spain were 41.86 €/MWh and 42.13 €/MWh, respectively [40]. The average daily maximum price was 71.06 €/MWh (observed on 10 October) and the average monthly maximum price was 58.91 €/MWh (observed in September). The highest market price was 110 €/MWh (observed in only 1 h on 17 February).

The year 2015 was characterized by an improvement of both the Portuguese and the Spanish economies, with growth figures of 1.5 and 3.2%, respectively. For this work, the most salient aspect of this economic scenario was a growth in the demand for electricity on the Iberian Peninsula: demand rose in Portugal by around 0.3% and in Spain by nearly 1.8%. The market prices were slightly higher than those in other European markets, mainly due to the increase in demand, along with

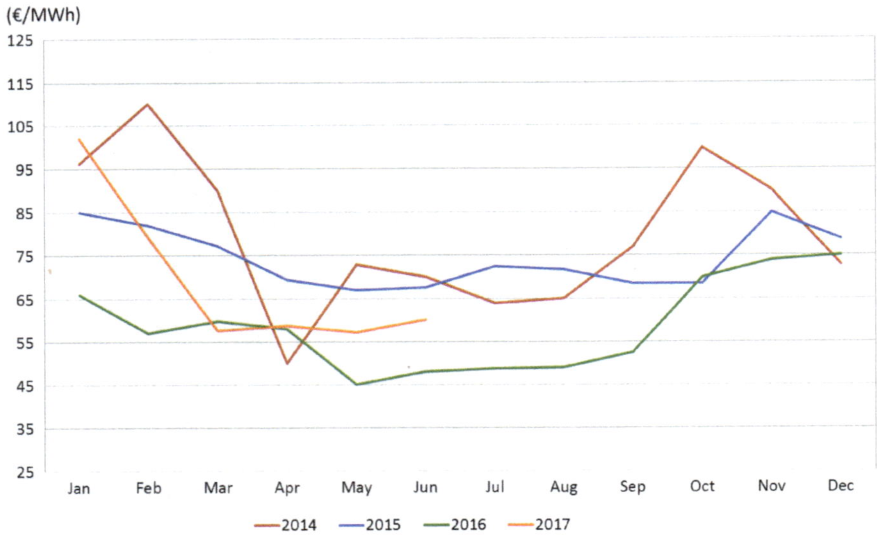

Fig. 10.3 Highest-priced hours per month in the period 2014–2017 (based on data published by OMIE [40–44])

lower levels of availability of both hydro and wind resources during the first part of the year. Specifically, the average price was 50.43 €/MWh in Portugal, and 50.32 €/MWh in Spain [41]. Other important figures are as follows: (i) the average daily maximum price was 67.12 €/MWh (observed on 2 December), (ii) the average monthly maximum price was 59.61 €/MWh (observed in July), and (iii) the highest market price was 85.05 €/MWh (observed in only 1 h on 7 January, although the price reached 85 €/MWh on 23 November at 8 p.m.).

In 2016, the Governments of Portugal and Spain appointed OMIE as the Nominated Electricity Market Operator (NEMO) for the day-ahead and intra-day markets meaning that it meets all requirements to act as a leading company in the project for creating the European internal energy market. The demand for electricity grew 0.8% in Spain, while the Portuguese increase stood at nearly 0.4%. The spot market maintained its levels of liquidity, with a trading volume of about 270 TWh, a figure slightly higher than that for 2015, involving an economic volume of 11.027 million €. The market prices decreased substantially and were 15% lower than the average prices for the past five years. In particular, the average price in Portugal was 39.44 €/MWh, while in Spain it was 39.67 €/MWh [42]. The average daily maximum price was 66.83 €/MWh (observed on 15 December) and the average monthly maximum price was 60.27 €/MWh (observed in December). The highest market price was 75 €/MWh (observed in only 1 h on 21 December).

The year 2017 marks the tenth anniversary of the operational launch of the spot market for the Iberian Peninsula, integrating the Portuguese and Spanish areas. In the first half of the year (i.e., from January 1, 2017 to June 30, 2017), the average monthly electricity prices varied between 71.52 €/MWh (in January) and 43.95 €/MWh,

Table 10.2 Number of hours per year with prices above a given threshold (or in a specific range) for the period between January 1, 2014 and June 30, 2017 (based on data published by OMIE [39])

Price (€/MWh)	Number of hours				
	2014	2015	2016	2017	Total
$P \geq 75$	70	28	1	328	427
$P \geq 80$	34	4	0	201	239
$P \geq 85$	23	2	0	139	164
$P \geq 90$	13	0	0	90	103
$P \geq 95$	6	0	0	36	42
$P \geq 100$	1	0	0	3	4
$75 \leq P < 80$	36	22	1	127	186
$80 \leq P < 85$	11	2	0	62	75
$85 \leq P < 90$	10	2	0	49	61
$90 \leq P < 95$	7	0	0	54	61
$95 \leq P < 100$	4	0	0	33	37
$100 \leq P < 105$	0	0	0	3	3
$105 \leq P \leq 110$	1	0	0	0	1

(in March). The market prices were extremely high in January [43]: the average daily price ranged between 40.80 €/MWh to a maximum of 101.99 €/MWh. The highest price (101.99 €/MWh) was observed on 25 January at 9 p.m. (although the prices rose above 75 €/MWh in all hours of the day). At the time of writing, it is also worth mentioning that 2017 is considered a decisive year for the implementation of the European platform for the continuous cross-border intra-day market, which will undoubtedly lead to a greater integration of renewable energies.

Table 10.2 shows the number of hours per year during which the Iberian market prices exceeded a given threshold.[21] The threshold starts at 75 €/MWh and is gradually increased by increments of 5 €/MWh until an upper bound of 100 €/MWh.[22] The prices rose above 75 €/MWh in 427 h. At higher price thresholds, this number

[21] A *price-duration curve* is a curve that shows the fraction of the number of hours of a year during which the electricity price was less than a given value. This curve appears commonly in the technical literature on energy markets (see, e.g., [45, 46]). It allows us to observe the percentage of hours where the price reached and/or exceeded a given value. To enhance readability, and also in the interests of completeness, we present this information in a table.

[22] The reader familiar with the Iberian market (and other European markets) might find the lower bound (75 €/MWh) and the upper bound (100 €/MWh) of the price threshold somewhat arbitrary (and even rather ad hoc). Nevertheless, the lower bound is considerably higher than the average price for each year of the observation period (see Table 10.1) and, therefore, suitable to the purposes of this work. Furthermore, the decision to set the upper bound at 100 €/MWh seems to be natural, intuitive and rather elegant.

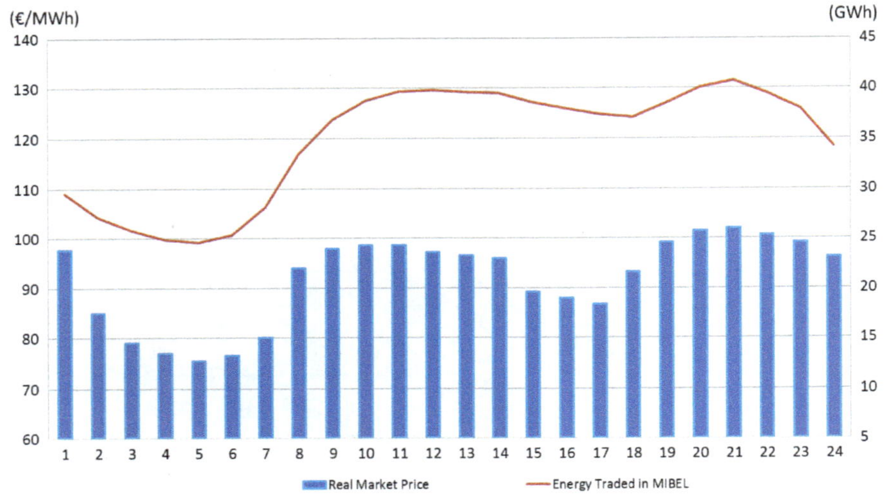

Fig. 10.4 Real market prices (column chart) and energy traded in MIBEL (line chart) on 25 January 2017 (based on data from [39])

decreases gradually. Specifically, the prices rose above 80 €/MWh in 239 h, exceeded 90 €/MWh in 103 h, and were greater than 100 €/MWh in 4 h.[23]

Table 10.2 also shows the number of hours per year during which the prices were in a given range. The lower bound of the first interval corresponds to a price of 75 €/MWh and the upper bound of the last interval to a price of 110 €/MWh (i.e., the highest price observed in the period under consideration). The prices were in the range between 75 and 80 €/MWh in 186 h. This number decreases gradually as the lower and upper bounds of the intervals increase (by increments of 5 €/MWh).

Now, a simple approach to analyze the influence of demand response on spot power prices is to investigate the interaction between a specific level of DR and the market price. Accordingly, we examine the Iberian market in a working day from the fifth week of January 2017, namely January 25 (a Wednesday), when the prices were at their highest. Figure 10.4 depicts the spot market prices (column chart) and the energy traded in the market (line chart) for each hour of this weekday. From the figure, we can observe the following:

- The price ranged between 75.51 €/MWh (at 5 a.m.) to a maximum of 101.99 €/MW (at 9 p.m.). The average price in Portugal was 91.91 €/MW. The price was in the range between 90 and 100 €/MW in 12 h and rose above 100 €/MW in 3 h.
- The total energy traded in the market varied between 24.5 GWh (at 5 a.m.) to a maximum of 40.7 GWh (at 9 p.m.). The energy traded was in the range between 30 and 40 GWh in 15 h and exceeded 40 GWh in 2 h.

[23]The market prices never exceeded 110 €/MWh during the observation period. The highest price (110 €/MWh) was observed on Monday 17 February 2014 at 10 p.m.

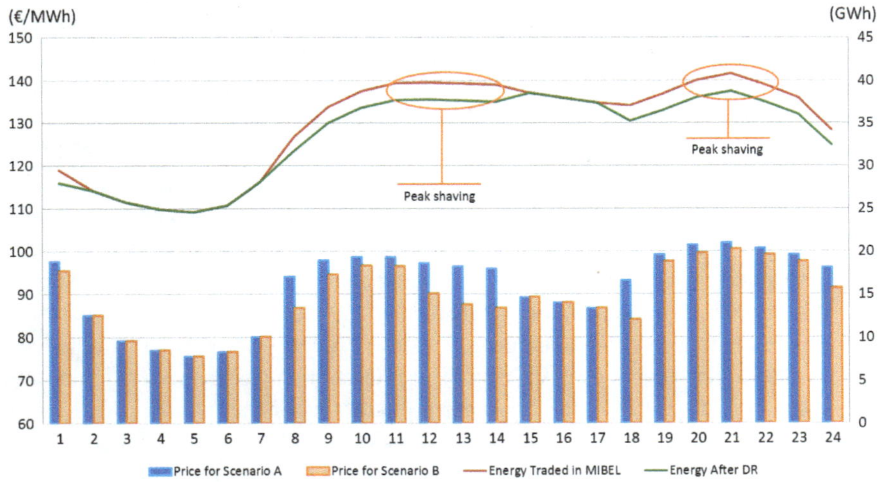

Fig. 10.5 Simulated market prices for scenarios A and B, actual energy traded in MIBEL, and energy quantities after DR on 25 January 2017 (based on data from [39])

- As one would expect, the price was lower during periods of low demand and increased considerably during periods of high demand.

The analysis involves the simulation of the Iberian market prices on Wednesday 25 January 2017. The simulation is carried out with the help of the agent-based system called MATREM. The system supports a day-ahead market and an intra-day market (see, e.g., [47, 48]) and a futures market. MATREM also supports a bilateral marketplace for negotiating the details of tailored (or customized) long-term bilateral contracts (see, e.g., [49–51]).[24]

The analysis considers a 3% load reduction and focuses on the highest-priced hours. Specifically, the following two scenarios are considered:

- Scenario A: the supply and demand curves are built from the bids and offers submitted to MIBEL. The hourly spot market prices are determined by using the MATREM system.
- Scenario B: the supply curve is built as in scenario A. However, to simulate what would have been the market prices in the presence of a certain level of DR, the value of the electricity demand is changed correspondingly. Specifically, DR is modeled as a 3% load reduction when prices rise above 90 €/MWh. Again, the prices are determined by using the MATREM system.

The aim is, therefore, to examine how a 3% load reduction in the hours with prices above 90 €/MWh can affect spot market prices.

Figure 10.5 shows the simulated prices for scenarios A and B. The prices for scenario A correctly fit those actually reported by OMIE (but see next subsection for

[24] *Chapter* 8 is entirely devoted to the agent-based simulation tool and the interested reader is referred to it for details.

a detailed description of the procedure adopted here for defining the bids and offers of the MATREM agents). For each of the highest-priced hours under consideration, the prices for scenario B are considerably lower than the prices computed for scenario A. Specifically, the prices for scenario A vary between 93.2 €/MWh (at 6 p.m.) and 101.98 €/MWh (at 9 p.m.), while the prices for scenario B range between 84 €/MWh (at 6 p.m.) to a maximum of 100.43 €/MWh (at 9 p.m.). The price reduction varies between 1.46 and 9.27 €/MWh, reaching the highest value at 2 p.m. and the lowest at 7 p.m. On average, a price reduction of about 4.27 €/MWh is estimated for the entire day, corresponding to a decline of 4.37%. The financial benefits of demand response under the simulated market conditions are around 4.9 million €, a remarkable result.

Now, recall that a key role of demand response in energy markets with high levels of renewable generation is peak shaving [4]: a reduction in peak consumption during tight system conditions in order to release pressure on generation and grid capacity needs. Figure 10.5 shows the actual energy traded in MIBEL on 25 January 2017 and the energy quantities after demand response, illustrating the power of DR to reduce load at times of high market prices (and during periods of high demand). This demand-side flexibility may facilitate the integration of variable generation into energy markets in an efficient and reliable manner (particularly when the wind and solar penetration levels are low).

10.4.2 Simulation-Based Study

This section presents an insightful simulation-based study to analyze the price-effect of demand response on the Iberian electricity market prices. The study makes use of data published by the managing entity of the Iberian daily electricity market (OMIE), as well as data reported by the managing entity of the Portuguese electrical grid (REN). Specifically, the following sources of data are considered:

- Hourly day-ahead (spot) prices and hourly energy quantities submitted to the Iberian market. Also, market-clearing prices and energy quantities traded in the Iberian market (data published by OMIE [39]).
- Market prices and traded energy quantities (data reported by REN [52]).

The observation period has the duration of 42 months: from January 1, 2014 to June 30, 2017 (a total of 30648 h).

The study involves the simulation of the Iberian market prices using the agent-based simulation tool called MATREM. To this end, there is a need to define the software agents that participate in the simulated day-ahead (spot) market—that is, the suppliers or sellers and the demanders or buyers—and also to specify their bids and offers. The method adopted here is a natural extension of the procedure described in Chap. 9 (see Sect. 9.4.2). Table 10.3 presents the various supplier and demander agents and summarizes their energy bids. To overcome the added computational

Table 10.3 Software agents and their energy bids/offers for a particular hour of the market horizon

Agent type	Energy quantity (MWh)	Energy price (€/MWh)
Supplier	Quantity of all bids at 0 €/MWh	0
Supplier	Quantity of all bids in the range]0, 5] €/MWh	5
Supplier	Quantity of all bids in the range]5, 10] €/MWh	10
. . .	. . .	. . .
Supplier	Quantity of all bids in the range]55, 60] €/MWh	60
Supplier	Actual quantity submitted to MIBEL	Actual price: $P_1 > 60$
Supplier	Actual quantity submitted to MIBEL	Actual price: $P_2 \geq P_1$
. . .	. . .	. . .
Supplier	Actual quantity submitted to MIBEL	Equilibrium price: P^*
Demander	Quantity associated with the equilibrium price	180

complexity of considering a very large number of agents, we make the following simplifying assumptions:

- The number of agents representing the electricity supply industry varies between 50 and 350. The agents submit bids to sell energy at prices in the range between 0 €/MWh and the market-clearing price.
- The demand for electrical energy is modeled by a price-inelastic demand curve. A single agent submits an offer to buy the entire electricity demand at 180 €/MWh.

Notice that the generation technologies associated with the bids submitted to MIBEL are not publicly available and, therefore, no attempt is made to associate each supplier agent with only a single technology.

The rationale for defining the agents that represent the electricity supply industry is as follows. First, consider the bids to buy energy at 0 €/MWh submitted to the Iberian electricity market. For simplicity, we associate a single agent to all bids at 0 €/MWh. This decision seems to be intuitive and natural.

Next, consider the bids to buy energy at any price between 0 and 60 €/MWh. Again, in the interests of simplicity, we decompose the interval]0, 60] into 12 sub-intervals, based on a somewhat arbitrary increment of 5 €/MWh—that is, we define the sub-intervals]0, 5],]5, 10], . . . ,]55, 60]—and associate a single agent to each sub-interval. In this way, a software agent playing the role of a supplier submits a bid to sell the quantity of energy corresponding to the sum of the quantities of all bids (submitted to MIBEL) in the range]0, 5] at 5 €/MWh. Another agent submits a bid

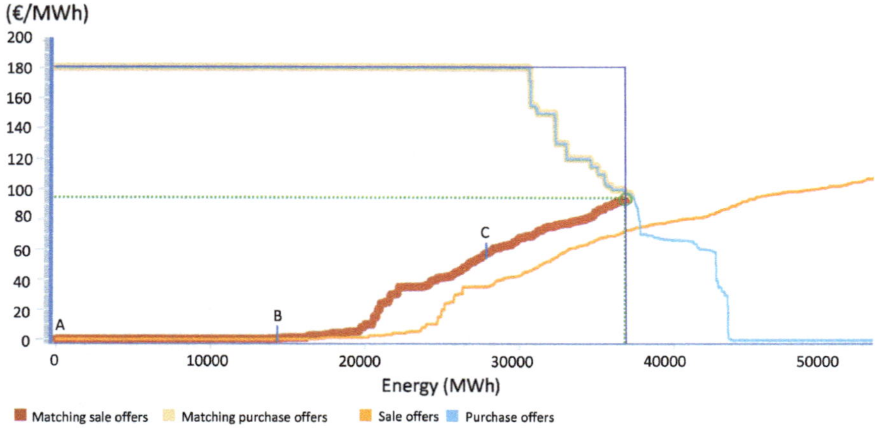

Fig. 10.6 Three (hypothetical) parts of the supply curve and the demand curve (dark blue curve) adopted in the simulation-based study on 20 January 2017 at 10 p.m. (adapted from [39])

to supply the sum of the quantities of all bids in the range]5, 10] at 10 €/MWh, and so on. This decision seems to be simple and satisfactory (yet somewhat ad hoc).[25]

Now, let P^* be the equilibrium price and consider the bids to buy energy at any price between 60 €/MWh and P^*. To perform computer simulations as close to the reality as possible, we associate a software agent to each supply bid submitted to MIBEL, and consider the actual values of the prices and quantities. This decision is relatively straightforward. The number of software agents involved in the simulation of this part of the supply curve ranges between 37 to a maximum of 337, a considerable number, although not (very) awkward to manage computationally.

Figure 10.6 shows the supply and demand curves published by OMIE [39] on Friday 20 January 2017 at 10 p.m. The intersection of the actual supply curve (red curve) and the actual demand curve (grey blue curve) determines the market-clearing price (about 95 €/MWh) and the equilibrium energy quantity (around 37206 MWh). The green point represents this market equilibrium.

Figure 10.6 also illustrates the three (hypothetical) parts of the actual supply curve mentioned above to define the supplier agents. The lower part corresponds to the segment $\overline{AB}$ and represents the supply bids at 0 €/MWh. It has a perfectly horizontal slope and is associated with a single software agent. The middle part of the curve represents the supply bids at any price between 0 €/MWh and 60 €/MWh and is modeled by 12 software agents. It slopes upward gradually until reaching a nearly, but not perfectly, vertical slope, then exhibits a nearly horizontal slope, and finally

[25] The reason for an upper bound of 60 €/MWh is as follows. The simulation-based study includes scenarios involving load reductions of 1, 3 and 5%, and price thresholds of 80, 90 and 100 €/MWh (see below). This means that the market-clearing price may drop below 80 €/MWh, depending on the scenario under consideration. However, for all scenarios, the lowest equilibrium price is always greater than 60 €/MWh.

ends with a considerable large slope. The upper part of the curve corresponds to the segment $\overline{CP^*}$ and represents the supply bids at any price between 60 €/MWh and the equilibrium price. It has a moderately upward slope and is modeled by a considerable number of software agents (ranging between 193 and 303).

Furthermore, Fig. 10.6 shows the price-inelastic demand curve (dark blue curve) adopted in the simulation-based study (for Friday 20 January 2017 at 10 p.m.). The curve is built by considering the instrumental purchase price or maximum price and the equilibrium energy quantity (around 37206 MWh). The inherent assumption—that is, the assumption of a perfectly inelastic demand in the short-term perspective of the day-ahead market—seems to be satisfactory with respect to the practical purpose of simulating the Iberian market prices (as close to the reality as possible). Also, this assumption is supported, at least in part, by the downward-sloping shape of the actual demand curve, falling rapidly from the maximum price to 0 €/MWh. This typical shape of the demand curve for the Portugal-Spain region suggests that the price elasticities in the Iberian energy sector are low in the short run.

The study involves 10 energy scenarios related to the Iberian daily electricity market. Scenario A considers the actual supply bids and demand offers submitted to MIBEL. Scenarios B1–B3, C1–C3, and D1–D3 consider load reductions when market prices rise above specific price thresholds. More concretely, the relevant characteristics of each scenario are as follows:

- Scenario A (base-case scenario): the supply and demand curves are built from the bids and offers submitted to MIBEL. In particular, the supply curve is built from the bids of the various supplier agents ranked in increasing order of price (as described above). The demand curve is built from the offer of a single agent. The hourly prices of electricity are determined using the MATREM system.
- Scenarios B1–B3: the supply curve is built as in the base-case scenario. However, to simulate what would have been the market prices in the presence of three different levels of DR, the values of the electricity demand are changed correspondingly. The three levels of DR are modeled as three distinct load reductions when prices exceed a threshold of 80 €/MWh, namely 1% load reduction for scenario B1, 3% load reduction for scenario B2, and 5% load reduction for scenario B3. The hourly prices are determined as in the base-case scenario.
- Scenarios C1–C3: scenarios very similar to scenarios B1–B3. However, the price threshold is set to 90 €/MWh. The load reductions are as follows: 1% for scenario C1, 3% for scenario C2, and 5% for scenario C3.
- Scenarios D1–D3: scenarios analogous to scenarios B1–B3. The price threshold is set to 100 €/MWh. The load reductions are 1% for scenario D1, 3% for scenario D2, and 5% for scenario D3.

Table 10.4 presents the various DR scenarios and summarizes the corresponding price thresholds and load reductions. The table also shows the number of demand response events considered in each scenario.[26]

[26]For the purposes of this work, a *demand response event* is any event that results in a load reduction when the spot power price rises above a given threshold. Such an event corresponds to a single hour

Table 10.4 Scenarios considered in the simulation-based study (period 2014–2017)

Scenario identifier	Price threshold (€/MWh)	Percentage of load reduction	Number of demand response events				
			2014	2015	2016	2017	Total
B1	80	1	34	4	0	201	239
B2	80	3	34	4	0	201	239
B3	80	5	34	4	0	201	239
C1	90	1	13	0	0	90	103
C2	90	3	13	0	0	90	103
C3	90	5	13	0	0	90	103
D1	100	1	1	0	0	3	4
D2	100	3	1	0	0	3	4
D3	100	5	1	0	0	3	4

To simulate the Iberian market in a correct and realistic fashion, the study accounts for the price differences between Portugal and Spain that result from market splitting in the daily horizon.[27] Specifically, the following two situations are considered to account for market splitting:

- For all hours in which there is no congestion in the Portugal-Spain interconnection, and thus there is a single price for Portugal and Spain, the MATREM simulations involve both the Portuguese and the Spanish regions.
- For the remaining hours, i.e., for all hours in which market splitting occurs, and thus different price areas are considered for Portugal and Spain, the MATREM simulations involve the Portuguese region only.

Overall, as pointed out throughout this section, the study involves a considerable number of software agents (up to 350 agents), 10 energy scenarios related to the Iberian electricity market, and a 42-month trading period. Accordingly, there is a need to perform 956 simulation runs, namely 239 simulation runs for scenario A and 717 for scenarios B1–B3 (since scenarios C1–C3 and D1–D3 may be analyzed by considering the results of the simulation runs carried out for scenarios B1–B3).

of operation, may occur at any time of the day, and is assumed to be related to a reduction in electricity usage by retail customers.

[27] *Market splitting* in the daily horizon involves basically the segmentation of the Iberian market into two independent markets due to congestion in the Portugal-Spain interconnection, typically leading to different prices for the Portuguese and Spanish areas, yet making possible to exhaust the available capacity safely. Chapter 9 presents a brief introduction to market splitting in the daily horizon and the interested reader is, therefore, referred to it for details (see also [53, 54]).

Each simulation run corresponds to a given hour of operation and involves basically the following:

1. Set agent-specific parameters.
2. Set scenario-specific parameters (notably, the price threshold and load reduction).
3. Obtain the hourly energy prices and quantities submitted to the daily Iberian market.
4. Analyze the occurrence of market splitting and consider either the Portuguese and the Spanish regions or the Portuguese region only.
5. Prepare the bids of all supplier agents (perform the procedures outlined above).
6. Prepare the offer of the demander agent.
7. Submit the supply bids and the demand offer to the day-ahead market (incorporated in the MATREM system).
8. Determine the market-clearing price and announce it to market participants.
9. Prepare a simulation report.

The capability of MATREM to produce realistic day-ahead (spot) market prices is very important to the successful completion of the study, and should be rigorously and thoroughly analyzed. Hence, in a calibration (and benchmarking) procedure, real hourly electricity prices published by OMIE are compared with simulated hourly prices generated by MATREM for scenario A. The simulated prices correctly fit those reported by OMIE, showing that MATREM is a reliable system for simulating the Iberian market prices (and also validating the procedure adopted here for defining the software agents and their bids and offers).

10.4.3 Results and Discussion

Tables 10.5, 10.6 and 10.7 summarize the results of the simulation-based study. Table 10.5 shows the average monthly differences between the simulated prices for scenario A and the simulated prices for any of the remaining scenarios (i.e., scenarios B1–B3, C1–C3 and D1–D3).[28] The results indicate that modest amounts of demand response—modeled as load reductions between 1 and 5% at times of high market prices, and (indirectly) associated with reductions in electricity usage by retail customers—have a relatively large effect on day-ahead (spot) market prices, providing a significant price relief and stability (i.e., considerably reducing market price volatility). In particular, the results link specific levels of load reduction to decreases in market prices, some of which indicate that the benefits may be quite significant.

The results for the year 2014 are striking. The average monthly reduction of the market price for scenarios B1, B2 and B3 varies between 3.57 €/MWh and 37.88 €/MWh. Under scenario B3, a modest load reduction of 5% would have mitigated the fairly high market prices—that is, the prices above 80 €/MWh—observed in

---

[28]The average monthly price reductions are computed by considering the hours of operation corresponding to demand response events only.

Table 10.5 Simulation results: average monthly price reductions

Scenario	Average price reduction (€/MWh)							
	2014					2015		2017
	Jan	Feb	Mar	Oct	Nov	Jan	Feb	Nov	Jan
B1	3.57	18.33	17.93	4.44	6.55	4.54	4.91	7.70	1.34
B2	12.10	29.84	27.20	9.96	13.05	9.24	10.91	10.40	3.67
B3	21.53	37.88	33.17	12.91	14.75	11.49	13.01	16.40	6.15
C1	6.70	17.75	15.45	9.75	7.00	–	–	–	1.33
C2	24.75	27.60	26.70	21.10	15.90	–	–	–	3.91
C3	38.85	35.95	32.90	25.05	18.20	–	–	–	7.38
D1	–	11.80	–	–	–	–	–	–	0.39
D2	–	23.60	–	–	–	–	–	–	1.66
D3	–	30.60	–	–	–	–	–	–	2.70

February and March by nearly one-third. The average monthly price reduction for scenarios C1, C2 and C3 ranges between 6.70 €/MWh to a maximum of 38.85 €/MWh, reaching the highest value in January. Under scenario C3, a load reduction of 5% would have mitigated the very high market prices—that is, the prices above 90 €/MWh—observed in the period between January and March by about one-third. Also, this slight reduction of the load would have mitigated the very high prices observed in October by more than one-quarter. The highest price of 110 €/MWh (observed on 17 February at 10 p.m.) is reduced by 11.80, 23.60 and 30.60 €/MWh under scenarios D1, D2 and D3, respectively. Accordingly, a load reduction of 5% would have mitigated this extremely high price by more than one-quarter.

At this stage, the simulation results for the year 2014, particularly for the period between January and March, raise the following question: why a small decrease in load reduces the market prices by one-quarter to one-third? To answer this question consider, for example, the supply and demand curves published by OMIE [39] on Friday 3 January 2014 at 10 p.m. (see Fig. 10.7). The market-clearing price (P^*) is set at 94.59 €/MWh and the equilibrium energy quantity (Q^*) at 32477 MWh. The figure also shows the price-inelastic demand curve adopted in the simulation-based study for scenario A (right-hand dark blue curve) as well as the demand curve adopted for scenario C3 (left-hand dark blue curve).[29] Recall that scenario C3 involves a load reduction of 5% and a price threshold of 90 €/MWh. This load reduction results in a lower market-clearing price, defined by the intersection of the supply curve with the left-hand demand curve. The "new" market price (P_{DR}) is set at 52 €/MWh and the equilibrium quantity (Q_{DR}) at 30853 MWh. Accordingly, the price is reduced by 42.59 €/MWh, which corresponds to a decline of nearly 45%.

[29] Strictly speaking, the left-hand dark blue curve of Fig. 10.7 is also the price-inelastic demand curve adopted for scenario B3.

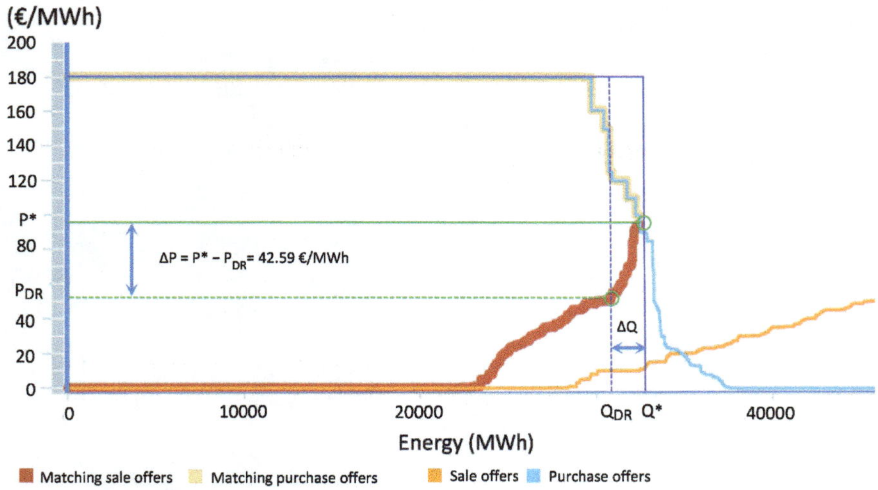

Fig. 10.7 Reduction of the market price for scenario C3 on 3 January 2014 at 10 p.m. (adapted from [39])

This (very) large reduction of the market price can be explained, at least in part, by the shape of the supply curve. The lower part of the curve has a perfectly horizontal slope, the middle part exhibits a considerable upward slope, and finally the upper part exhibits a nearly, but not perfectly, vertical slope. Clearly, this shape is consistent with the common hypothesis that high market prices are associated with supply conditions that are very sensitive to changes in demand, where even a small amount of DR can have a significant impact on prices. In other words, the fact that the supply curve increases very steeply at its upper end means that the marginal cost of electricity becomes very sensitive to changes in demand—that is, a small reduction in demand causes a large decrease in price.

Returning to the simulation results presented in Table 10.5, it is particularly noteworthy that the magnitude of the average monthly price reduction decreases gradually from 2014 to 2017. Indeed, in the year 2015, the price reduction varies between 4.54 €/MWh and 16.40 €/MWh, reaching the highest value in November (under scenario B3). In the year 2017, the price reduction is still lower, ranging between 0.39 €/MWh to a maximum of 7.38 €/MWh (under scenario C3).[30]

Again, such lower, albeit relatively considerable, reductions of the market price can be explained by the shape of the supply curve. To this end, compare, for instance, the supply curve published by OMIE on Friday 3 January 2014 at 10 p.m. (see Fig. 10.7) with the supply curve published by OMIE on Friday 20 January 2017 at 10 p.m. (see Fig. 10.6). As noted above, the upper part of former curve has a nearly vertical slope, meaning that the marginal cost of electricity is very sensitive to changes in demand. In contrast, the upper part of the latter curve slopes upward

[30]Notice that the year 2017 involved the highest number of demand response events (namely, 201 DR events). All events occurred in January.

Table 10.6 Simulation results: annual price reductions

Scenario	Average price reduction (€/MWh)			Average price reduction (%)		
	2014	2015	2017	2014	2015	2017
B1	8.10	5.42	1.34	9.45	6.51	1.57
B2	16.75	9.95	3.67	19.53	11.94	4.45
B3	23.61	13.10	6.15	27.52	15.72	6.95
C1	11.95	–	1.33	13.30	–	1.49
C2	24.78	–	3.91	27.58	–	4.56
C3	33.56	–	7.38	37.35	–	7.87
D1	11.80	–	0.39	13.95	–	0.38
D2	23.60	–	1.66	27.90	–	1.63
D3	30.60	–	2.70	36.17	–	2.67

gradually, exhibiting a considerable large but finite slope, meaning that the marginal cost of electricity is less sensitive to changes in demand (although still considerably sensitive). Put another way, the steeply sloping former curve means that a small reduction in demand causes a large decrease in price. The moderately sloping latter curve means that a small reduction in demand causes a substantial decrease in price (to be concrete, under scenario C3, the price is reduced by about 10 €/MWh).

Table 10.6 shows the average annual differences between the simulated prices for scenario A and the simulated prices for any of the remaining scenarios. The first column indicates the DR scenario, the three adjacent columns in the middle specify the annual price reductions (expressed in €/MWh), and the rightmost 3 columns present the corresponding percentages. In the year 2014, the prices are reduced by 8.10–33.56 €/MWh, which corresponds to a decline of 9.45–37.35%. In the year 2017, the prices are reduced by 0.39–7.38 €/MWh (a decline of 0.38–7.87%), highlighting what appears to be an inherent tendency for the magnitude of the price reduction to decrease from 2014 to 2017.

Table 10.7 presents the market value of the electricity traded in the spot market during the hours of operation corresponding to DR events (see the three adjacent columns in the middle), and also quantifies, in monetary terms, the potential benefit of demand response for each of the DR scenarios (see the rightmost 3 columns). The market value of the energy is undoubtedly very important for the discussion of the financial benefit of DR. This value may be roughly calculated by multiplying the estimated hourly day-ahead (spot) prices by the corresponding hourly energy quantities.[31] Naturally, it decreases from the scenarios involving a load reduction of 1% to the scenarios involving a load reduction of 5%. To be concrete, in the year 2014, the market value of the energy ranges between 1.53 million € (under scenario D3)

[31] As stated in *Chap.* 9, a considerable volume of electricity is traded via bilateral contracts, whose price may deviate from the spot market price (see, e.g., [37]). Nevertheless, the market price is the leading price indicator for all electricity trades and, therefore, the price paid through bilateral contracts is based, to some extent, upon that price (but see, e.g., [55].

Table 10.7 Simulation results: financial benefits of demand response

Scenario	Market value of energy (million €)			Demand response benefit (million €)		
	2014	2015	2017	2014	2015	2017
B1	76.44	10.64	501.97	11.27	0.85	17.28
B2	66.55	9.82	472.55	21.16	1.67	46.70
B3	58.71	9.20	442.63	29.00	2.29	76.62
C1	29.21	–	309.42	7.07	–	6.79
C2	23.80	–	294.81	12.48	–	21.40
C3	20.03	–	277.16	16.25	–	39.05
D1	2.15	–	11.98	1.13	–	0.18
D2	1.77	–	11.59	1.51	–	0.57
D3	1.53	–	11.23	1.75	–	0.93

to 76.44 million € (under scenario B1). Interestingly, although not too surprising, the data shows a remarkable increase in the market value of the energy in the year 2017, reaching 501.97 million €.

The financial benefit of demand response for a given scenario S ∈ {B1, ..., D3} is computed by subtracting the market value of the energy for scenario A from the market value of the energy for scenario S. Specifically, the financial benefit of DR—or the economic impact of DR—is estimated by the following equation:

$$B = \sum_{h=1}^{t} (P \times Q) - (P_{DR} \times Q_{DR}) \tag{10.1}$$

where t is the number of hours of operation corresponding to the occurrence of DR events (under scenario S), P is the hourly spot market price estimated for scenario A (€/MWh), Q is the actual energy traded in the market (MWh), P_{DR} is the hourly spot market price estimated for scenario S (€/MWh), Q_{DR} is the hourly energy quantity after the (small) reduction specified in scenario S (MWh), and B represents the financial benefit of demand response (€).

The results are striking. They indicate that modest amounts of demand response, modeled as load reductions in the range of 1–5% at times of high market prices, can provide significant benefits to market participants (and eventually to most retail customers). Indeed, in the year 2014, the financial benefit of DR for scenarios B1, B2 and B3 ranges between 11.27 and 29 million €. This benefit is also substantial for scenarios C1, C2 and C3, varying between 7.07 and 16.25 million €. Furthermore, under scenarios D1, D2 and D3, the financial benefit from reducing the highest price of 110 €/MWh (observed on 17 February at 10 p.m.) is still relatively high, varying between 1.13 and 1.75 million €.

Especially noteworthy is the remarkable increase of the financial benefit of DR during the year 2017.[32] In fact, a modest load reduction of 5% at times of fairly high market prices—that is, prices above 80 €/MWh—can yield the very large benefit of 76.62 million €. Also, the same decrease in load at times of very high market prices—that is, prices above 90 €/MWh—can provide the (still large) benefit of 39.05 million €. Furthermore, a load reduction of 5% when prices exceed 100 €/MWh can generate almost 1 million € in savings.

Overall, the financial benefit of DR reaches the considerable value of 76.62 million € in the year 2017. However, in addition to this quite significant benefit, another aspect related to the short-term market impact of DR should be taken into account, namely the fact that DR averts the need to use some of the most costly-to-run power plants, which may interact with the climate policy and the EU emission trading system (ETS). Put another way, demand response may push out of the energy market fossil fuel fired power plants, which, in turn, may reduce the CO_2 emissions and, consequently, the demand on the emissions trading market (or carbon market). This typically leads to a reduction of the market price of allowances (or carbon price), creating savings for the different entities that take part in the ETS. Although very interesting, an estimation of the CO_2 savings is a highly complex task and, therefore, deferred to future work.[33]

10.5 Conclusion

This chapter has investigated the price effect of demand response—modeled as modest load reductions at times of high market prices—on the Iberian electricity market (MIBEL) and analyzed the potential benefits that result to retail customers. The first part of the chapter introduced the concept of demand response, described two key categories of DR programs (i.e., incentive-based and price-based programs), and presented a brief overview of DR in Spain and Portugal.

The second part introduced the Iberian electricity market, notably the day-ahead market, and discussed the short-term market impact of demand response. In particular, it described how demand response may avert the need to use the most costly-to-run power plants during periods of high demand, thus driving market prices down. The description was illustrated with a practical example using data published by the managing entity of the Iberian market (OMIE).

The third part examined, in detail, the hourly prices published by OMIE in the period between January 1, 2014 and June 30, 2017. It also investigated the interaction between a specific level of demand response and the market prices in a working day

[32]As noted above, the magnitude of the price reduction decreases gradually from 2014 to 2017. In direct contrast, the financial benefit of DR is much greater in 2017 than in 2014. The main reason for this increase is naturally related to the number of DR events considered in both years, namely 34 events in 2014 and 201 in 2017.

[33]In this chapter, and throughout the book, the term "environmental benefit of DR" is used to denote the CO_2 savings resulting from demand response.

from the fifth week of January 2017, namely January 25 (a Wednesday), when the prices were at their highest. Two scenarios were considered: (i) the hourly spot market prices were simulated by using an agent-based simulation tool, called MATREM, and (ii) DR was modeled as a 3% load reduction when prices rose above 90 €/MWh (the corresponding market prices were simulated as in the previous scenario). On average, a price reduction of about 4.27 €/MWh was estimated for the entire day, corresponding to a decline of 4.37%. The financial benefits of demand response were around 4.9 million €, a remarkable result.

Following this introductory material, the chapter presented a study to investigate the impact of demand response on the Iberian market prices—that is, the price effect of DR on MIBEL—during the period 2014–2017. The main results generated by the MATREM system are as follows:

- The prices were reduced by 8.10–33.56 €/MWh in the year 2014, which corresponds to a decline of 9.45–37.35%.
- The price reduction ranged between 5.42 and 13.10 €/MWh in 2015 (a decline of 6.51–15.72%).
- In 2017, the price reduction was substantially lower, ranging between 0.39 €/MWh to a maximum of 7.38 €/MWh (a decline of 0.38–7.87%), highlighting what appears to be an inherent tendency for the magnitude of the price reduction to decrease from 2014 to 2017.
- The financial benefit of DR reached the considerable value of 29 million € in the year 2014.
- The financial benefit of DR increased considerably in 2017. A modest load reduction of 5% when prices rose above 80 €/MWh yielded the very large benefit of 76.62 million €. Also, the same decrease in load at times of very high market prices—that is, prices above 90 €/MWh—provided the (still large) benefit of 39.05 million €. Furthermore, a load reduction of 5% when prices exceeded 100 €/MWh generated almost 1 million € in savings.

The results are, therefore, striking. They indicate that modest amounts of demand response—modeled as load reductions between 1 and 5% when prices rise above a threshold between 80 and 100 €/MWh, and (indirectly) associated with reductions in electricity usage by retail customers—have a relatively large effect on market prices, providing a significant price relief and stability (i.e., considerably reducing market price volatility), and therefore creating substantial benefits to market participants (and eventually to most retail customers).

The results lead inevitably to the following conclusions:

> Demand response should be a key element of the Iberian market design. The near-universal sentiment that encouraging demand response is a necessary element of effective wholesale market design is undeniable. The current lack of demand response may lead to a number of problems in the wholesale market, including very high market prices at times of system constraints. The optimal amount of demand response should be determined by the market

rather than specified a priori. State institutions should work with system operators and market participants to examine how much demand response is needed to improve the efficiency and reliability of the wholesale and retail markets.

Since 2014, Spain has carried out public annual auctions to assign interruptible load to large industrial customers. Also, the VPSC tariff for residential and small business customers came into force in 2014 (in Spain). In direct contrast, Portugal may be considered "closed" to the participation of incentive-based DR in the market, and also "passive" in relation to the implementation of dynamic pricing options for retail customers. This is largely due to a lack of regulatory structures defining roles and responsibilities, baselining, payments and all other technical aspects required for implementing demand response programs. Accordingly, the following recommendations seem to be pertinent to foster demand response in the country through both incentive-based and price-based programs:

1. *Strengthening the analysis and valuation of demand response*: additional work should be carried out to quantify the costs and benefits of demand response, including the savings that are passed through to retail customers, clarifying the link that DR provides between the wholesale and retail markets.
2. *Improving incentive-based demand response*: incentive-based programs, such as interruptible contracts, should be adopted/maintained, expanded and eventually adapted to new market situations and circumstances.
3. *Fostering price-based demand response*: time-varying tariffs, such as RTP and CPP, should be offered as default tariffs to retail customers, letting them to take control of their electricity costs. State institutions, system operators, electric utilities and other market participants should consider providing education, outreach, and technical assistance to retail customers to maximize the effectiveness of such tariffs.
4. *Adopting enabling technologies*: state legislatures and institutions should consider the development of new rules that encourage the large-scale deployment of DR enabling technologies, such as advanced metering systems.

Acknowledgements This work was performed under the project MAN-REM (FCOMP-01-0124-FEDER-020397), supported by FEDER funds, through the program COMPETE ("Programa Operacional Temático Factores de Competividade"), and also National funds, through FCT ("Fundação para a Ciência e a Tecnologia"). Hugo Algarvio was funded by FCT (PD/BD/105863/2014). The authors wish to thank João Santana, from INESC-ID and also the Technical University of Lisbon (IST), and João Martins and Anabela Pronto, from the NOVA University of Lisbon, for their valuable comments and helpful suggestions to improve the chapter.

References

1. Braithwait, S., Eakin, K., Inc., Laurits R. Christensen A.: The role of demand response in electric power market design. Technical Report, Edison Electric Institute, Washington, D.C. (October 2002)

2. DOE: Benefits of Demand Response in Electricity Markets and Recommendations for Achieving Them. A Report to the United States Congress Pursuant to Section 1252 of the Energy Policy Act of 2005, US Department of Energy (February 2006)
3. FERC: Standardized Transmission Service and Wholesale Electric Market Design. Working Paper of the Federal Energy Regulatory Commission Staff, Washington, D.C. (March 2002)
4. IEA: Re-powering Markets: Market Design and Regulation during the Transition to Low-carbon Power Systems. International Energy Agency, Paris, France (2016)
5. European Union: Directive 2012/27/EU of the European Parliament and of the Council on energy efficiency, amending Directives 2009/125/EC and 2010/30/EU and repealing Directives 2004/8/EC and 2006/32/EC (25 October 2012) http://eur-lex.europa.eu/LexUriServ/LexUriServ.do?uri=OJ:L:2012:315:0001:0056:en:PDF (Cited on 17 June 2017)
6. European Commission: Delivering the Internal Electricity Market and Making the Most of Public Intervention. Communication from the Commission (5 November 2013) https://ec.europa.eu/energy/sites/ener/files/documents/com_2013_public_intervention_en_1.pdf (Cited on 17 June 2017)
7. European Commission: Proposal for a Regulation of the European Parliament and of the Council on the internal market for electricity. 2016/0379 (COD) (30 November 2016) http://www.ipex.eu/IPEXL-WEB/dossier/document/COM20168611FIN.do (Cited on 17 June 2017)
8. Fernández, J., Payán, M., Santos, J., García, A.: The voluntary price for the small consumer: real-time pricing in spain. Energy Policy **102**, 41–51 (2017)
9. IRC: Harnessing the Power of Demand – How ISOs and RTOs are Integrating Demand Response into Wholesale Electricity Markets. Report prepared by the Markets Committee of the ISO/RTO Council, United States (16 October 2007)
10. Brown, T., Newell, S., Oates, D., Spees, K.: International Review of Demand Response Mechanisms. Report prepared by the Brattle Group for the Australian Energy market Commission, Cambridge (October 2015)
11. Hirst, E., Kirby, B.: Retail-Load Participation in Competitive Wholesale Electricity Markets. Technical Report, Edison Electric Institute, Washington D.C. and Project for Sustainable FERC Energy Policy, Alexandria, VA (2001)
12. IEA: The Power to Choose: Demand Response in Liberalised Electricity Markets. International Energy Agency, Paris, France (2003)
13. Hogan, W.: Demand Response Pricing in Organized Wholesale Markets. Report Prepared for ISO/RTO Council (13 May 2010) www.hks.harvard.edu/fs/whogan/Hogan_IRC_DR_051310.pdf (Cited on 17 June 2017)
14. FERC: National Action Plan on Demand Response. Report of the Federal Energy Regulatory Commission Staff, Docket N. AD09-10, Washington, D.C. (June 2010)
15. FERC: Assessment of Demand Response and Advance Metering. Staff Report of the Federal Energy Regulatory Commission, Docket AD06-2-000, Washington, D.C. (August 2006, revised December 2008)
16. FERC: Assessment of Demand Response and Advance Metering. Staff Report of the Federal Energy Regulatory Commission, Washington, D.C. (December 2015)
17. FERC: Assessment of Demand Response and Advance Metering. Staff Report of the Federal Energy Regulatory Commission, Washington, D.C. (December 2016)
18. SEDC: Mapping Demand Response in Europe Today. Report of the Smart Energy Demand Coalition, Brussels, Belgium (2015)
19. NYISO: Day-Ahead Demand Reduction Program Manual. New York Independent System Operator, New York (July 2003)
20. NEMS: Implementing Demand Response in the National Electricity Market of Singapore. Energy Market Authority, Singapore (October 2013)
21. KEMA-XENERGY: Protocol Development for Demand Response Calculation: Findings and Recommendations. Consultant Report prepared by KEMA-XENERGY for the California Energy Commission, CA (February 2003)
22. FERC: Demand Response Compensation in Organized Wholesale Energy Markets. Order N. 745, Docket N. RM10-17-000, Washington, D.C. (March 2011)

23. Bertoldi, P., Zancanella, P., Boza-Kiss, B.: Demand Response status in EU Member States. Science for Policy report by the Joint Research Centre, EUR 27998 EN, European Union, Brussels, Belgium (2016)
24. SEDC: Explicit Demand Response in Europe: Mapping the Markets 2017. Report of the Smart Energy Demand Coalition, Brussels, Belgium (2017)
25. CNMC: Spanish Energy Regulator's National Report To The European Commission 2017. Report of the Comisión Nacional de Los mercados y la Competencia, Madrid, Spain (July 2017)
26. ERSE: Annual Report on the Electricity and Natural Gas Markets in 2015 in Portugal. Entidade Reguladora dos Serviços Energéticos, Lisboa, Portugal (July 2016)
27. ERSE: Annual Report on the Electricity and Natural Gas Markets in 2016 in Portugal. Entidade Reguladora dos Serviços Energéticos, Lisboa, Portugal (July 2017)
28. EPRI: A Taxonomy of Retail Electricity Products. Report of the Electric Power Research Institute, Palo Alto, California (December 2002)
29. ComED: Hourly Pricing Program Guide 2016–2017. Commonwealth Edison Company, Chicago, Illinois (2017) https://hourlypricing.comed.com/wp-content/uploads/2017/10/2017-HP-program-guide-v13.pdf (Cited on 24 June, 2017)
30. Penn State Extension: Renewable and Alternative Energy Fact Sheet – Real-time Pricing for Electricity. The Pennsylvania State University, PA (2013) https://extension.psu.edu/real-time-pricing-for-electricity (Cited on 24 June, 2017)
31. Eakin, K., Faruqui, A.: Pricing retail electricity - making money selling a commodity. In: Faruqui, A., Eakin, K. (eds.) Pricing in Competitive Electricity Markets, pp. 5–31. Springer Science+Business Media, New York (2000)
32. Huso, S.: Real time pricing - a unified rate design approach. In: Faruqui, A., Eakin, K. (eds.) Pricing in Competitive Electricity Markets, pp. 307–312. Springer Science+Business Media, New York (2000)
33. O'Skeasy, M.: How to buy low and sell high. Electr. J. **11**(1), 24–529 (1998)
34. EURELECTRIC: Dynamic pricing in electricity supply. Union of the Electricity Industry Position Paper, Brussels, Belgium (February 2017)
35. CNMC: Informe Sobre el Cumplimiento del Primer Hito del Plan de Substitución de Contadores. Report of the Comisión Nacional de Los mercados y la Competencia, INF/DE/006/15, Madrid, Spain (May 2015)
36. CNMC: Informe Sobre la Efectiva Integración de los Contadores con Telemedida y Telegestión Eléctricos con Potencia Contratada Inferior a 15 KW a Finales del Primer Semestre de 2016. Report of the Comisión Nacional de Los mercados y la Competencia, IS/DE/002/16, Madrid, Spain (February 2017)
37. MIBEL Regulatory Council: Description of the Operation of MIBEL (November 2009) http://www.erse.pt/eng/electricity/MIBEL/Documents/Description_Operation_MIBEL.pdf (Cited on 24 June, 2017)
38. OMIE: Daily and Intraday Electricity Market Operating Rules (May 2014) http://www.omie.es/files/20140509_reglas_v11_ingles.pdf (Cited on 24 June, 2017)
39. OMIE: "Operador del Mercado Ibérico de Energía (Spanish Electricity Market Operator)." Market Results (online data) http://www.omie.es/files/flash/ResultadosMercado.swf (Cited on 24 June, 2017)
40. OMIE: Market Report (2014) http://www.omie.es/en/files/informe_mercado_ingles.pdf (Cited on 24 June, 2017)
41. OMIE: Price Report (2015) http://www.omie.es/en/files/omie_informe_precios_2015_english_0.pdf (Cited on 24 June, 2017)
42. OMIE: Price Report (2016) http://www.omie.es/files/informe_precios_ing_navegable.pdf (Cited on 24 June, 2017)
43. OMIE: Informe Mensual (January 2017) http://www.omie.es/files/informe_mensual_enero_2017.pdf (Cited on 24 June, 2017)
44. OMIE: Informe Mensual (July 2017) http://www.omie.es/files/informe_mensual_julio_2017.pdf (Cited on 24 June, 2017)

45. Stoft, S.: Power Systyem Economis – Designing Markets for Electricity. IEEE Press and Wiley Interscience (2002)
46. Kirschen, D., Strbac, G.: Fundamentals of Power System Economics. Wiley, Chichester (2004)
47. Vidigal, D., Lopes, F., Pronto, A., Santana, J.: Agent-based Simulation of Wholesale Energy Markets: a Case Study on Renewable Generation. In: Spies, M., Wagner, R., Tjoa, A. (eds.) 26th Database and Expert Systems Applications (DEXA 2015), pp. 81–85. IEEE (2015)
48. Algarvio, H., Couto, A., Lopes, F., Estanqueiro, A., Santana, J.: Multi-agent energy markets with high levels of renewable generation: a case-study on forecast uncertainty and market closing time. In: Omatu, S., et al. (eds.) 13th International Conference on Distributed Computing and Artificial Intelligence, pp. 339–347. Springer International Publishing (2016)
49. Lopes, F., Rodrigues, T., Sousa, J.: Negotiating bilateral contracts in a multi-agent electricity market: a case study. In: Hameurlain, A., Tjoa, A., Wagner, R. (eds.) 23rd Database and Expert Systems Applications (DEXA 2012), pp. 326–330. IEEE (2012)
50. Sousa, F., Lopes, F., Santana, J.: Contracts for difference and risk management in multi-agent energy markets. In: Demazeau, Y., Decker, K., Pérez, J., De la Prieta, F. (eds.) Advances in Practical Applications of Agents, Multi-Agent Systems, and Sustainability: The PAAMS Collection (PAAMS 2015), pp. 339–347. Springer International Publishing (2015)
51. Lopes, F., Mamede, N., Novais, A.Q., Coelho H.: Negotiation in a multi-agent supply chain system. In: Third International Workshop of the IFIP WG 5.7 Special Interest Group on Advanced Techniques in Production Planning and Control, pp. 153–168. Firenze University Press (2002)
52. REN: Redes Energéticas Nacionais, Preços Mercado Spot, Portugal e Espanha. http://www.mercado.ren.pt/PT/Electr/InfoMercado/InfOp/MercOmel/Paginas/Precos.aspx (Cited on 25 June, 2017)
53. ERSE and CMVM: Proposta de Mecanismo de Gestão conjunta da interligação Espanha-Portugal. Entidade Reguladora dos Serviços Energéticos and Comisssão do Mercado de Valores Mobiliários (March 2006). http://www.erse.pt/pt/mibel/compatibilizacaoregulatoria/Documents/PROPOSTADEMECANISMODEGEST%C3%83OCONJUNTADAINTERLIGA%C3%87%C3%83OESPANHAPORTUGAL.pdf (Cited on 15 May, 2017)
54. ERSE, CMVM, CNE and CNMV: Proposta do Conselho de Reguladores para a Repartição da Capacidade de interligação entre os Mecanismos de "Market Splitting" e Leilões Explícitos de Capacidade no Âmbito do MIBEL. Entidade Reguladora dos Serviços Energéticos, Comisssão do Mercado de Valores Mobiliários, Comísión Nacional de Energía and Comísión Nacional del Mercado de Valores (May 2007). http://www.erse.pt/pt/mibel/compatibilizacaoregulatoria/Documents/PropostaCRreparticaodacapacidadedeinterligacao.pdf (Cited on 25 June, 2017)
55. OMIP: General Contractual Terms: MIBEL PTEL Base Load Physical Futures Contracts. MIBEL Derivatives Market, Lisbon, Portugal (2016)

Chapter 11
Multi-agent Electricity Markets and Smart Grids Simulation with Connection to Real Physical Resources

Tiago Pinto, Zita Vale, Isabel Praça, Luis Gomes and Pedro Faria

Abstract The increasing penetration of distributed energy sources, mainly based on renewable generation, calls for an urgent emergence of novel advanced methods to deal with the associated problems. The consensus behind smart grids (SGs) as one of the most promising solutions for the massive integration of renewable energy sources in power systems has led to the development of several prototypes that aim at testing and validating SG methodologies. The urgent need to accommodate such resources require alternative solutions. This chapter presents a multi-agent based SG simulation platform connected to physical resources, so that realistic scenarios can be simulated. The SG simulator is also connected to the Multi-Agent Simulator of Competitive Electricity Markets, which provides a solid framework for the simulation of electricity markets. The cooperation between the two simulation platforms provides huge studying opportunities under different perspectives, resulting in an important contribution to the fields of transactive energy, electricity markets, and SGs. A case study is presented, showing the potentialities for interaction between players of the two ecosystems: a SG operator, which manages the internal resources of a SG, is able to participate in electricity market negotiations to trade the necessary amounts of power to fulfill the needs of SG consumers.

T. Pinto (✉) · Z. Vale · I. Praça · L. Gomes · P. Faria
GECAD-Research Group on Intelligent Engineering and Computing for Advanced Innovation and Development, Institute of Engineering, Polytechnic of Porto, R. Dr. António Bernardino de Almeida 431, Porto, Portugal
e-mail: tmcfp@isep.ipp.pt

Z. Vale
e-mail: zav@isep.ipp.pt

I. Praça
e-mail: icp@isep.ipp.pt

L. Gomes
e-mail: lufog@isep.ipp.pt

P. Faria
e-mail: pnfar@isep.ipp.pt

F. Lopes and H. Coelho (eds.), *Electricity Markets with Increasing Levels of Renewable Generation: Structure, Operation, Agent-based Simulation, and Emerging Designs*, Studies in Systems, Decision and Control 144, https://doi.org/10.1007/978-3-319-74263-2_11

11.1 Introduction

The use of renewable energy sources (RES) has increased significantly, stimulated by policies and incentive programs aiming at decreasing the dependency on fossil fuels and avoiding environmental damages. The European Union (EU) has defined the well-known "20-20-20" targets [1]. Such targets will enable the EU as a whole to reach 20% energy consumption from renewable energy sources in 2020, more than doubling the 2010 level of 9.8%. In October 2014, a commitment has been achieved to reduce EU domestic greenhouse gas emissions by at least 40% below the 1990 level by 2030 [2].

Renewable energy sources, such as wind and solar power, poses new challenges to the power sector and electricity markets (EMs). Many different market approaches have been experimented around the world, and all have been subject to multiple revisions. The primary focus is on adapting electricity markets to deliver the intended economic efficiency and reliability outcomes under the new paradigm of a growing share of renewable energy sources [3].

One of the main EU priorities concerns the formation of a pan-European energy market. The majority of European countries have already joined together into common market operators, resulting in joint regional electricity markets composed of several countries. Additionally, in early 2015, several of these regional European electricity markets have been coupled in a common market platform, operating on a day-ahead basis [4]. This achievement has been enabled by the multi-regional coupling (MRC), a pan-European initiative dedicated to the integration of power spot markets in Europe. The common market platform has resulted from an initiative of seven European power exchanges, called price coupling of regions (PCR) [4], which have joined efforts to develop a single price coupling solution used to calculate electricity prices across Europe and to allocate cross-border capacity on a day-ahead basis. This is a crucial step to achieve the overall EU target of a harmonized European electricity market.

The centralized top-down approach of electricity markets has proven to be insufficient to take full advantage from the participation of small players, both consumers and distributed generation (DG). Electricity markets still do not allow to integrate the required amount and diversity of DG and put serious limitations to the participation of small and medium size resources [5]. Moreover, the tentative reforms of retail markets are not being able to achieve the envisaged goals as they are being built under the same top-bottom principles as wholesale markets. Electricity prices for smaller consumers still do not reflect the market prices and the introduction of flexible, innovative tariffs adapted to consumers' needs and behaviours, able to promote and fairly remunerate their contribution towards an increasingly efficient energy system are still distant targets. New approaches that are able to bring a closer connection between small consumers, DG and the wholesale electricity market are required promptly. A pioneer solution to overcome these problems is currently being implemented in the New York electricity market, in the US. The creation of local electricity markets as part of the regional electricity market is being put into practice, enabling smaller portions of the power network (microgrids) to participate in the electricity market as aggregators of the resources that are part of the portion of the grid. This way, resources can be managed at a local level, enhancing the potential of smaller sized resources, and their participation in electricity markets is facilitated by microgrid operators. This provides an important incentive for the development of adequate methods to manage resources at lower levels and make their connection with higher, wholesale electricity markets, levels more effective.

One of the main achievements of the power and energy sector in recent years is the common acceptance by the involved stakeholders that power systems require major changes to accommodate in an efficient and secure way an intensive use of renewable based and DG [6, 7]. The conclusion that the so-called smart grids (SGs) are required is a crucial foundation for the work to be done in the coming years towards the modernization and restructuring of the power sector according to the new paradigms [8]. Huge investments have already been made in projects concerning smart grids [9], including research and development projects, pilot installations, and roll-out of smart metering. A list with 459 projects related to smart grids involving Member States is included in [9].

The large number of smart grid related projects is resulting in important advances in the field, namely concerning demonstration pilots and management and control methodologies. However, the quick emergence process of smart grids is not entirely free of problems. A large number of practical applications, although very expensive, are enabling solutions that present serious limitations and provide little return of investment. It is not clear that the rolled out equipment is sufficiently open and flexible to be useful for the next generation of smart grid solutions that should appear in the coming years. Additionally, although important contributions are being achieved, these still remain as solutions for partial problems. In highly dynamic and co-dependent areas, such as power networks, smart grids and electricity markets, the cooperation between different systems becomes essential in order to look at the global problem as a whole. Most of the smart grid related works consist in practical implementations, highly industry driven, and involving almost exclusively large stakeholders in the field, such as regulators, operators and utilities, resulting in an almost complete focus on achieving fast ways to overcome present problems.

A closer attention should also be given to the demand side and especially to its interaction with the new methods for the operation and management of smart grids. The demand side role is recognized as very important in many documents (as in [10]). However, most projects are not considering this matter or are considering it in a very shallow way. Demand response (DR) is a high value resource with low cost, when compared with the other available substituting resources [11]. It has already been proved that DR is able to adequately prevent and/or solve emergency situations [12]. DR use in Europe is still very incipient and even in the US, where the integration of DR is much more mature, the way it should be implemented is still a focus of controversy, as exemplified by the FERC Order 745 saga [13]. The potential of DR is still highly unexplored, and the delay in implementing adequate measures to take full advantage of its benefits is causing significant drawbacks. Suitable models and solutions to explore the full potential of DR are urgently needed.

Simulation combined with distributed artificial intelligence techniques is growing as an adequate form to study the evolution of electricity markets and the coordination with smart grids, in order to accommodate the integration of the growing DG penetration [8, 9]. Modelling the smart grid environment with multi-agent systems enables model enlargements to include new players and allows studying and analyzing the individual and internal performance of each distinct player, as well as the global and specific interactions between all involved players [14].

MASGriP (Multi-Agent Smart Grid simulation Platform [7, 14]) is a multi-agent system that models the internal operation of SGs. The system considers all the typical players, which are modeled by software agents with the capability to represent and simulate their actions. Additionally, some agents, namely the ones representing small players, are directly connected to physical installations, providing the means for an automatic management of the associated resources. MASGriP uses real-time simulation [15] to complement simulations with the analysis of the impact of the methods in the energy flows and transmission lines.

The Multi-Agent Simulator of Competitive Electricity Markets (MASCEM) [16, 17] is a modelling and simulation tool that has been developed to study complex electricity markets' operation. It provides market players with simulation and decision-support resources, being able to provide them a competitive advantage in the market. MASCEM includes the market models of several real electricity markets, especially from EU operators. Simulations in MASCEM are based on real data, extracted in real-time from the websites of several market operators.

The presented work considers the integration between MASCEM and MASGriP, providing the means for a joint simulation of electricity markets and smart grids. The participation of smart grid players in electricity markets, in a controlled, simulated environment, brings huge studying opportunities, with the aim of bringing DG participation in the market a step closer to reality. It also provides invaluable studying opportunities for the solidification of the smart grid concept, and smart grid participation in competitive electricity negotiation environments. Additionally, the connection to real physical installations, the real-time simulation capabilities, and the use of real data to generate simulation scenarios, brings huge advantages for the validation of the achieved results, and consequent projection of simulated results into the reality.

This chapter consists of six sections. After this introductory section, Sect. 11.2 presents the MASCEM electricity market simulator. Section 11.3 is dedicated to the MASGriP smart grid simulator, including the description of this simulator's connection to physical infrastructures. Section 11.4 describes the energy resource management methodology that is applied by the smart grid operator agent to manage the smart grid resources before and after participating in electricity market negotiations using the MASCEM simulator. Section 11.5 presents a case study concerning the joint simulation of electricity markets and smart grids. Besides describing the simulation process, this section also presents the simulation results, providing a discussion on the influence of the joint simulation in the management perspective of the smart grid operator. Finally, Sect. 11.6 presents a discussion of the most relevant conclusions and future work.

11.2 MASCEM Electricity Market Simulator

MASCEM [16, 17] provides a simulation platform for the study of complex electricity markets. MASCEM considers the most important entities and their decision support features, allowing the definition of bids and strategies, granting them a competitive advantage in the market. Players are provided with biding strategic behavior so that they are able to achieve the best possible results depending on the market context. MASCEM players include: market operator agents, independent system operator agents (ISO), market facilitator agents, buyer agents, seller agents, virtual power player (VPP) [18] agents, and VPP facilitators.

MASCEM allows the simulation of the main market models: day-ahead pool (asymmetric or symmetric, with or without complex conditions), bilateral contracts, balancing market, forward markets and ancillary services. Hybrid simulations are also possible by combining the market models mentioned above. Also, the possibility of defining different specifications for the market mechanisms, such as multiple offers per period per agent, block offers, flexible offers, or complex conditions, as part of some countries' market models, is also available. Some of the most relevant market models that are fully supported by MASCEM are those of the Iberian electricity market (MIBEL), central European market (EPEX), and northern European market (Nord Pool).

Simulation scenarios in MASCEM are automatically defined using the realistic scenario generator (RealScen) [19]. RealScen uses real data available online, usually in market operators' websites. The gathered data concerns market proposals, including quantities and prices, accepted proposals and established market prices, details of proposals, execution of physical bilateral contracts, statement outages accumulated by unit type and technology, among others. By combining real extracted data with the data resulting from simulations, RealScen offers the possibility of generating scenarios for different types of electricity markets. Taking advantage on MASCEM's ability to simulate a broad range of different market mechanisms, this framework enables users to consider scenarios that are the representation of real markets of a specific region, or even consider different configurations, to test the operation of the same players under changed, thoroughly defined scenarios [19]. When summarized, yet still realistic scenarios are desired (in order to decrease simulations' execution time or facilitate the interpretation of results), data mining techniques are applied to define the players that act in each market. Real players are grouped according to their characteristics' similarity, resulting in a diversity of agent types that represent real market participants.

In order to allow players to automatically adapt their strategic behavior according to the operation context and with their own goals, a decision support system has been integrated with MASCEM. This platform is ALBidS (Adaptive Learning Strategic Bidding System) [17], which provides agents with the capability of analyzing contexts of negotiation, allowing players to automatically adapt their strategic behavior according to their current situation. In order to choose the most adequate strategy for each context, ALBidS uses reinforcement learning algorithms (RLA), and the

Bayesian theorem of probability. The contextualization is provided by means of a context definition methodology, which analyzes similar contexts of negotiation (e.g. similar situations in the past concerning wind speed values, solar intensity, consumption profiles, energy market prices, and types of days and periods, i.e. business days vs. weekends, peak or off-peak hours of consumption, etc.). This contextualization allows RLAs to provide the most adequate strategic support to market players depending on each current context. ALBidS strategies include: artificial neural networks, data mining approaches, statistical approaches, machine learning algorithms, game theory, and competitor players' actions prediction, among others. Figure 11.1 shows the connection between MASCEM and ALBidS, including the diverse modules that compose both systems.

ALBidS is implemented as a multi-agent system itself, in which each agent is responsible for an algorithm, allowing the execution of various algorithms simultaneously, increasing the performance of the platform. It was also necessary to build a suitable mechanism to manage the algorithms efficiency in order to guarantee the minimum degradation of MASCEM execution time. For this purpose, a methodology to manage the efficiency/effectiveness (2E) balance of ALBidS has been developed [17].

All communications between agents are carried out through the exchange of messages [20]. FIPA suggests the agent communication language (ACL) as a standard for communications between agents. Its content includes the content language, specifying the syntax, and the ontology, which provides the semantics of the message assuring a correct interpretation [21]. MASCEM agents use ontologies to allow the interoperability with other systems that intend to participate in the available electricity markets, as is the case of MASGriP [22]. These ontologies are also used to facilitate the interoperability with ALBidS, and they open the possibility for interaction with agents from external systems [22].

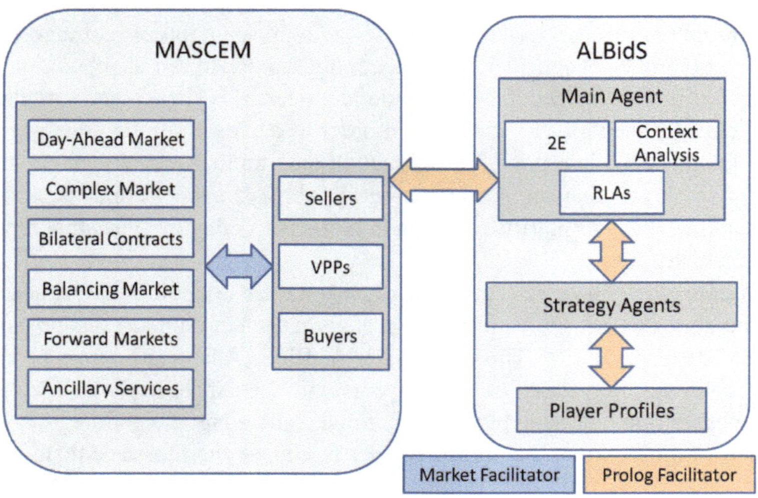

Fig. 11.1 Integration of ALBidS and MASCEM

11.3 MASGriP Smart Grid Simulation Platform

MASGriP simulates, manages and controls the most relevant players acting in a smart grid environment [14]. The proposed system includes fully simulated players, which interact with software agents that control real hardware. This enables the development of a complex system capable of performing simulations with an agent society that contains both real infrastructures and simulated players, providing the means to test alternative approaches, such as energy resource management (ERM) algorithms, DR models, negotiation procedures, among many others, in a realistic simulation environment [7].

11.3.1 Multi-agent Model

MASGriP provides a simulation platform that allows the experimentation and analysis of different types of models, namely energy resource management methodologies, contract negotiation methods, energy transaction models, and diverse types of DR programs and events. Among the many alternative DR models that are supported by MASGriP, both price-based and incentive-based models are considered [12], regarding three types of actions: load curtailment, reduce, and shift. Direct load control [12] is also included.

The simulated players in MASGriP have been implemented to reflect the real world. These players include some operators, such as the Distribution System Operator and the Independent System Operator (ISO). However, the majority of players represents energy resources, such as several types of consumers (e.g. industrial, commercial, residential), different types of producers (e.g. wind farms, solar plants, co-generation units), electric vehicles (EVs) with vehicle-to-grid capabilities, among others.

Aggregators present an important role in the future power systems management and operation. Some examples are: (i) VPPs [18], which can aggregate any other resource, including other aggregators, (ii) curtailment service providers (CSP) [23], which aggregate consumers that participate in DR programs, and (iii) SG operators, which manage the players that are contained in a specific SG.

The communications are implemented through the JADE platform, compliant with FIPA specifications [20]. To facilitate the exchange of information, our own ontology has been developed, where each event has its own predicates and characteristics [22]. By using FIPA-ACL, external developed players and resources will also be able to participate in simulations within this system by using the implemented communication system. The interface with software agents that allow the interaction with real players (humans) and with real hardware (loads, generators units, storage systems, protections, etc.) is achieved using an interface agent that allows the communication with hardware. Communications are performed using the Internet Protocol to communicate with a Programmable Logic Controller and RS-485 to communicate with soft-starters, measurement units, etc.

11.3.2 Connection to Physical Resources

MASGriP agents that represent physical players detain all the information concerning the physical installation, including its geographic coordinates and the electric characteristics. Concerning the type of player, the business model and the contracts being used, each agent has the necessary information to share with other agents. The sharing rules can be modified according to negotiations between the players and the aggregators, making MASGriP a dynamic system.

MASGriP is also used to control real physical installations through its integration with the SCADA (Supervisory Control and Data Acquisition) Office Intelligent Context Awareness Management (SOICAM). SOICAM was developed in GECAD (Knowledge Engineering and Decision Support Research Group) under the "GID-Microrede" project. The physical installations consist of four main spaces. Three of these are campus buildings where GECAD operates. These buildings include several offices, classrooms, kitchens, and bathrooms. The fourth place is a laboratory controlled house. SCADA-House [24] is located in a GECAD laboratory, and contains a large set of different loads, normally used in a common house. These loads are connected to a SCADA management system, which is controlled by a software agent. Some resources are not available in our lab, making their physical integration in the system impossible. In order to overcome this limitation, OPAL-RT [25] is used to simulate resources that are not physically available. The integration between OPAL-RT and the remainder of the system is done through the Java API of OPAL-RT. Among many other resources, the OPAL-RT platform simulates wind generators making possible to obtain outputs according to their electrical models, which can also be validated by using the platform capabilities. Additionally, OPAL-RT is also able to perform real time simulations of the components, loads and facilities that cannot be used or simulated in conventional systems. The integration of real loads in OPAL-RT is possible through the connection to software agents that represent different players in the electricity market (e.g. large consumers, large producers, and virtual players) and players connected to the distribution network (such as facilities and microgrids) [24]. This merge allows the use of different methods for management and control of the distribution network while the real time simulator analyses the impact of the methods in the energy flows and transmission lines.

The GECAD buildings where SOICAM is implemented cover more than 30 researchers. SOICAM was implemented in June, 2014. The system monitors all the consumption and generation of GECAD. The generation data (namely solar and wind based) is stored individually every 10 s. The consumption data is divided by three main types (Fig. 11.2): illumination, heating, ventilation and air conditioning (HVAC), and electrical sockets. The consumption data is also stored every 10 s. All data is stored in a structured query language (SQL) Server database, allowing the study of consumption and generation in GECAD.

SOICAM is also able to control HVAC systems. This functionality is only available for one building, affecting 19 researchers. The possible control is only on/off (for now). New hardware is being developed and implemented to allow individual load management and control.

Fig. 11.2 Monitored loads (blue: illumination; red: HVAC; green: electrical sockets)

SOICAM uses five switchboards to incorporate the energy analysers and the HVAC control system. These switchboards communicate with two main communication switchboards (one for each building) via RS485. The data acquired by SOICAM is used to test and validate the participation of SOICAM as a SG player. Additionally, the use of MASGriP for real-time control enables the simulation of real scenarios with visible outcomes on the loads.

The inclusion of a large set of different players, the combination of technical and economic treatment of future power systems, the inclusion of both real and simulated players, and the facilitation in adding or testing alternative algorithms, such as energy resource management methods, forecasting methodologies, DR models, and negotiation procedures, are characteristics that distinguish the proposed simulation platform from other existing simulators. The integration with MASCEM enables the simulation platform to go a step further yet, by including electricity market simulation capabilities to the joint simulations.

11.4 Day-Ahead and Hour-Ahead Energy Resource Management

The mathematical formulation of the ERM platform is classified as a mixed-integer non-linear programming problem. The SG operator can maximize the profits or minimize the costs to supply the required energy in both phases of the proposed methodology. The XA index refers to each phase of the methodology, namely day-ahead (DA) and hour-ahead (HA) scheduling. To maximize profits, the SG operator uses the cheaper resources, i.e. minimize the cost, and maximize the income (In).

$$Minimize\ f = C \tag{11.1}$$

$$Maximize\ f = In - C \tag{11.2}$$

The intention of the SG operator is to obtain profits from ERM. The ERM platform allows that consumers attempt to use more energy in lower price periods and avoid energy use in higher price periods. To determine the SG operator's income (Eq. 11.3), we consider the revenues from supplying the demand power to consumers, $P_{Load_{XA}(L,t)}$, the selling energy to the electricity market, $P_{Sell_{XA}(t)}$, the charging process of storage units, $P_{Ch_{XA}(ST,t)}$, and the charging process of EVs, $P_{Ch_{XA}(V,t)}$. To limit the charge of the EV, a weight λ is applied in order to charge the essential energy to make the return trip. The SP_X terms refer to SG prices and the index X refers to the type of energy resources used in the income.

$$In = \sum_{t=1}^{T}\left(\sum_{L=1}^{N_L} SP_{Load(L,t)} \times P_{Load_{XA}(L,t)} + SP_{Sell(t)} \times P_{Sell_{XA}(t)} + \sum_{ST=1}^{N_{ST}} SP_{Ch(ST,t)} \times P_{Ch_{XA}(ST,t)} + \lambda \times \sum_{V=1}^{N_V} SP_{Ch(V,t)} \times P_{Ch(V,t)}\right) \tag{11.3}$$

For the operation cost (Eq. 11.4) of the resources managed by the SG operator, we consider the cost of all the available resources, namely the DG cost, $P_{DG_{XA}(DG,t)}$, the cost with the energy bought to external suppliers, $P_{SP_{XA}(SP,t)}$, the cost of energy discharged by the storage systems, $P_{Dch_{XA}(St,t)}$, the cost of energy discharged by the EVs, $P_{Dch_{XA}(V,t)}$, the DR events from the system operator (load curtailment, $P_{Cut_{XA}(L,t)}$, and load reduction, $P_{Red_{XA}(L,t)}$), the non-supplied demand, $P_{NSD_{XA}(L,t)}$, and penalization with generation curtailment, $P_{GCP_{XA}(DG,t)}$, considering the "take-or-pay" contracts.

$$C = \sum_{t=1}^{T}\left(\sum_{DG=1}^{N_{DG}} C_{DG(DG,t)} \times P_{DG_{XA}(DG,t)} + \sum_{SP=1}^{N_{SP}} C_{SP(SP,t)} \times P_{SP_{XA}(SP,t)} + \sum_{ST=1}^{N_{ST}} C_{Dch(ST,t)} \times P_{Dch_{XA}(ST,t)} + \sum_{V=1}^{N_V} C_{Dch(V,t)} \times P_{Dch_{XA}(V,t)} + \sum_{L=1}^{N_L} C_{Cut(L,t)} \times P_{Cut_{XA}(L,t)} + \sum_{L=1}^{N_L} C_{Red(L,t)} \times P_{Red_{XA}(L,t)} + \sum_{L=1}^{N_L} C_{NSD(L,t)} \times P_{NSD_{XA}(L,t)} + \sum_{DG=1}^{N_{DG}} C_{GCP(DG,t)} \times P_{GCP_{XA}(DG,t)}\right) \tag{11.4}$$

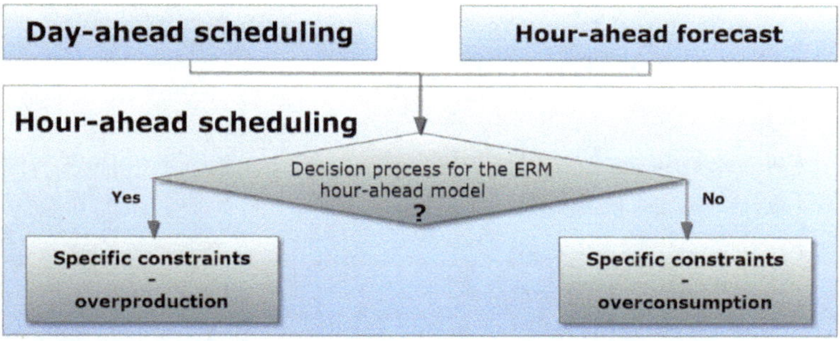

Fig. 11.3 Decision process for the hour-ahead scheduling methodology

Problem constraints of the ERM platform include both technical and economic aspects, such as the Kirchhoff's Law, voltage limits, line thermal limits, the maximum capacity considering the available resources, the storage resources, and DR power limits. A detailed description of all the constraints used is presented in [8].

The decision process for the ERM hour-ahead model (Eq. 11.5) determines whether there is surplus or shortage of generated energy between the newer forecast (hour-ahead forecast) and the day-ahead scheduling, for each period t. Equation 11.5 considers the hour-ahead forecast of the demand power, $P_{Load_{FHA}}$, the hourly forecasts of the renewable energy sources, $P_{DG_{FHA}}$, and the results of day-ahead scheduling, particularly the generation, $P_{DG_{DA}}$, and the load, $P_{Load_{DA}}$.

$$\left(\sum_{DG=1}^{N_{DG}} P_{DG_{FHA}(DG,t)} - \sum_{DG=1}^{N_{DG}} P_{DG_{DA}(DG,t)}\right) \geq \left(\sum_{L=1}^{N_L} P_{Load_{FHA}(L,t)} - \sum_{L=1}^{N_L} P_{Load_{DA}(L,t)}\right) \tag{11.5}$$

The set of constraints is divided into two groups: overproduction (Eq. 11.6), when there is surplus of generated energy, and over-consumption (Eq. 11.7), when there is shortage of generated energy. Figure 11.3 shows the decision process for the ERM hour-ahead model.

For the specific constraints of overproduction (Eq. 11.6) and over-consumption (Eq. 11.7), we consider the technical limits of the distributed energy resources for each period t, where $X_{Cut_{XA}}$ represents the binary variable of DR curtailment of load in the XA phase of the methodology proposed. The P_{XMin} term refers to the minimum active power and the index X refers to the type of energy resources used. Also, the P_{XMax} term refers to maximum active power and the index X refers to the used type of energy resources.

$$
\begin{aligned}
&P_{DGMin(DG,t)} \leq P_{DG_{HA}(DG,t)} \leq P_{DG_{DA}(DG,t)} \\
&P_{SPMin(SP,t)} \leq P_{SP_{HA}(SP,t)} \leq P_{SP_{DA}(SP,t)} \\
&P_{Ch_{DA}(ST,t)} \leq P_{Ch_{HA}(ST,t)} \leq P_{ChMax(ST,t)} \\
&P_{Dch_{HA}(ST,t)} \leq P_{Dch_{DA}(ST,t)} \\
&P_{Ch_{DA}(V,t)} \leq P_{Ch_{HA}(V,t)} \leq P_{ChMax(V,t)} \\
&P_{Dch_{HA}(V,t)} \leq P_{Dch_{DA}(V,t)} \\
&P_{Red_{HA}(L,t)} \leq P_{Red_{DA}(L,t)} \\
&P_{Cut_{HA}(L,t)} = \begin{cases} 0 & \text{, if } X_{Cut_{DA}(L,t)} = 0 \\ P_{MAxCut(L,t)} \times X_{Cut_{HA}(L,t)} & \text{, if } X_{Cut_{DA}(L,t)} = 1 \end{cases} \\
&P_{Sell_{DA}(t)} \leq P_{Sell_{HA}(t)} \leq P_{SellMax(t)}
\end{aligned} \tag{11.6}
$$

$$
\begin{aligned}
&P_{DG_{DA}(DG,t)} \leq P_{DG_{HA}(DG,t)} \leq P_{DGMax(DG,t)} \\
&P_{SP_{DA}(SP,t)} \leq P_{SP_{HA}(SP,t)} \leq P_{SPMax(SP,t)} \\
&P_{Ch_{HA}(ST,t)} = P_{Ch_{DA}(ST,t)} \\
&P_{Dch_{DA}(ST,t)} \leq P_{Dch_{HA}(ST,t)} \leq P_{DchLimit(ST,t)} \\
&P_{Ch_{HA}(V,t)} = P_{Ch_{DA}(V,t)} \\
&P_{Dch_{DA}(V,t)} \leq P_{Dch_{HA}(V,t)} \leq P_{DchLimit(V,t)} \\
&P_{Red_{DA}(L,t)} \leq P_{Red_{HA}(L,t)} \leq P_{MaxRed(L,t)} \\
&P_{Cut_{HA}(L,t)} = \begin{cases} P_{MAxCut(L,t)} \times X_{Cut_{HA}(L,t)} & \text{, if } X_{Cut_{DA}(L,t)} = 0 \\ P_{Cut_{DA}(L,t)} & \text{, if } X_{Cut_{DA}(L,t)} = 1 \end{cases} \\
&P_{Sell_{HA}(t)} \leq P_{Sell_{DA}(t)}
\end{aligned} \tag{11.7}
$$

11.5 Case Study

The potential of the joint simulation of SGs and EMs using MASGriP and MASCEM is demonstrated with a case study using real data, which includes several players that control physical installations. The interface between the two environments is done by the SG operator, which is responsible for managing the internal resources of the SG using the ERM methodology presented in Sect. 11.4, and also for participating in the EM in order to purchase the required amount of power to fulfill the SG needs when the local generation is low, and sell eventual surplus power when the generation is high. The ERM is performed on a day-ahead basis, including DG, consumption and market price forecasts for the following day. The market transacted power and resulting prices are used by the SG operator to adapt its management results, namely by performing a new ERM on an hour-ahead basis for each hour of the objective day.

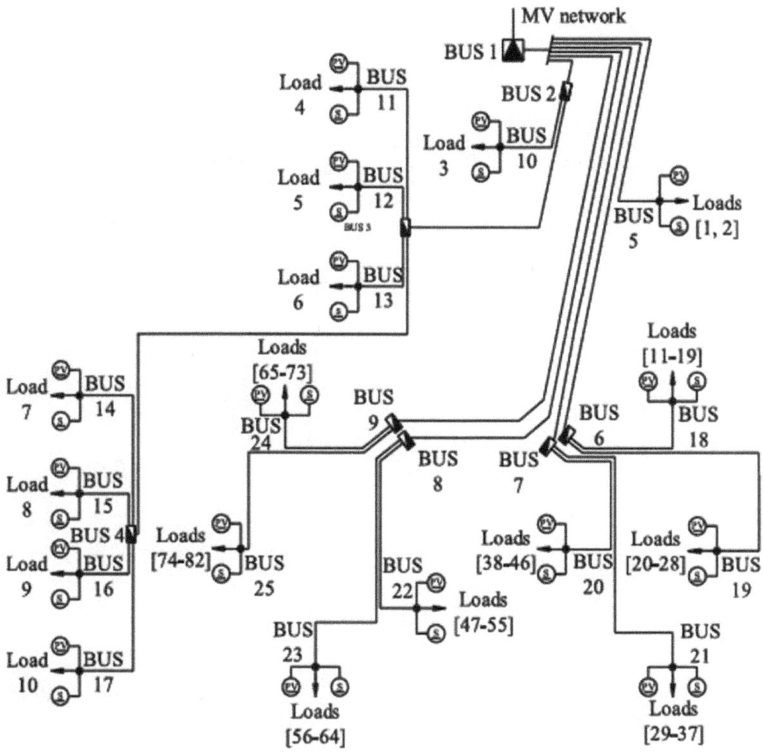

Fig. 11.4 Distribution network used for the SG simulation

11.5.1 Case Study Characterization

The simulated scenario considers a SG that is modeled using MASGriP, involving a real distribution network located in Portugal, with 25 buses (Fig. 11.4). The private distribution network is connected to the main grid trough a MV/LV transformer. The SG accommodates distributed generation (photovoltaic and wind based generation) and storage units, which are integrated in the consumption buildings (8 residential houses, 8 residential buildings, and 1 commercial building). The two loads connected to Bus 5 are physical installations, namely Buildings I and N of GECAD. The simulation results have a direct impact on the real loads of these two GECAD buildings. The photovoltaic generation, wind based generation and storage units are related to the installed consumption power, according to the current legislation in Portugal. Further details on the considered distributed network can be found in [26].

The case study considers a simulation day during the summer time in Portugal, namely September 4th, 2014. In this context, the photovoltaic generation reaches high values, especially during the mid-hours of the day. The sequence of events for the case study is presented in Fig. 11.5, considering the actions of the SG operator

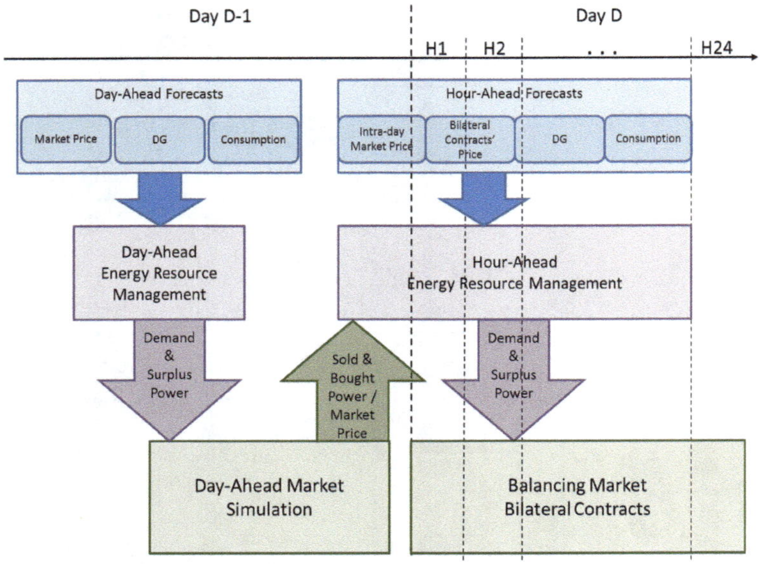

Fig. 11.5 Simulation sequence process

agent when managing the energy resources, in the scope of MASGriP, and when participating in the EM, using MASCEM, in order to sell the surplus of generated power of the SG or buying the required power to fulfill the requirements of the SG consumers when necessary.

From Fig. 11.5, it is visible that the SG operator agent starts by executing some preliminary forecasts, considering the expected market price of the following day, the expected DG (including the forecast of the wind speed, solar intensity, and temperature in order to model the expected generation), and also the expected consumption of all the consumers that are part of the SG. These forecasts are performed using ALBidS, as presented in [26–29], and are used to perform a day-ahead ERM. From this, results the optimal hourly schedule of generation, consumption, application of demand response programs, charge and discharge of EVs' batteries, and also the total hourly needs for power that must be bought from outside the SG, and hourly surplus power that can be sold in order to improve the incomes of the SG operator. Using the results of the day-ahead ERM, the SG operator agent participates in the EM simulation, using MASCEM, in order to negotiate the amounts of power that must be sold or purchased.

Finally, the achieved market results are used to execute new, adapted, hour-head ERMs, considering the deals that have been established in the market, and new, updated, hour-ahead forecasts of DG and consumption in the SG. From the hour-ahead ERMs result the final scheduling of the SG resources, and eventually new amounts of power (to sell or buy) that should be negotiated with external entities by means of bilateral contracts or by participating in near real-time markets, such as balancing markets. These negotiations are the last resource to achieve the required power to fulfill the SG consumers' needs, or a final opportunity to sell extra power to increase the incomes of the SG operator.

Real-time simulation using the connection to OPAL-RT is executed after each ERM, in order to analyze the impact of the scheduled actions in the power network, and validate if the results are suitable to be implemented, from the network standpoint. The present case study allows, in all the loads, the use of incentive-based demand response programs to pay participating customers to reduce their load at a maximum until 30% of the initial load. Moreover, it allows energy shifting in commercial building located at bus 5 (loads 1 and 2), at a maximum of 60% of the initial load [30]. The use of DR resources can be seen both in the simulated environment by analyzing the outcomes of the software agents, and also in the real resources, by verifying the implication of load curtailment, reduce, and shift, in the physical installations.

The ERM methodology has been developed in TOMLAB Optimization with CPLEX solver using MATLAB R2014a 64 bits software. The simulations presented in this case study have been executed in a machine with one Intel® Xeon® E5-2620v2–2.10 GHz processor, with 12 cores, 16GB of Random-Access-Memory and Windows 8.1 Professional.

11.5.2 Results

Figure 11.6 presents the scheduling results of the day-ahead ERM, including DG, consumption and day-ahead market price forecasts. The results concern the generated power, the initial expected load and the final load resulting from the application of DR, the charge and discharge of batteries, and the amount of power that needs to be transacted outside the SG.

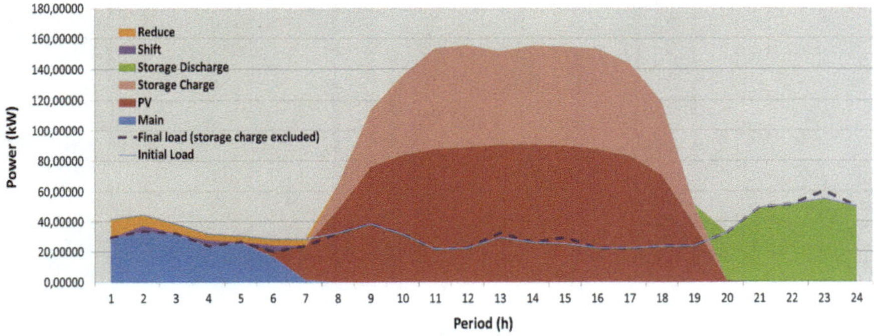

Fig. 11.6 Scheduling results of the day-ahead ERM

From Fig. 11.6, it is visible that the SG achieves large volumes of generated power during most hours, due to the high value of photovoltaic generation during the day. The corresponding amount is used to charge the batteries, so that the consumption can be guaranteed in the last periods of the day (when the generation decreases), namely from periods 20 to 24. During the first hours of the day, since the batteries have not been charged yet (all batteries started the simulation completely empty), and there is still no photovoltaic generation, the consumption has to be assured by external sources. In this case, from power bought in the electricity market. Additionally, DR programs are used to lower the consumption during the first hours of the day, so that the cost of purchasing power externally is minimized. The reduction of load has been applied during hours 1–7, and the shifting of load has also been used, namely in periods 2, 4 and 6. Figure 11.7 details the load shifting process in the GECAD buildings (Bus 5).

From Fig. 11.7, as can be confirmed by the comparison between the dashed and the solid lines of Fig. 11.6, some loads (referring to GECAD's buildings) have been moved from hours 2, 4 and 6 to hours 13, 15 and 23. This DR process allows taking load away from periods when the demand is higher, when compared to the generated production, and incentivizing consumers to make this consumption in hours that are more convenient to the network (due to lower consumption, higher generation, or lower energy prices from external sources).

After the execution of the day-ahead ERM, a real-time simulation using OPAL-RT is executed, so that the impact of the optimal scheduling results on the power network can be evaluated. Figures 11.8, 11.9 and 11.10 present the comparison between the data resulting from the ERM and the values resulting from the OPAL-RT simulation.

From Figs. 11.8, 11.9 and 11.10, it is visible that the output from the OPAL-RT simulation is almost identical to that of the day-ahead ERM, especially for the cases of photovoltaic generation (Fig. 11.9) and batteries (Fig. 11.10). This occurs because the model has been built with flow sources based on Three-Phase Programmable

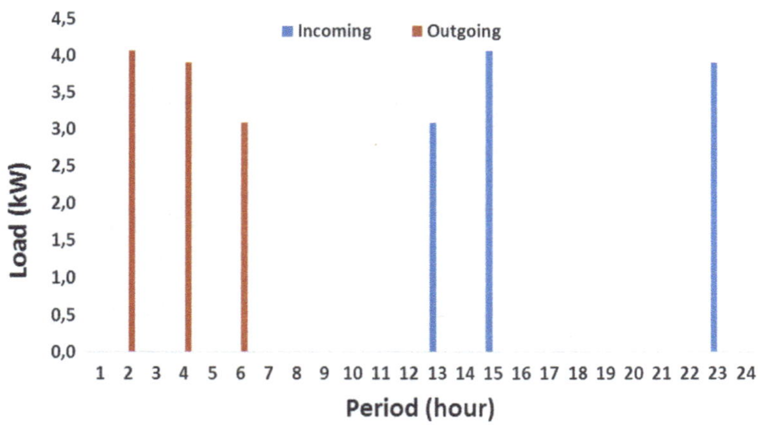

Fig. 11.7 Load shifting in GECAD's buildings

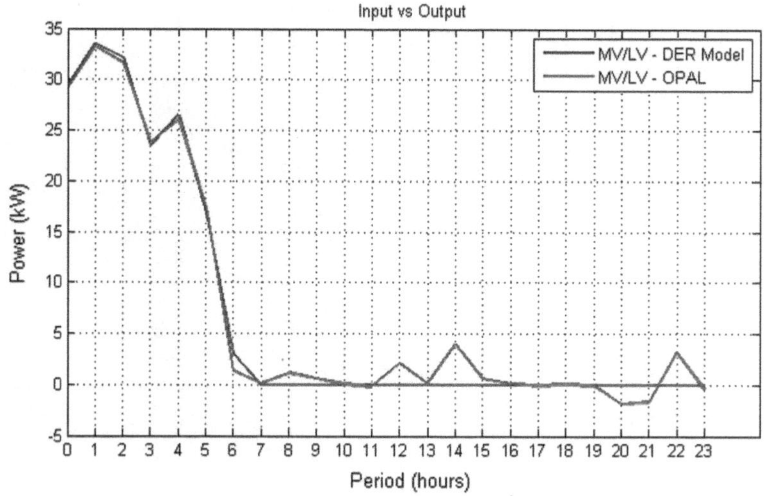

Fig. 11.8 ERM and OPAL-RT results comparison, regarding the energy traded by the SG with the external network in MV

Source (PLL). The most notorious difference is verified in the interaction with the external network, in bus 1 (Fig. 11.10), due to the response time of the physical components. The synchronization is not instantaneous, and for this reason some discrepancies occur. This can be better visualized in Fig. 11.11, which shows the behavior of the loads, comparing the expected behavior that results from the ERM and the actual behavior that results from the real time OPAL-RT simulation.

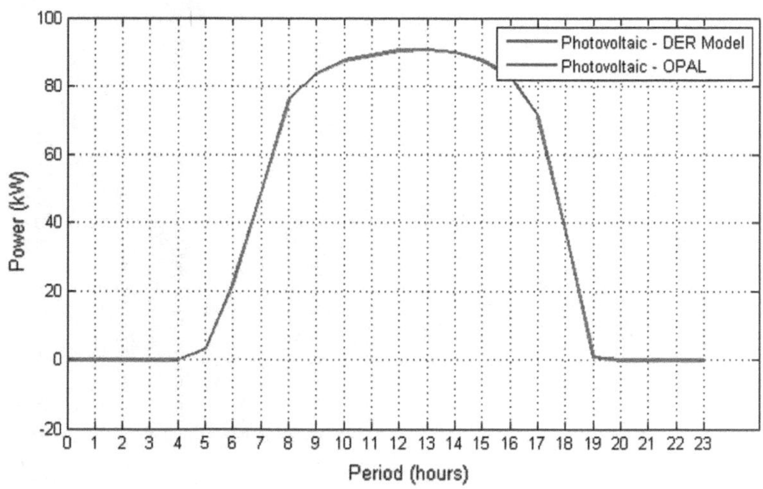

Fig. 11.9 ERM and OPAL-RT results comparison, regarding the total photovoltaic generation of the SG

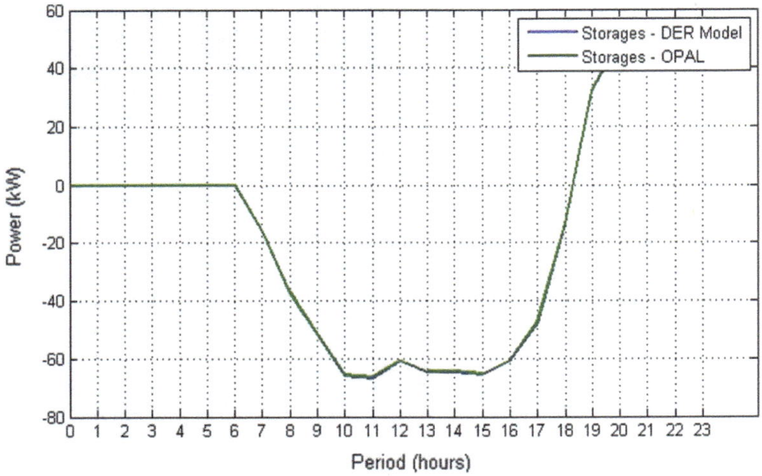

Fig. 11.10 ERM and OPAL-RT results comparison, regarding the total SG storage charge and discharge

From Fig. 11.11, it is visible that the real-time simulation results are very similar to the expected ones. The larger discrepancies are verified in the results of the real buildings (GECAD building N and I). Since these loads have different response times, which require a larger synchronization process, the results are more unstable when compared to the simulated loads.

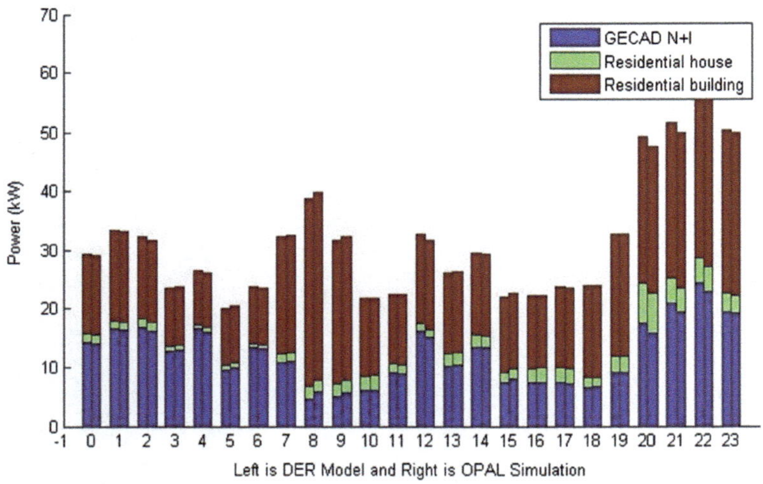

Fig. 11.11 Comparison of the loads as result from the day-ahead ERM scheduling and the OPAL-RT real time simulation

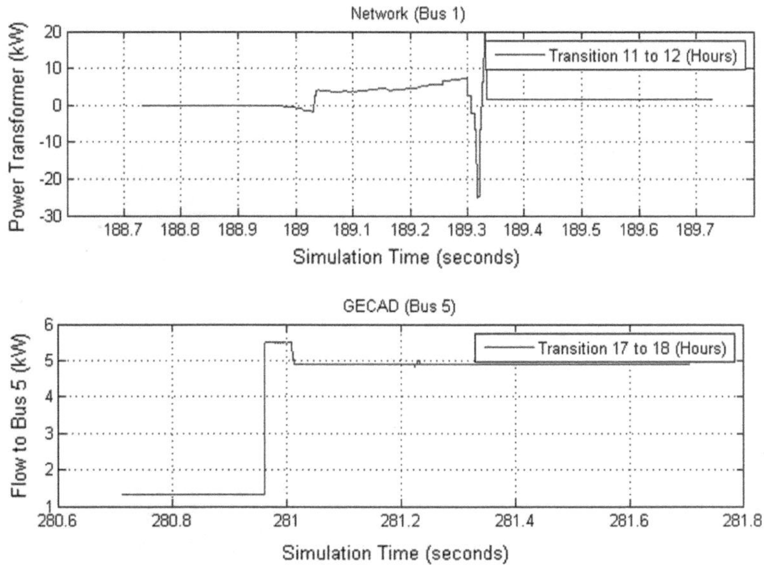

Fig. 11.12 Active power synchronization during hour transition in: Bus 1 – connection with the external MV network (top), and Bus 5 – GECAD buildings (bottom)

The synchronization process can be visualized in Fig. 11.12, which presents the active power synchronization during the transition from one hour to the following, in two different buses, namely Bus 1 (power transformer, top), and Bus 5 (GECAD buildings, bottom).

From Fig. 11.12 (top), one can see that the synchronization in Bus 1 takes approximately 306 ms. This time corresponds to the time that the MASGriP software agent takes to send the variable values from the ERM to the OPAL-RT simulation. The total number of variables that are sent in each hour transition is of 116. In Fig. 11.12 (bottom), it is visible that the synchronization regarding Bus 5 is much smoother, since the number of variables referring to a single Bus is smaller.

Considering the results of the day-ahead planning, the SG operator needs to purchase some power in the day-ahead market in order to fulfill the consumption needs of the SG, namely from hours 1 to 7. Figure 11.13 presents the market results achieved from MASCEM, concerning the participation of the SG operator in the EM to buy the required power during the first hours of the simulation day.

Figure 11.13 shows that the smart grid operator was able to purchase the required amount of power from the market in hours 3–6. This occurred due to the bid prices that the SG operator has considered in the market, which are superior to the market price during these hourly periods. The higher values reflect the maximum value that the SG operator agent is willing to pay for the purchased power, taking into account the values of the optimization performed in the day-ahead ERM, using the day-ahead EM price forecasts as a basis. However, in hours 1, 2 and 7, the SG operator was not

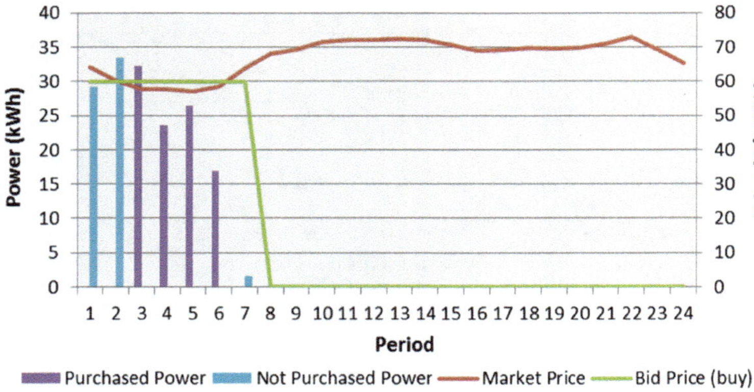

Fig. 11.13 EM results of the SG operator agent when participating in the day-ahead market in MASCEM, with the objective of buying the demanded power

able to purchase the required amount of power. This occurs due to bid price from the SG operator, which is inferior to the established market price during these hours.

This means that the corresponding amount of power will have to be ensured by another way, either by participating in other types of negotiations (bilateral contracts with nearby SGs or neighbor players, or in balancing markets), or by applying further DR programs. How this amount of power will be achieved is determined by the hour-ahead ERM process, which already considers the real values of day-ahead market results, and more up-to-date forecasts of both demand and generation. The hour-ahead ERM runs independently for each hour, one hour before it occurs. Thus, it is able to include up-to-date, hour-ahead forecasts of demand, consumption and generation. Figure 11.14 presents the results of the final, adapted hour-ahead energy resource scheduling plan.

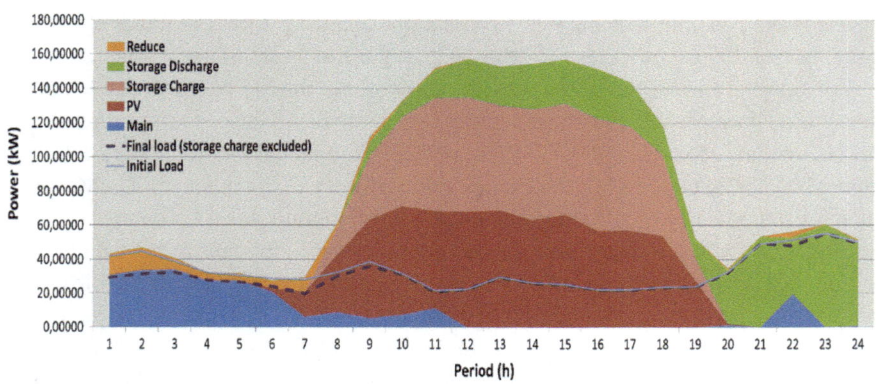

Fig. 11.14 Scheduling results of the hour-ahead ERM

Figure 11.14 shows that considering the already transacted power from the day-ahead EM and the updated forecasts, the hour-ahead ERM results include the need for further energy transactions with external entities, in order to deal with the changes of expected consumption and generation throughout the day. For this reason, in addition to the amount of power that were already required to be bought (the amount from hours 1, 2 and 7 that could not be transacted in the day-ahead EM), there is now the identification of a further need in some other hours. These amounts need to be bought from alternative market opportunities (e.g. bilateral contracts with nearby SGs or near real-time balancing markets). Additionally, further DR is also required, in order to face the changes that are expected from the day-ahead planning to a more up-to-date hour-ahead plan. Further load reduction is verified for hours 1 to 7, and in hour 22. The need for load reduction in hour 22 is verified because the energy stored in the batteries during the day is not enough to supply all the load in the final hours of the day, as expected in the day-ahead planning.

11.6 Conclusion

The practical implementation of SGs is, nowadays, a reality. Several pilot implementations have been experimented and full scale tests and validations are being conducted in order to draw conclusions. With the worldwide implementation of SGs, management and negotiation mechanisms need to be robust in order to take full advantage of the potential of DG and local control of demand.

This chapter described the integration between two complementary multi-agent simulators, MASCEM and MASGriP, which together provide the means to create realistic simulation environments, involving SGs and EMs. The integration makes possible to simulate the participation of SG players in EMs, to reach conclusions on the steps that are necessary to enable the full participation of DG in markets, and also to validate potential alternatives for a competitive SG market environment.

A case study based on real data was presented, which includes a smart grid composed of a simulated distribution network involving several loads. It also includes the participation of a software agent (the SG operator) in both simulators simultaneously, by managing the internal resources of the SG using day-ahead and hour-ahead ERM methodologies. And also by participating in the EM to transact the required power to fulfill the needs of the internal SG resources. Additionally, real-time simulation capabilities provided by the integration of MASGriP with OPAL-RT have provided the means to test the impact of the planned actions in the power network.

The multi-agent platform presented in the chapter opens important opportunities under different perspectives, which result in important advances in the fields of transactive energy, EMs, and SGs. The contributions that this work provides includes: multi-agent simulation of SG environments, multi-agent simulation of EMs, joint simulation of EMs and SGs by interconnecting MASGriP and MASCEM, participation of a SG operator in multi-agent EM simulations, and adaptation of SG operator's ERM based on the results achieved in the EM.

As future work, the participation of SG players in alternative EMs, e.g., bilateral contracts and balancing markets, can bring added value. Additionally, the simultaneous management of several SGs and the participation of different SG operators in the same EM environment are also relevant. Regarding the used methodologies, the integration of further near real-time ERM methods are important, in order to provide a closer adaptation to the reality. As seen from the simulation results, changes from the day-ahead to the hour-planning were significant. Thus, a further step that approximates the scheduling plans towards real time should also brings additional benefits, by adapting the plans so that they can be better prepared to deal with unexpected changes.

Acknowledgements This work has received funding from the European Union's Horizon 2020 research and innovation programme under the Marie Sklodowska-Curie grant agreement N. 641794 (project DREAM-GO). It has also received FEDER Funds through the COMPETE program and National Funds through FCT under the project UID/EEA/00760/2013. The authors gratefully acknowledge the valuable contribution of Bruno Canizes, Daniel Paiva, Gabriel Santos and Marco Silva to the work presented in the chapter.

References

1. European Commission: The 2020 climate and energy package (2009)
2. European Commission: 2030 Framework for climate and energy policies (2014). https://ec.europa.eu/clima/policies/strategies/2030_en. Accessed 10 September 2017
3. Sioshansi, P.: Evolution of Global Electricity Markets. New paradigms, New Challenges, New Approaches. Academic Press, Oxford (2013)
4. PCR: EUPHEMIA public description: PCR market coupling algorithm. Price coupling of regions (2014). https://www.apxgroup.com/wp-content/uploads/Euphemia-public-description-Nov-20131.pdf. Accessed 15 September 2017
5. Shahidehpour, M., Yamin, H., Li, Z.: Market Operations in Electric Power Systems: Forecasting, Scheduling, and Risk Management. Wiley, New York (2002)
6. Saber, Y., Venayagamoorthy, K.: Resource scheduling under uncertainty in a smart grid with renewables and plug-in vehicles. IEEE Syst. J. **4**, 103–109 (2012)
7. Gomes, L., Faria, P., Morais, H., Vale, Z., Ramos, C.: Distributed, agent-based intelligent system for demand response program simulation in smart grids. IEEE Intell. Syst. **29**, 56–65 (2014)
8. Borlase, S.: Smart Grids: Infrastructure, Technology, and Solutions. CRC Press, New york (2013)
9. Covrig, C., Ardelean, M., Vasiljevska, J., Mengolini, J., Fuli, G., Amoiralis, E.: Smart Grid Projects Outlook 2014. Science and Policy Report by the Joint Research Centre of the European Commission, Luxembourg (2014)
10. European Commission: Incorporing Demand Side Flexibility, in Particular Demand Response, in Electricity Markets. Commission Staff Working Document (2013)
11. DOE: Benefits of Demand Response in Electricity Markets and Recommendations for Achieving Them. A Report to the United States Congress Pursuant to Section 1252 of the Energy Policy Act of 2005, US Department of Energy (2006)
12. Faria, P., Vale, Z., Baptista, J.: Constrained consumption shifting management in the distributed energy resources scheduling considering demand response. Energy Convers. Manag. **93**, 309–320 (2015)

13. Walton, R.: 2014 for Demand Response: The Best of Times, the Worst of Times. Utility Dive (2014)
14. Oliveira, P., Pinto, T., Morais, H., Vale, Z.: MASGriP - a multi-agent smart grid simulation platform. In: IEEE Power and Energy Society General Meeting, pp. 1–5. IEEE Press (2012)
15. Fernandes, F., Silva, M., Faria, P., Vale, Z., Ramos, C., Morais, H.: Real-time simulation of energy management in a domestic consumer. In: IEEE/PES Innovative Smart Grid Technologies Europe (ISGT EUROPE), pp. 1–5. IEEE Press (2013)
16. Praça, I., Ramos, C., Vale, Z., Cordeiro, M.: MASCEM: a multiagent system that simulates competitive electricity markets. IEEE Intell. Syst. **18**(6), 54–60 (2003)
17. Pinto, T., Vale, Z., Sousa, T., Praça, I., Santos, G., Morais, H.: Adaptive learning in agents behaviour: a framework for electricity markets simulation. Integr. Comput. Aided Eng. **21**(4), 399–415 (2014)
18. Pinto, T., Morais, H., Oliveira, P., Vale, Z., Praça, I., Ramos, C.: A new approach for multi-agent coalition formation and management in the scope of electricity markets. Energy **36**(8), 5004–5015 (2011)
19. Teixeira, B., Silva, F., Pinto, T., Praça, I., Santos, G., Vale, Z.: Data mining approach to support the generation of realistic scenarios for multi-agent simulation of electricity markets. In: IEEE Symposium on Intelligent Agents (IA), pp. 1–5. IEEE Press (2014)
20. FIPA: Agent management specification. Foundation for intelligent physical agents, Document number SC00023K (2004). http://www.fipa.org/specs/fipa00023/SC00023K.pdf. Accessed 15 September 2017
21. FIPA: FIPA ACL message structure specification. Foundation for intelligent physical agents, Document number SC00061G (2002). http://www.fipa.org/specs/fipa00061/SC00061G.pdf. Accessed 15 September 2017
22. Santos, G., Pinto, T., Morais, H., Sousa, T., Pereira, I., Fernandes, R., Praça, I., Vale, Z.: Multi-agent simulation of competitive electricity markets: autonomous systems cooperation for European market modelling. Energy Convers. Manag. **99**, 387–399 (2015)
23. Moran, D., Suzuki, J.: Curtailment service providers: they bring the horse to water... do we care if it drinks? In: ACEEE Summer Study on Energy Efficiency in Buildings, pp. 287-298. ACEEE Publications (2010)
24. Fernandes, F., Morais, H., Vale, Z., Ramos, C.: Dynamic load management in a smart home to participate in demand response events. Energy Build. **82**, 59–606 (2014)
25. OPAL-RT: OP5600 Off-the-shelf Hardware-in-the-Loop (HIL) simulator. OPAL-RT Technologies, Inc., Quέbec, Canada. www.opal-rt.com. Accessed 15 September 2017
26. Pinto, T., Sousa, T., Vale, Z.: Dynamic artificial neural network for electricity market prices forecast. In: IEEE International Conference on Intelligent Engineering Systems (INES), pp. 1–5. IEEE Press (2012)
27. Pinto, T., Ramos, S., Sousa, T., Vale, Z.: Short-term wind speed forecasting using support vector machines. In: IEEE Symposium on Computational Intelligence in Dynamic and Uncertain Environments (CIDUE), pp. 1–4. IEEE Press (2014)
28. Marques, L., Pinto, T., Sousa, T., Praça, I., Vale, Z., Abreu, S.: Solar intensity forecasting using artificial neural networks and support vector machines. In: 2nd ELECON Workshop – Consumer control in Smart Grids, pp. 83–93. ELECON Press(2014)
29. Ramos, S., Soares, J., Vale, Z., Ramos, S.: Short-term load forecasting based on load profiling. In: IEEE Power and Energy Society General Meeting (PES), pp. 1–5. IEEE Press (2013)
30. Mitra, J., Suryanarayanan, S.: System Analytics for Smart Microgrids. In: IEEE Power and Energy Society General Meeting (PES), pp. 1–4. IEEE Press (2010)

FSC
www.fsc.org
MIX
Papier | Fördert
gute Waldnutzung
FSC® C083411

Kolibri 360 GmbH
Ferdinand-Jühlke-Straße 7
99095 Erfurt, Deutschland
produktsicherheit@kolibri360.de